T0206244

Records in Stone
Papers in memory of Alexander Thom

Alexander Thom.

RECORDS IN STONE

Papers in memory of Alexander Thom

Edited by

C. L. N. RUGGLES
University of Leicester

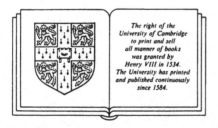

The right of the
University of Cambridge
to print and sell
all manner of books
was granted by
Henry VIII in 1534.
The University has printed
and published continuously
since 1584.

CAMBRIDGE UNIVERSITY PRESS

Cambridge

New York   New Rochelle

Melbourne   Sydney

PUBLISHED BY THE PRESS SYNDICATE OF THE UNIVERSITY OF CAMBRIDGE
The Pitt Building, Trumpington Street, Cambridge, United Kingdom

CAMBRIDGE UNIVERSITY PRESS
The Edinburgh Building, Cambridge CB2 2RU, UK
40 West 20th Street, New York NY 10011–4211, USA
477 Williamstown Road, Port Melbourne, VIC 3207, Australia
Ruiz de Alarcón 13, 28014 Madrid, Spain
Dock House, The Waterfront, Cape Town 8001, South Africa

http://www.cambridge.org

First published 1988
First paperback edition 2002

A catalogue record for this book is available from the British Library

Library of Congress cataloguing in publication data
Records in stone: papers in memory of Alexander Thom / edited by
C. L. N. Ruggles.
p.   cm.
Bibliography: p.
ISBN 0 521 33381 4 hardback
1. Megalithic monuments.   2. Leys.   3. Astronomy, Prehistoric.
4. Thom, A. (Alexander)   I. Thom, A, (Alexander)
II. Ruggles, C. L. N. (Clive L. N.)
GN790.R43   1988
936.1-dc19   88-1723   CIP

ISBN 0 521 33381 4 hardback
ISBN 0 521 53130 6 paperback

Contents

Foreword

Alexander Thom died on 7 November, 1985, aged 91. He had had a distinguished academic career, holding the Chair of Engineering Science at the University of Oxford from 1945 to 1961. He had also developed, quite independently, a deep and active interest in the prehistoric megalithic sites of his native Scotland, and upon his retirement this spare-time interest became his principal one. Ironically perhaps, it is for his contribution to archaeology that he will be best remembered by many.

The volume opens with appreciations from two people who knew Alexander Thom both as an engineer and as a field worker passionately interested in megalithic sites. One of these, his son Archie, was his devoted helper and collaborator for many years.

Between the 1930s and the 1970s Thom visited and surveyed hundreds of megalithic sites in Britain and Brittany. These sites - stone rings, stone rows and single standing stones, together with burial monuments such as chambered tombs and cairns - were erected in considerable numbers in the British Isles and north-western France during the third and second millennia BC. Thom's surveys, which accurately record sites many of which are in a continuing state of deterioration, give students of British prehistory an invaluable corpus of field data which is of lasting value. His field notebooks and most of his original plans are now deposited with The National Monuments Record of Scotland, in Edinburgh, and the centrepiece of the current volume is the catalogue for this collection which has been prepared by Mrs. Lesley Ferguson of the Royal Commission on the Ancient and Historical Monuments of Scotland.

Thom's interpretations of his field data have led to widespread interest and debate in three areas: geometry (the methods used to set out the megalithic rings, many of which are clearly non-circular), mensuration (the possible use of 'standard' units of measurement in setting out megalithic rings and rows), and astronomy (the possible alignment of structures upon the horizon rising and setting positions of certain celestial bodies). All three topics are controversial, and continue to arouse strong and even impassioned debate. It is fitting that two of them are covered in Alexander Thom's last paper, written together with his son Archie, which is published here for the first time.

Many advances have been made since Thom's work first became widely recognised. It is now generally realised, for instance, that studies of prehistoric astronomy and geometry are meaningless if they take place in a cultural vacuum, and discussions about these topics have moved beyond their original narrow confines. What then of Thom's contribution to archaeological research? It would be short-sighted indeed to attempt to measure this simply by comparing his own conclusions with current or future consensus amongst prehistorians. Instead, we must look to his success in opening up new fields of enquiry which have important implications for the study of the people who, through a long and changing period of north-west European prehistory, constructed and used monuments in stone.

Thom's conclusions, particularly concerning astronomy, have also been the catalyst for a number of new areas of investigation stretching far beyond the British megalithic sites that were of such interest to Thom himself. Recent years have seen the rise of 'archaeoastronomy' - the study of astronomical practice in past societies - as a recognised field of enquiry in its own right. The *Archaeoastronomy* supplement to *Journal for the History of Astronomy* was created in 1979 and the first 'World Archaeoastronomy Symposium' was held in Oxford in 1981. This provided a forum for interaction between European and American archaeoastronomers, who had developed different methodologies for dealing with the study of astronomical practice in very different cultural contexts. The second such symposium was held in Merida, Mexico in 1986, and underlined the interdisciplinary involvement which is now prevalent in the field. The third World Archaeoastronomy Symposium is planned for 1990 and promises to be truly global both in content and participation. All these developments can be traced back to the impetus given by Alexander Thom's work.

The larger part of this volume contains material which, it is hoped, will serve as a memorial to Thom's contribution by illustrating current work in areas which it has inspired, both directly and indirectly. It is hoped that the contributions represent a fair cross-section of current opinion and the contributors a representative selection of the people actively involved. Most of the contributions relate to the megalithic sites of Britain and Brittany; the final two papers from American contributors concern world archaeo-astronomy.

It has become clear from the many points of view expressed over the years about Thom's ideas that the nature of admissible evidence, and of the conclusions that can reasonably be derived from that evidence, seems very different for a person trained in, say, the physical sciences from that for a colleague trained in a discipline such as social anthropology. Yet the astronomer, the statistician and the anthropologist each have a point of view which is undeniably relevant to the study of the nature of prehistoric astronomy. How then should their evidence be collated and what conclusions should be reached?

After many initial misunderstandings and fruitless exchanges, the wheels of interdisciplinary communication and collaboration necessary to confront this question have at last been set in motion. In many archaeoastronomical papers today one sees attempts to consider the astronomical and statistical evidence alongside, and on equal terms with, the anthropological and the ethnohistoric. There is far to go: but in drawing attention to problem areas where such collaboration is necessary, Thom's work may prove in the longer term to have opened up the interdisciplinary arena for a fascinating exchange of views across the 'two cultures' which could have methodological consequences far beyond the mere study of megalithic remains and archaeoastronomy. In the longer term this may well prove to be the most significant benefit of all to be derived from the work of Alexander Thom.

In the meantime his contribution to megalithic studies bears witness to his own remarkable range of skills, both theoretical and practical, as well as to his sheer enthusiasm and determination. It is hoped that this volume, as well as commemorating Thom's endeavours, will contribute significantly to their continuation.

I wish to record my sincerest thanks to those without whose help *Records in Stone* would simply not have been possible: to Archie Thom, who acted as advisor and consultant throughout; to Susan Kruse, who acted as editorial assistant; to the referees, who gave me invaluable help in selecting a worthy and representative cross-section of papers for inclusion; to Jackie Macklin, who re-typed those manuscripts (the great majority) which were not received in 'soft' form; and to Paul Warren, who helped considerably in the conversion of these manuscripts to a format suitable for the laser printer. The manuscript was prepared using the facilities of the Computing Studies Department at Leicester University.

The catalogue of the Alexander Thom archive (Chapter 4) and the article by Ritchie on Brodgar (Chapter 15) are published by courtesy of the Commissioners and with the assistance of a grant from the Royal Commission on the Ancient and Historical Monuments of Scotland. Shorter versions of the personal note by Archie Thom (Chapter 1) and the list of publications (Chapter 3) have appeared in *Archaeoastronomy*, the journal of the Center for Archaeoastronomy, Maryland, USA, and the relevant passages are reproduced with permission. The Astronomical Society of Australia have given permission to reproduce parts of the review by Norris, which were first published in their Proceedings.

I should like to thank the following for their kind permission to publish or reproduce illustrations: the British Library Board (Ritchie, Figs. 2 & 3); the Landsbókasafn Islands, Reykjavík (Ritchie, Fig. 4); Oxford University Press (Thom & Thom, Figs. 1 & 2; Myatt, Figs. 3, 6, 8 & 9); Science History Publications Ltd. (Thom & Thom, Fig. 4); the Photographic Unit of the University of Glasgow (Thom & Thom, Fig. 6); the Historic Monuments and Buildings Branch, Department of the Environment for Northern Ireland (Burl, Fig. 5); the Royal Commission on the Ancient and Historical Monuments of Scotland (Ritchie, Figs. 1 & 5-8); the Scottish Development Department, Historic Buildings and Monuments Branch (Curtis, Fig. 9); the Royal Statistical Society (Myatt, Fig. 4); the National Museums of Scotland (Ponting, Figs. 9 & 11); the Society of Antiquaries of Scotland (Ponting, Fig. 10); and the Danish National Museum (Ponting, Figs. 12 & 13).

Clive Ruggles

Notes about contributors

Anthony Aveni is Charles A. Dana Professor of Astronomy and Anthropology at Colgate University, Hamilton, New York. Since 1970 he has worked in Mesoamerican and Andean archaeoastronomy and is the author/editor of eight texts and numerous articles on the subject. Most recently he has edited *World Archaeoastronomy*, the proceedings of the 2nd Oxford International Conference on Archaeoastronomy, and *The Lines of Nazca*, contributing two articles to both.

Aubrey Burl, formerly Principal Lecturer in Archaeology at Hull College of Higher Education but now retired, is an authority on prehistoric stone circles. As well as many papers and articles he has written *The Stone Circles of the British Isles*, *Prehistoric Avebury* and *Megalithic Brittany: A Guide*. His most recent book, *The Stonehenge People*, was published in 1987.

Thaddeus M. Cowan teaches in the Department of Psychology at Kansas State University. He received his doctorate in experimental psychology from the University of Connecticut in 1965. His interests in archaeology extend over seventeen years. They include megalithic design analysis and the astronomical significance of the effigy mounds of North America.

G. Ronald Curtis is a chartered civil engineer with the North of Scotland Hydro-Electric Board in Edinburgh. For over fifteen years, both privately and on behalf of the Scottish Development Department, he has organised and undertaken field surveys, archaeological excavations and restoration work on historic masonry bridges and old roads in the Highlands. He has also made precise surveys of many megalithic sites in the Outer Hebrides.

Alan Davis teaches Physics and Mathematics at Lancaster Royal Grammar School. He graduated in Physics at Sheffield University in 1968, and received an M.Sc. in Radio Astronomy at Manchester University in 1970. Since 1980 he has worked, in his spare time, on various aspects of the Thoms' metrological hypotheses. He is currently attempting a reassessment of their work on the stone rows in Britain and Brittany.

Lesley Feguson graduated from Edinburgh University and is a member of the curatorial staff of the Archaeology Section of the National Monuments Record of Scotland, Royal Commission on the Ancient and Historical Monuments of Scotland.

David Fraser, after studying geography at Aberdeen University, was awarded his Ph.D from Glasgow University for a study of the Neolithic monuments of Orkney and their surrounding landscape. He is now an Inspector of Ancient Monuments with the Historic Buildings and Monuments Commission for England.

Peter Freeman is Professor of Statistics at the University of Leicester and has a long-standing interest in the application of statistical ideas to archaeology in general and the data of Professor Thom in particular.

Pierre-Roland Giot is Professor of Archaeology at the University of Rennes. A leading authority on the archaeology of Brittany for over forty years, he has worked on all periods from the Palaeolithic to the Medieval. His special interest is in the use of geological methods in archaeological science.

Chris Jennings is a freelance artist who lives and works in Oxford. He studied fine art at Hornsey College of Art, London. His work - photographs, prints, drawings and sculpture - has been exhibited widely in this country and abroad. He has recently completed a major sculpture commission for the main courtyard at Southampton General Hospital.

Ed Krupp is an astronomer and the Director of the Griffith Observatory in Los Angeles. He is editor/co-author and author of several books on ancient and prehistoric astronomy and also writes astronomy books for children.

Euan MacKie is Senior Curator in Archaeology and Anthropology at the Hunterian Museum, Glasgow.

Hans Motz is an Emeritus Professor of Engineering at Oxford University and an Emeritus Fellow of St. John's College and of St. Catherine's College, Oxford. He was born in Vienna, Austria, in 1909 and is also an Honorary Professor at the Technical University of Vienna.

Leslie Myatt is Head of the Engineering and Building Department at Thurso Technical College. A graduate of London University and a Chartered Electrical Engineer, he has lived in Caithness for the past twenty-two years. He has been inspired by the research of Professor Thom to continue the work on the stone settings in the north of Scotland.

Ray Norris is a Senior Research Scientist at CSIRO Division of Radiophysics, Australia, where he is studying the astrophysics of active galaxies, and is involved in the design and software of the Australia telescope. He was previously at Jodrell Bank, University of Manchester, where he spent much of his spare time surveying megalithic sites.

Jon Patrick is Senior Lecturer in Computing at Deakin University, Victoria, Australia. As a qualified surveyor, he has produced much valuable data on Irish megalithic sites and made the first survey of the astronomical alignment of the great passage grave at Newgrange. His Ph.D. thesis applied the concepts of minimum message length to evaluate Professor Thom's hypotheses about the shapes of stone rings.

Margaret Ponting lives at Callanish, and has acquired a detailed knowledge of the archaeological sites in the area. She has taken part in and directed excavations at some of the local stone rings.

Graham Ritchie has worked since 1965 surveying in Argyll as an archaeological investigator with the Royal Commission on the Ancient and Historical Monuments of Scotland, Edinburgh. He has undertaken excavations on stone circles including Stenness in Orkney and Balbirnie in Fife.

Clive Ruggles has recently been appointed editor of *Archaeoastronomy*, the supplement to *Journal for the History of Astronomy*. He has worked for several years reassessing the ideas of Professor Thom, and has himself surveyed some three hundred Scottish megalithic sites. More recently he has undertaken archaeoastronomical fieldwork at Teotihuacan in Mexico and Nazca in Peru. He is currently Lecturer in Computing Studies at the University of Leicester.

Archibald S. Thom is Honorary Senior Research Fellow in the Department of Aeronautics and Fluid Mechanics in the University of Glasgow, having retired in 1979. A Chartered Civil and Mechanical Engineer, he has specialised in the measurement of the hydraulic efficiency of hydro-electric turbines and pumps. He often helped his father, Alexander Thom, with fieldwork and, in later years, he co-authored his father's books and papers on archaeoastronomy.

Part 1

ALEXANDER THOM'S
LIFE AND WORK

1

A personal note about my late father, Alexander Thom

ARCHIE THOM

When my father was born, Queen Victoria was on the throne; the law of the land required a mechanically propelled vehicle to have a man walking in front with a red flag; powered flight was still a dream; the Yukon Trail was known to only a few men; and later feats such as the moon walk and sending probes to visit Halley's comet were mere fantasy.

My father was born at The Mains Farm, Carradale, on 26 March, 1894. His father was a dairy farmer; his mother was the daughter of a Glasgow muslin manufacturer. The family left Carradale when my father was seven years old, and about that time his brother was born. He remembered Carradale for the rest of his life, having absorbed much of the atmosphere of that active fishing village. He always had a soft place in his heart for his native county, and especially for the peninsula of Kintyre. In later years he really enjoyed his frequent survey trips to the many prehistoric sites in Argyllshire.

His boyhood was spent at Dunlop, on the family farm acquired in 1901. With access to all of his father's tools (his father ought to have been an engineer) the growing boy had a marvellous time, building his own playthings - an electrically driven pendulum clock, model aircraft, model boats, canoes, kites, bows and arrows - even a glider. On two pram wheels and pulled by a rope against the wind, I believe that it lifted him two feet off the ground. He attended Dunlop School where he acquired an excellent basic education. One amusing highlight occurred on the dark night when he flew a box kite carrying a storm lantern over Dunlop village. About 1907 or 1908 he was sent to Kilmarnock Academy, some eight miles distant. Upon leaving that school he took a compressed course at Skerries College and sat the Preliminary Examinations for entrance to University. He attended engineering classes at the Technical College for three sessions, gaining the qualification of Associateship of the Royal Technical College in 1914. He was awarded the degree of B.Sc. in engineering in 1915 after a further year attending courses at the University of Glasgow. During that year he also studied astronomy under Professor L. Becker, who was Head of the Department of Astronomy.

By this time his many interests emerged. He extended the use of a waterwheel, which his father had built for threshing oats, by making it drive a dynamo for charging a battery to light his parents' house. This, I am told, was the first house to be lit by electricity in the Parish of Dunlop. In this period he built a powerful windmill to help to boost the house-lighting battery.

He became the assistant Scout Master in the local troop. He used a camera successfully. He was obviously very interested in things astronomical and was reading voraciously. One of his uncles presented him with a three-inch refractor telescope, and a special removable section of the roof in the attic of the house was built. From that relatively sheltered position he did some observation work which was published. One paper which many years later he was to remember reading was that report by B. Somerville on Callanish (Somerville 1912).

In 1917 my father married Jeanie Boyd Kirkwood. I was born in 1918. With the help of my mother, during the 1922 summer vacation from Glasgow University, he built, on the farm at Dunlop, a small cottage called Thalassa, from which the sea can be seen. My late brother Alan and my sister Beryl were born there in 1923 and 1926 respectively. Deep ponds usually take two nights' frost to freeze over for safe skating, and A.T. built a flat concrete area which could be flooded easily and which gave us a fine skating area after even a light frost. In summer the area served as a model sailing boat pond. Many flat-bottomed models were made and raced.

We were on holiday on the Island of Arran once, about 1930 or 1931, and it was there that I recall first hearing my father talking about a standing stone. Until then the only standing stones I knew about were the six upright 'rubbing stones' in the fields on the farm here in Dunlop. I can remember him theorising at the nine ft-high stone in South Glen Sannox, Arran, explaining how it might be possible that prehistoric people watched the sun setting on the high mountain ridge to the south west, viewing it from the big stone. He was thinking then of what he was later to call a 'backsight'. He surveyed Sannox years later and observed that from the stone, at Martinmas and Candlemas, the sun set behind the mountain ridge and reappeared momentarily an hour later in the col.

In the early 1920s he took up colour photography, and became skilful in producing the large positive glass slides for display. In 1938 I remember him struggling for two days to attempt to produce a coloured print from a coloured positive, using a trichrome printing method. This was years before the arrival of the present type of colour printing.

In the area of Dunlop, fields are separated by stone dykes, fences or hedges. Itchy cattle love to rub themselves against solid objects and it was customary to have an upright stone, about five ft high, in each field, so that the boundary dykes, fences, hedges and so on would not be subjected to too much wear and tear from livestock. These are the 'rubbing stones' referred to above. Needless to say, operators of modern agricultural machinery do not like any obstruction in their path. Once, about 1945, my father made a tenant re-erect a six ft-high rubbing stone which he, the tenant, had felled and then buried below ploughshare depth, because it was too heavy to drag to the edge of the field. (Thirty years later, at the Hall of Clestran on Orkney, we were to find an eight ft-long menhir recently removed from the middle of the field and lying away against the stone dyke. 'Remove not the ancient landmarks' (Proverbs, 23:28).)

My second memory of my father's interest in standing stones recalls a sailing cruise in 1933. Whenever possible, he arranged a summer cruise in the Hebrides. His sailing experience spanned the years from 1909 to 1974, when he became too frail for the sport.

During his lifetime he skippered yachts of lengths ranging from 25 to 66 ft. His knowledge of the waters and islands of the west coast of Scotland was vast. He often thought seriously about owning and maintaining a yacht of his own, but he never owned a sailing vessel after 1921, when he had sailed an open boat on The Solent. He had found out as the years passed that two to four weeks of hard sailing each season did him for the rest of the year. I know because I sailed with him regularly for half a century.

On this particular cruise in 1933 we had been having a long hard sail from the Sound of Harris northward, in the open North Atlantic, not a peaceful ocean, and in the evening had entered the sheltered waters of East Loch Roag, Isle of Lewis. A quiet anchorage, where the ship's company can have a good night's sleep, is always sought for while cruising, and this time the Skipper navigated his way carefully as far in from the open sea as possible, chose a little bay and dropped anchor. As we stowed sail, I well remember looking up and seeing the full moon rising over the low land and there, silhouetted against the orb, were the Stones of Callanish. We were within a biscuit's toss of The Stonehenge of Scotland. I do not know to this day whether the Skipper had come here on purpose, but the six of us all went ashore in the moonlight immediately after dinner and explored the site. The date would be about 9 August 1933, with sunset about 10.07 p.m.

While exploring the site in the moonlight, the thought was running through his mind that the Megalithic Builders had lived here on Lewis as well as on the mainland of Britain, and his respect for them rose considerably. 'The Boys' as he often called them later, had had to cross The Minch too, like ourselves. He saw that one of the rows of stones pointed to the Pole of the heavens, and he remembered Somerville's report.

He has written that this was the time when he decided to collect more information about as many megalithic sites as possible. I cannot say exactly when he began surveying: my own first memory of site surveying with him was in 1938. As the years passed he made many forays into inaccessible areas, usually with a companion, and I suppose that I myself have visited more sites with him than any one other person. I was living nearby, and it was natural that I should help. My two children also became involved as they grew up. When my son Alasdair and daughter Susan became old enough they occasionally accompanied their grandfather. Once Susan was with him at Kintraw on the day that he decided to cross the deep valley to the northeast and investigate the ground. She had just sat down on the stone at 'the platform' and her grandfather was setting up the theodolite. He asked if there were any stones about and she evidently said, 'What about this one here?' It was called The Susan Stone thereafter. When she had been younger she said once 'What does grandfather do when he stands on the stones?'

As the time passed he found it necessary to go back frequently to sites. Perhaps he wanted a sun shot, a revision of the horizon profile, or other additional site particulars. My mother, who perforce had to adjust to living with her husband, once likened him to the mole-catcher - reputed always to leave a pair to breed so that he would always be asked back. When possible, he liked to take an azimuth on two separate days.

Over the decades it came about that as he expounded his theories and hypotheses I acted as a sounding board for his ideas.

He began to realise that he was dealing with the work of men who had an advanced knowledge of geometry. When he understood that these people were not only intelligent but scientific in their approach, he was able to get ahead. Metaphorically speaking, he put himself in their position and 'wore their moccasins'. His lack of knowledge of archaeology was in some ways an advantage because, as he said, he had no prejudices to overcome.

After I retired in 1979, I could of course spend much more time with him. Not much more fresh survey work was done however.

He knew that Halley's Comet had been seen through powerful telescopes in 1985; he could recall having seen it in 1910. He maintained, however, that the comet which passed by in 1911 was much more spectacular.

Some photographs of a comet, taken in December 1916 at Winchester, Mass., attracted his attention, and he began to calculate the orbit. I have a file of his figures with a letter dated 8 August 1919, addressed to Professor Becker of the Department of Astronomy at The University of Glasgow. The young enthusiast almost certainly completed the calculations between April and August 1919, in the interval between two jobs. He always found something to work at. Professor Becker, who had taught him Astronomy before the Kaiser's war (my father's terminology) and had been interned in the U.K. during hostilities, returned to Germany. Quite a friendship must have existed between them, because in 1936 Professor Becker wrote him a testimonial.

During Hitler's war (again my father's terminology), while living at Fleet, Hants, he had little time to do field work, but by that time he had quite an amount of standing stone data to mull over. He must have relaxed on occasion from the demanding work at the Royal Aircraft Establishment, 'The Factory' as it was called, by poring over one-inch maps and making marginal notes of the S.S. (standing stones) positions almost always shown.

At the R.A.E. he was initially involved in work on the small low-speed wind tunnels but after a year he was appointed to be the man in charge of the High Speed, pressurised Wind Tunnel. The task before him and his team was to get the tunnel working as soon as possible, with fans, force balances and so on, and then put it to use. He gained a considerable reputation and was advanced from Senior Scientific Officer to Principal Scientific Officer (covering, in his own words, a very wide ground). To quote Professor Austin Mair:

> A.T.'s energy, enthusiasm and drive led to the completion of the High Speed Tunnel in 1942 and later in that year the first model aircraft (a Spitfire) was tested in the tunnel. From that time until after the end of the war the tunnel was in continuous use, normally seven days a week, and there were no breakdowns. The intensive use of the H.S.T. from 1942 to 1945 led to greatly improved understanding of the problems of high speed flight. Among the most valuable achievements were understanding the importance of thin wings and the causes of longitudinal trim changes at high Mach numbers.

Models of the De Haviland Vampire and the Gloster Meteor were tested in the High Speed Tunnel.

In the early summer of 1945 he was one of a team of British Engineers sent to Germany to find out how far advanced the Germans were in their technology.

In May 1944 he sucessfully applied for the Chair of Engineering Science at the University of Oxford, taking up his duties there more than a year later in the autumn of 1945. Under his guidance, both the teaching of undergraduates and postgraduate research work flourished. Soon the Engineering Laboratory was too small for the ever-increasing numbers of students. As the years passed, its deficiencies became more and more obvious. Architects were called in, and action taken. The laboratory extention was finally named, in his honour, The Thom Building.

On retiring, he and my mother returned to The Hill, to live in Thalassa, the cottage they had built in 1922 'to last for five years'. It was modernised to suit their requirements, and an excellent workshop was fitted out. He was soon contentedly engaged in making himself a twelve-inch Cassegrainian telescope, both mirrors for which he carefully ground and polished himself. At that time I am sure that he was Scotland's leading amateur mirror grinder. My mother was greatly dismayed with the jeweller's rouge transferred from the seat of his pants to the cushions of the house furniture in Thalassa.

He built an observatory for the telescope, complete with clock drive and adjustable seat. I have a photograph of part of the moon's limb taken with this telescope, but his eyesight was failing and no more work was done with it. About this time he built a 4.25 inch refractor telescope, the object glass of which was a lens from a submarine periscope - a German one, I think.

From 1946 onwards he devoted much of his spare time and energy to what was later called archaeoastronomy. Field work was done mostly during the Easter and Summer vacations from Oxford. Winter was used for calculation work. The yacht cruises were continued. On board would be his smallest theodolite, specially reduced in weight by himself in the workshop. He carefully machined excess brass from it, gleefully saving and weighing the resulting swarf.

Many islands were visited from the yachts, islands which are often almost inaccessible by public transport. A tent was carried on many Easter and summer expeditions; I remember awakening one March morning to find two inches of snow on the tent. Camping allowed us on occasion to visit up to a dozen sites in a day. Azimuths were obtained from sun observations whenever possible, and a radio was always carried to monitor the rate of the watch, used for timing each shot of the sun.

He began to write up and seek to publish his findings in the early 1950s. Astronomers (Thom 1954) and statisticians (Thom 1955) were the only people interested, his work being looked upon with contumely by the archaeological establishment. A glance at the list of his publications in this volume will show how slowly his work came to be recognised after 1954.

At one time he sent a paper to the U.S.A. which was returned to him after some six months, not accepted. This was too long for a senior citizen to wait and he proceeded to write a book. The rest is history. He did, at least, live long enough to enjoy the general acceptance of his efforts. Is it perhaps indicative of established thought that the last paper which he wrote (Thom 1984) was published by The Prehistoric Society? That article mentioned neither archaeoastronomy, calendars, metrology nor statistical analysis, but gave the engineer's thoughts on how men could have moved huge menhirs.

At one stage, during the Christmas and Easter vacation as well as in the summer, he made dozens of horizontal observations from Thalassa to enable him with reasonable certainty to know the effect of horizontal refraction. From The Hill the sun sets into the sea for about 30 days in winter, but he never had the good luck to observe the sunset on the sea horizon and measure its altitude in clear atmosphere.

After retiring, he spent much more time on archaeoastronomy and began to write about it more and more. He coined specialised terms such as major and minor lunar standstill, graze effect, lunar band, megalithic inch, megalithic yard and megalithic rod, as well as backsight and foresight.

His work extended from the Shetland Archipelago to Cornwall, from the Hebrides to Wales, Stonehenge, Avebury and Carnac. He did practically no field work in Ireland, making only one trip as far as County Tyrone.

To facilitate work in his study he built himself an index using notched cards with holes which allowed him to extract circles, flattened rings, eggs, ellipses and so on with ease, using a knitting needle.

Informed at a public meeting by Glyn Daniel in 1969 that he, Glyn Daniel, was in the process of inviting A.T. to Brittany in 1970 to look over the Carnac Alignments, my father spent the rest of the winter planning and arranging for a team of helpers. In all, five survey expeditions were made, one in July 1970 and the remaining four in the spring of each year from 1971 to 1974 inclusive. Much field work was necessary; who better to lead us than this old man who kept us regaled with yarns of his experiences in 1913 when he had spent the summer working as a chainman on Canadian Pacific Railway construction? He trained his teams well, and often, in the quiet Breton countryside, the humorists were heard to say (quoting the Skipper) 'Come on, "This is how it was done on the C.P.R."'

The plotting of these surveys alone took a long time; Kermario, for instance, ended up on an eight ft-long roll of paper. It alone took several months to plot. Main survey points or hubs were always carefully referenced in, each year, for future use, and by the end he knew the relative co-ordinates of, say, two menhirs two miles apart, to within ± 4 ft or better.

I put it on record that my father had a very good aural memory. He could recite the whole of Tam O'Shanter, the epic poem by Robert Burns, and The Pied Piper of Hamelin, but he seemed to have difficulty in learning school French. About 1929 he learned German, using the Pelman Method. Later, in the 1970s and 1980s he read many German novels for relaxation. 'It makes the story longer'. In spite of having spent 27 years of his life working in England, south of the Border, he still maintained the use of his native Scottish tongue, both the dialect and the accent. Towards the end of his tenure of the Oxford chair my mother once said 'It has been a long war this time!'

He was always seeking new horizons. During the early 1920s, having seen the need for power-assisted control of flying boat rudders and the like, he had invented a pneumatic servo-mechanism. In 1939 this device went to the R.A.E. with him for the experts to examine. There the backroom team he led helped to add 20 m.p.h. to the top speed of British fighter aircraft. Later, he designed and had made for him a beautiful pantograph which he needed for reducing or enlarging his survey plans. He designed a rain gauge suitable for use on board ship and on sloping mountain rain gauge sites (wind tunnel-tested of course). He performed experiments on conical models of hailstones in his 24 in x 24 in tunnel in Glasgow University. I would like to draw attention here to the work he did at Glasgow University between 1925 and 1934 on investigating the side forces brought into play by the flow of fluid past rotating cylinders. This work was done in his wind tunnel, and in his water and oil channel. I was never told, but Barnes Wallace must have referred back to all of this pure academic research work while inventing the spinning bomb, to be used so successfully in 'busting' the Möhne Dam in Germany in the 1939-45 war.

In the 1918-24 period he used two motor cycles. An epic four-day journey from Sussex to Dunlop in 1919 with my mother and me was, to them, a memorable occasion. Later about 1931 he took up pedal cycling seriously and he and my mother must have cycled several thousand miles on their tandem before the second war came. He cycled daily to the R.A.E. from Fleet for the six years of the war, then daily to the Oxford University Engineering Laboratory for another 16 years. By this time he had learned to look after himself; he had the habit of taking a cold plunge early each morning and saw to it that he took enough exercise.

In 1916 one of his uncles gave him an Albion motor car, which was requisitioned and sold to The Norman Thomson Flight Company (his employers) for conversion into a works truck. On this car, to the amazement of another uncle, he found out by himself how to 'double de-clutch' while shifting gear. In 1938-39 he owned a 4.25 litre Bentley touring car and used it on survey trips, but he did not drive regularly until after the war.

A.T. loved intricate woodwork, carving, lathe turning, screw-cutting and so on, and he became expert in moulding small aluminium castings. Pen-and-ink sketching was easy for him, and some of his water-colour paintings show his talent. He could sing reasonably well. I can remember him once or twice playing an old fiddle, but he was no musician. In the early 1920s when 'wireless' came in, he built his own receiver sets. At one stage he proudly claimed to be receiving broadcasts from the eastern U.S. This involved sitting up late at night. He once heard a programme, noted the broadcaster and the time, wrote and received a certificate that he had heard 'The Voice from Way Down East'. He built only receiving sets. I well remember the two very high aerial masts needed in those days for reception. The two-volt lead acid battery could always be charged by the water wheel-driven dynamo.

The theory of probability has to be taught to surveying students and when at Glasgow University A.T. invented a device to demonstrate a Gaussian error curve. A long horizontal row of about 200 one mm-diameter holes was drilled in a vertically-positioned brass sheet about seven inches high. Each year each student in the class was given ten balls; he then had to stand back a certain distance, decide where he estimated the centre to be, and then step forward and insert each ball in turn in the estimated hole. That was a good teaching device, as the theory could be applied thereafter, with the distribution curve as a basis for discussion.

His open-air activities included hill walking, a little ski-ing on 1926 vintage Norwegian skis, outdoor and indoor badminton, outdoor skating, gardening, greenhouse tending, roses, shrubs, maintenance of The Hill farm (90 acres), and some fairly arduous rock climbing on the soft granite of the ridges on the mountains of Arran. He possessed an ice axe, used in The Austrian Tyrol in 1929-30.

To quote my father, he tried to do unimportant things gracefully, and I think he succeeded. He certainly grew old gracefully, coping first with double cataract, then with a broken thigh, and later with retinal decay. At the end he was registered as a blind person, and he made great use of the talking book service.

Preferring to find out a thing for himself, rather than be told how to do it, he loved to explore and to extend the frontiers of the continents of knowledge.

Like his uncle, Sam Fulton Strang, my father had the type of personality which allowed him to get people to work for him, and with him. Leading an expedition of 12 or 14 volunteer helpers to make accurate plans of the thousands of standing stones at Carnac in Brittany was no mean task. Nor was it easy to skipper a crew of six or sometimes ten on sailing cruises. Once, to quote one of his crew, a member of his staff in Oxford, his approach to a situation was referred to as 'a superb piece of man-management'. He was always able easily to admit co-existence with people of all types.*

He was a first class seaman, and never in all the time I knew him did he have a misadventure at sea. He always reduced sail in time and took great care of his ships and crews. A highlight in his sailing experience was his seven-day voyage in a 12 m yacht under sail from Brixham to the Clyde.

In the late 1920s my father began to work on the arithmetical solution of problems in steady two-dimensional flow. His publications indicate that he was an expert in this branch of applied mathematics. He used Brunsvega calculators for the long, slow, time-consuming arithmetic needed. Later he used Curta pocket calculators, before the advent of the modern electronic machines. He had seen the changes in calculating machines: ten-inch slide rules, cylindrical drum-type devices, digital calculators, the Curta machine and finally the electronic machines. There was a set of Napier's Bones in his study, but I never saw him use them.

In the mid 1930s he became interested in studying the zodiacal light and spent much time in library work noting the dates of its having been seen and recorded throughout the last few centuries. His final decision was that the intensity of zodiacal light was in phase with the sunspot cycle.

At home in the early 1930s he constructed yet another device to generate electricity, a small jet-type turbine. This worked successfully and intermittently for ten or twelve years. The Hill is on a hill, and there are not more than 30 acres of catchment area available at the actual farmstead itself: the head was 17 ft.

---

*   By 'admit co-existence' I do not mean 'co-exist'. We all co-exist, but there are some people so insensitive to others that they can walk past acquaintances in the village street without acknowledging their presence. Such people are so puffed up by their self-importance that it is not considered necessary to *let on* that they co-exist with others. My father wasn't like that: he had the ability to enjoy a fine morning, for example, along with someone - the postman, say, or the scout in Brasenose, or the labourer making hay. He was willing and pleased to admit co-existence with anyone.

In the mid 1970s much time and effort was expended not only on the field work at Carnac, but also on the very necessary follow-up calculations. One minor triumph worth recording was the occasion when the distance between two trig. stations as calculated from our traverse work was found not to fit the distance calculated from the given grid co-ordinates. A.T. spent several weeks combing over all of the figures, but to no avail. The two stations were about 12000 ft apart, and he trusted the work of his helpers. He was not happy. It turned out in the end that the officially supplied co-ordinates on the Breton grid were not correct, some numbers having been transposed inadvertently, and he was greatly relieved. A small example like this shows the importance he put on being correct in detail.

Having lived a very full and interesting life, he slipped away peacefully on 7 November, 1985.

Vale, Pater.

2

A personal appreciation of Professor Alexander Thom

HANS MOTZ

Professor Thom came to Oxford in 1945 to take up the Chair of Engineering vacated some time earlier by Sir Richard Southwell. I met him at that time, when I was a departmental demonstrator who had come to Oxford in an unusual way. Having been interned as an Austrian National in 1941 and released in the same year, I was directed to Oxford by Dr C. P. Snow, the novelist who was then in charge of deployment of scientific manpower at the Ministry of Labour. The Reader in Electrical Engineering had left on war work and I was to take over the teaching of Electricity in the Engineering school. It was very small at the time; we had about 15 undergraduates per year. By the time Professor Thom retired in 1961 it was a big department and he had designed the ten-floor building named after him which was needed to house it.

Southwell had developed numerical methods for the solution of problems of Mechanics and Elasticity. They are known as Southwell's Relaxation method. The term is somewhat misleading, because it involved weeks and months of tedious work turning the handle of a mechanical calculator, an occupation which was not exactly relaxing. I know it because I became a member of the team, but I was lucky enough to acquire an assistant (Dr Laura Klanfer, an able and patient lady paid by the Admiralty) who helped me with my research on Radar. I had adapted Southwell's ideas to the solution of problems of electromagnetism, in particular microwave problems in Radar research.

Professor Thom, not entirely by chance, was the independent originator of another method for dealing with the same class of problems. In solving partial differential equations according to both methods, values of the function were calculated at the points of a discrete net. Partial differential equations approximately valid at these points were used. Starting from arbitrary values, the errors at every point were calculated and successively removed. In the case of the relaxation method this was done by intelligent and purposeful actions by the operator; in the case of Thom's method it was achieved by a process which could eventually be automated. This is why, with the advent of electronic digital computers, the Thom method was used and is still employed to this day. Professor Thom may therefore be regarded as a pioneer of modern computing methods. His contributions are by no means confined to this field. He did important work in fluid dynamics, and he designed one of the first large wind tunnels for testing high-speed aircraft.

When he came to the department, after an interregnum, I had regrettably already made my decision to leave, largely because I anticipated a certain resentment at Oxford (in the College) for having been planted there by Government intervention in war-time and because I did not want to block a position needed by young people returning from the war. I went to California, but not before making friends with Professor Thom and his family. I met Mrs Thom and Beryl, his daughter, and I was often a guest in their home. Mrs Thom, the kindliest of persons, had an unpretentious charm and directness with a heartfelt warmth; one simply had to love her.

I had some extremely fruitful years at Stanford. But somehow I felt catapulted into the twenty-first century and I was already too old to 'Americanise'. I must have spoken about this when I visited the Thoms at Oxford on the occasion of a half-year mission to Cambridge when I was on leave from Stanford. A year or so later, Professor Thom wrote to me to say that the Donald Pollock Readership was vacant, because E.B. Moullin, the holder, had accepted a chair at Cambridge, and he suggested that I might apply. It was Professor Moullin's teaching which I had taken over during my war-time stay at Oxford. I applied, and to my great joy was elected and returned to Oxford.

My work hardly overlapped with Professor Thom's, but there were points of contact between his fluid dynamics and my electricity. He took a great interest in the 1953 tidal surge which had devastated the East Coast of the country and he had constructed a model of the North Sea which correctly represented important features. He had a theory about the resonance of the wind driving the sea with tidal waves which resembled one which occurred in my work on electronic amplifiers.

At Oxford he built up a loyal and devoted staff; harmonious collaboration was essential in a school of non-specialised general engineering, and he made some excellent appointments. Compared with present standards he was rather cautious with money: his rule was one of enlightened autocracy. His method of discreetly consulting people was, in my opinion, preferable to government by committee voting.

He became the great old man of Brasenose College. He computed corrections to a sun-dial to be applied during the different months of the year. They can be found on a plaque on the opposite side of the quadrangle, recessed in a porch, and serve as a permanent memorial. College servants as well as laboratory staff who go back to his time still enquire after him.

I now come to my association with Professor Thom's archeological fieldwork. I went with him, his son Archibald Stevenson Thom (Archie) and the other camp followers to Stonehenge, Avebury, the Orkneys and Carnac. The ventures left me with some of the best memories of my life and I only regret that I did not come with him more often. I suppose the trouble was that he was so unassuming when he vaguely suggested that I might come that I had hardly noticed the hint.

The parties consisted of the professor, Archie, Archie's wife and their son Alasdair Strang, and Beryl Austin, the Professor's daughter. I remember Robert L. Merritt, an American benefactor, his son Ethan and his daughter Elizabeth at Stonehenge, and a young Frenchman Jean-Luc Quinio at Carnac; there may have been others.

Professor Thom was a Grand Master of the theodolite. Whoever has seen engineers use it to survey street corners would probably not suspect that, in his hands, it could become a delicate precision instrument, capable of measuring seconds of arc. In the department he laid great stress on the art of surveying, always in danger of being supplanted by some new-born engineering subjects. He believed that working carefully with a theodolite is the best introduction to experimental science. I am afraid that on these outings he never gave me a chance of acquiring the gentle art. It is true that I could never have emulated him successfully: I had to be content with taking turn with others in taking down the readings in a carefully-laid-out book.

Surveying was the alpha and omega of his observational art. Theorising came later. On a site, whatever its nature, and even on territory separating points of importance, exact measurement of positions and distances was the order of the day. Exacting measurements were needed to determine the right ascension or declination of stars, the sun or the moon; perhaps the most delicate measurements I remember pertained to events of the lunar cycle as evidenced by the setting or rising moon at spots marked by special features of the landscape at the horizon.

Professor Thom was a tall man with a pleasant somewhat gaunt face, a small moustache and a kindly expression. He kept his manly bearing into old age and there was something about him that commanded respect wherever he went. He had a natural aristocratic gentility; he was a Laird, perhaps, by nature. When we put up somewhere, the landlord would know that he needed to produce the best food or wine. Yet Professor Thom did not give himself any airs, not in the least; he never sent a bottle back; one just knew. His natural authority extended, of course, to his family and it was a pleasure to be with them, to watch them love, tease, spoil and respect him.

I found the surveying at Avebury particularly impressive. It is hard to describe the beauty and charm of the site and I can strongly recommend a pilgrimage. It had been surveyed a long time before by Alexander Keiller, the marmalade king, and his helpers. His researches are recorded in a book written by Dr I.F. Smith, who had been Keiller's secretary, and published by Oxford University Press (Smith 1965). On the occasion when I was there we were entertained to tea by Dr Smith, by then an old lady, who had settled down in a cottage near the museum. Her book is a most impressive document. Unfortunately the Press kept no copies, so that it is now available in libraries only. In parentheses, I wish to state my disappointment with the practice of publishers to dispose of their stock of important records, thus not showing a responsibility transcending their mere commercial interests.

There were many uncertainties left by the earlier survey which Professor Thom wished to clear up by his own. The Ministry of Works had constructed a little museum in the village and the permanent custodians were very helpful to the professor and his team when we went there. They put a room, which is part of their office, at our disposal for making drawings and carrying out calculations on the spot. Some ten years later I visited Avebury and was recognized by a custodian who remembered Professor Thom and spoke highly of him.

I was a modest acolyte helping with taking down the Professor's readings of the theodolite dials. The site is not altogether easy to survey. Several traverses had to be made, and we went back and forth and round and round until we came back to the initial point of the survey. When the work has been done correctly, the final point must lie near the initial point. On that occasion Thom was out by no more than six inches which is extraordinarily good.

The visit to the Brogar Ring in the Orkneys in 1974 started at the professor's home at 'The Hill', a seventeenth or eighteenth century steading near Dunlop in Ayrshire. At that time Archie and his family lived in a large house on the farm and Professor Thom, with his wife, in a wooden building which they had built with their own hands. In the courtyard there was a bathtub filled from a continuously-running private water supply, in which Thom had a dip every morning: this probably explained his long-lasting state of health. In the bungalow, there was the professor's study, equipped with drawing table, pantograph, his files and manuscripts. He had a home-made telescope, with a mirror which he had ground himself in countless hours of patient work. With this he watched the stars and when he found that a tree cut out a portion of the sky he resolutely cut it down.

The journey to the Orkneys was undertaken in Archie's Range Rover. If a person has never ridden this marvel of engineering he or she does not know what comfort of travel can be. I could have written this contribution by hand, no less legible than usual, during such a journey. The beauty of the mountain ranges, the heather and the castles was enhanced by the explanations and anecdotes of the professor. We traversed the whole of Scotland right up to Sutherland and Caithness, to Thurso and finally Scrabster where we crossed on the car ferry to the mainland of Orkney. The Standing Stones Hotel, near the Brogar Ring, offered every comfort one might desire.

I believe that the fierce wind which we experienced is a constant feature of the climate in the Orkneys. Here the wisdom of the aerodynamicist Thom came in useful. He made us lie down flat in the hollows of the ground, in the quiet boundary layer where one could get respite from the elements.

I can not go into the details of the work (*see* Thom & Thom 1973) but I do want to give an account of a most impressive episode. The Brogar Ring is surrounded by foresights, both in the shape of cairns and in features of the not-too-distant hills such as the Kame of Corrigall, a steep part of which has the same slope as that of the moon at maximum declination at a major standstill. This had been investigated by the team on previous occasions, but this time many things needed to be checked. In particular the azimuthal angle under which the feature of the Kame of Corrigall appears, when viewed from the circle, had to be measured with great accuracy to identify it with the azimuth expected at a major standstill during a night observation by neolithic man.

The skies were too cloudy for optical observation in daytime and the following strategy was employed. The Range Rover was driven up as high as possible near the Kame. The headlights were then screwed off and carried, together with the car battery, up to a cairn which stands at the site. The Thoms had not examined the site before, but inspection revealed a platform to which large stones had been brought, presumably from a location nearby which could be clearly identified as a quarry for large stones. If this platform could be identified as being of megalithic origin, this would strongly support Thom's hypothesis, that the Brogar Ring had amongst other functions that of a lunar observatory. We were equipped with walkie-talkies and it had been agreed that a car light should be lit in the darkness of the night, when the professor had set the hair-line of the theodolite precisely at the azimuthal setting predicted by the astronomical theory. I was with him when the signal was given and the light spot appeared, exactly on the hair-line. There is, in validating science, always a difference between accurate prediction and comparison with existing data. Witnessing confirmation of a prediction is indeed very impressive.

I shall always regret that I did not take part in any of Professor Thom's expeditions to Carnac prior to 1973. When I joined him there in that year he was already well known in Carnac by some city notables. He had booked into a very good hotel where the owner was also the chef. His choice of wines was superb (perhaps this is not too difficult in France) and the quality of the sea food, caught locally, was excellent.

At lunchtime the fare was much simpler. We were fed from the back of the Range Rover with beverages prepared by Beryl and also by the professor himself. To the casual onlooker we must have looked more like a bunch of Stonehenge hippies than a party of respectable archaeologists.

According to Thom's theory, the sites at Le Ménec and Kermario served as gigantic computers to carry out interpolations between lunar measurements. The centrepiece of the theory is Er Grah, sometimes known as Le Grand Menhir Brisé, which now lies broken in four pieces near Locmariquer. Its length must have been at least 67 ft and from its cubic content its weight may be estimated to be over 340 tons. The great mystery is how the megalithic engineers managed to transport and erect it. It might have been observed from various places: Le Moustoir, Kerran, Trevas, Petit Mont, or from Quiberon. The names of places like Kerlescan, Kervilor, and Kermario still ring in my ears.

There were problems connected with the line of sight from these places to Er Grah. Professor Thom was ready to assume that runners from observation posts might have carried information to the man busy at the 'computer' because they did not yet have walkie-talkies, despite their advanced skills in engineering. But it was of great interest whether territory, now heavily wooded but which was almost certainly bare in their days, allowed direct sighting. I remember most clambering through the woods, traversing when the terrain was almost impassable, and measuring changes of elevation as we made our arduous way. The professor would issue his usual 'Let's get on with the job' when we showed signs of tiring.

After a good dinner we would reassemble again to work out the results. Formulae of spherical trigonometry had to be evaluated in order to relate the repeated sun observations to our measurement net. A certain water tower from which one could observe Er Grah figured prominently as a reference point.

For his computations the professor used a gadget (a Curta calculator) looking like a cross between a Tibetan Prayer Wheel and a Turkish Coffee Grinder with a rotating handle. I had brought along a Texas Instruments programmable calculator, but he did not altogether trust it and checked everything.

Professor Thom was regaling us all the time with anecdotes, stories and jokes, sometimes quite bawdy, but none the worse for that. All this was with his inimitable Scots accent, in a matter-of-fact dry manner. He was jolly good company.

3

The career and publications of Alexander Thom

Compiled by ARCHIE THOM

Notes

In this list, A.S. Thom and Alexander Strang Thom designate respectively Alexander Thom's son Archibald Stevenson Thom and his late grandson Alasdair Strang Thom.

Abbreviations for the titles of journals which are used in the general bibliography are also used here. The names of other journals are given in full.

Alexander Thom 1894-1985

1914	A.R.C.S.T. Associate of the Royal College of Science and Technology, formerly the Royal Technical College
1915	B.Sc., University of Glasgow
1926	Ph.D., University of Glasgow
1929	D.Sc., University of Glasgow
1930-1935	Carnegie Teaching Fellow
1945	M.A., Honorary degree, University of Oxford
1960	Hon. LL.D., University of Glasgow
1961	Emeritus Fellow, Brasenose College

Member of the British Astronomical Association
Fellow of the Royal Astronomical Society

List of Publications

1.   Thom, A. (1916). Variable stars; some features of light curves. *JBAA* **26**, 162-4.

2.   Thom, A. (1919). Determination of the scale error of a sextant. *Nautical Magazine* **102**, 148-50.

3.   Thom. A. (1920). The correction of aerofoil characteristics for scale effect. *Flight* **12**, 1042-3.

4.   Thom, A. (1921). Some points in flying boat design. *Proceedings of the Institution of Engineers and Shipbuilders in Scotland* **64**, 1-23.

5.   Thom, A. (1921). An empirical method of predicting the aerodynamics of an aerofoil. RMARC no. 837.

6.   Thom, A. (1925). Experiments on the air forces on rotating cylinders. *RMARC* no. 1018.

7.   Thom, A. (1925). The aerodynamics of the rotating cylinder. *Proceedings of the Institution of Engineers and Shipbuilders in Scotland* **68**, 1-27.

8.   Thom, A. & Small, J. (1925). Velocity of wind in conical ducts. *The Engineer* **40**, 3.

9.   Thom, A. (1926). Windchannel experiments. *Ph.D. Thesis, University of Glasgow.*

10.   Thom, A. (1926). The pressure round a cylinder rotating in an air current. *RMARC* no. 1082.

11.   Thom, A. (1928). The boundary layer of the front portion of a cylinder. *RMARC* no. 1176.

12.   Thom, A. (1928). An investigation of fluid flow in two dimensions. *RMARC* no. 1194.

13.   Thom, A. (1929). Experimental and theoretical investigation on fluid flow in two dimensions. *D.Sc. Thesis, University of Glasgow.*

14.   Thom, A. (1929). Some studies of the flow past cylinders. In A. Gilles, L. Hopf & Th. v. Karman (eds.), *Sonderdruck aus Vorträge aus dem Gebiete der Aerodynamik und verwandter Gebiete*, Berlin: Springer, 58-63.

15.   Thom, A. (1930). Eddies behind a circular cylinder. *RMARC* no. 1373.

16.   Thom, A. (1930). The pressure on the front generator of a cylinder. *RMARC* no. 1389.

17. Thom, A. & Orr, J. (1931). The solution of the torsion problem for circular shafts of varying radius. *PRS* **A131**, 30-7.

18. Thom, A. (1931). Experiments on the flow past a rotating cylinder. *RMARC* no. 1410.

19. Thom, A. (1931). Experiments on cylinders oscillating in a stream of water. *Philosophical Magazine* **12** (7th series), 490-503.

20. Thom, A. (1932). Arithmetical solution of problems in steady viscous flow. *RMARC* no. 1475.

21. Thom, A. (1932). Flow past circular cylinders at low speeds. *RMARC* no. 1539 (abstract only).

22. Thom, A. and Sengupta, S.R. (1932). Air torque on a cylinder rotating in an air stream. *RMARC* no. 1520.

23. Thom, A. (1933). The flow past circular cylinders at low speeds. *PRS* **A141**, 651-69.

24. Thom, A. (1933). Arithmetical solution of equations of the type $\nabla^4\psi = $ constant. *RMARC* no. 1604.

25. Thom, A. (1934). Effect of discs on the air forces on a rotating cylinder. *RMARC* no. 1623.

26. Thom, A. (1935). *Standard tables and formulae for setting out road spirals*. London: Pitman. 28 pp.

27. Thom, A. (1938). Correspondence on 'The flow of water in short channels' by C.F.J. Leslie. *Journal of the Institution of Civil Engineers* **9**, 427-8.

28. Thom, A. (1939). The Zodiacal Light. *JBAA* **49**, 103-13.

29. Thom, A. & Swart, P. (1940). The forces on an aerofoil at very low speeds. *Journal of the Royal Aeronautical Society* **44**, 761-70.

30. Thom, A. (1941). On suitability of Ben Nevis summit as a station for studying icing. *R.A.E. Aeronautical Departmental Note - High Speed Tunnel* no. 22.

31. Thom, A. (1941). Note on the arithmetical solution of the two-dimensional flow of a compressible fluid. *R.A.E. Aeronautical Departmental Note - High Speed Tunnel* no. 28.

32. Thom, A. (1941). Squares method of solution as applied to compressible flow. *R.A.E. Aeronautical Departmental Note - High Speed Tunnel* no. 30.

33. Thom, A. (1942). Tables for rapid arithmetical solution of Laplace's equation with certain boundaries. *R.A.E. Aeronautical Departmental Note - High Speed Tunnel* no. 32.

34. Thom, A. (1943). Blockage corrections in a closed high speed tunnel. *RMARC* no. 2033.

35. Thom, A. & Owen, P.R. (1943). Velocities and speeds through a shock wave. *RMARC* no. 2044.

36. Thom, A. (1945). Pressure ratio across a shock wave. *ARCR* no. 1191.

37. Thom, A. *et al*. (1945). Arithmetical solution of compressible flow past a symmetrical aerofoil. *R.A.E. Technical Note - Aeronautics* no. 1592.

38. Thom, A. (1945). The effect of sweepback at high speed. *R.A.E. Technical Note - Aeronautics* no. 1640.

39. Thom, A. & Douglas, G.P. (1945). High speed tunnels and other research in Germany. *ARCR* no. 8907 (this no. is not on the 41-page restricted report) & *Combined Intelligence Objectives Sub-Committee*, item no. 25.

40. Thom, A. (1945). Notes on tunnel blockage at high speeds. *R.A.E. Report - Aeronautics* no. 2020.

41. Thom, A. & Jones, M. (1945). Tunnel blockage near the choking condition. *RMARC* no. 2385. Published 1952 by HMSO, London.

42. Thom, A. & Klanfer, L. (1946). The method of influence factors in arithmetical solutions of certain field problems. *RMARC* no. 2440 (9854, 11,010). Published 1953 by HMSO, London.

43. Thom, A. (1947). Tunnel wall effect from mass flow considerations. *RMARC* no. 2442 (11,004). Published 1952 by HMSO, London.

44. Thom, A. (1947). Some arithmetical studies of the compressible flow past a body in a channel. *ARCR*, no. unknown (Fluid Motion Sub-Committee, no. 968a).

45. Thom, A. (1948). Note on C_L max. at high speed. *ARCR* no. 11,197.

46. Thom, A. & Perring, W.G.A. (1948). The design and work of the Farnborough High Speed Tunnel. *Journal of the Royal Aeronautical Society* **52**, 205-50.

47. Thom, A. (1949). Flow of a compressible fluid past a symmetrical aerofoil in a wind tunnel and in free air. *ARCR* no. 12,435 (Fluid Motion Sub-Committee, no. 1368).

48. Thom, A. (undated). A cambered aerofoil treated by squares method. *ARCR* no. 11, 652.

49. Thom, A. (1950). Shape of a slot for a given wall velocity. *ARCR* no. 12, 953.

50. Woods, L.C. & Thom, A. (1950). A new relaxational treatment of the compressible two-dimensional flow about an aerofoil with circulation. *RMARC* no. 2727 (13,034). Published 1953 by HMSO, London.

51. Thom, A. (1950). Treatment of the stagnation point in arithmetical methods. *RMARC* no. 2807. Published 1954 by HMSO, London.

52. Thom, A. & Klanfer, L. (1951). Tunnel wall effect on an aerofoil at subsonic speeds. *RMARC* no. 2851. Published 1957 by HMSO, London.

53. Thom, A. & Klanfer, L. (1952). Designing a slot for a given wall velocity. *ARCR* no. 13,604 (Current Paper no. 76).

54. Thom, A. (1953). The arithmetic of field equations. *ARCR* no. 15,419; and *Aeronautical Quarterly* **4**, 205-50.

55. Thom, A. (1954). Some refraction measurements at low altitudes. *Journal of the Institute of Navigation* **7**, 301-4.

56. Thom, A. (1954). The flow at the mouth of a Stantop pitot. *RMARC* no. 2984. Published 1956 by HMSO, London.

57. Thom, A. (1954). The solar observatories of Megalithic Man. *JBAA* **64**, 396-404.

58. Thom, A. (1955). A statistical examination of the megalithic sites in Britain. *JRSS* **A118**, 275-95.

59. Thom, A. *et al.* (1956). Efficiency of 3-speed bicycle gears. *Engineering* **182**, 78-9.

60. Thom, A. & Apelt, C.J. (1956). Convergence of numerical solutions of the Navier-Stokes equations. *RMARC* no. 3061. Published 1958 by HMSO, London.

61. Thom, A. (1957). Circum-meridian altitudes. *Empire Survey Review* **14**, 170-75.

62. Thom, A. & Apelt, C.J. (1958). Pressure in a two-dimensional static hole at low Reynolds' number. *ARCR* no. 18,798; and *RMARC* no. 3090.

63. Thom, A. (1958). Boundary troubles in arithmetical solutions of the Navier-Stokes equations. *ARCR* no. 20,289.

64. Thom, A. (1958). An empirical investigation of atmospheric refraction. *Empire Survey Review* **14**, 248-62.

65. Thom, A. (1960). A tidal model of the North Sea. *New Scientist* **8**, 1130-32.

66. Thom, A. (1961). The geometry of Megalithic Man. *Mathematical Gazette* **45**, 83-93.

67. Thom, A. (1961). Egg-shaped standing stone rings of Britain. *Archives Internationales d'Histoire des Sciences* **14**, 291-303.

68. Thom, A. & Apelt, C.J. (1961). *Field computations in engineering and physics*. London: Van Nostrand. 165 pp.

69. Thom, A. (1962). The forces on a cylinder in shear flow. *RMARC* no. 3343.

70. Thom, A. (1962). The megalithic unit of length. *JRSS* **A125**, 243-51.

71. Thom, A. (1964). Megalithic geometry in standing stones. *New Scientist* **21**, 690-1.

72. Thom, A. (1964). Observatories in ancient Britain. *New Scientist* **23**, 17-19.

73. Thom, A. (1964). The larger units of length of Megalithic Man. *JRSS* **A127**, 527-33.

74. Thom, A. (1965). Testing a hyperboloidal mirror for a Cassegrainian telescope. *JBAA* **75**, 322-27.

75. Thom, A. (1966). Megalithic astronomy: Indications in standing stones. *VA* **7**, 1-57.

76. Thom, A. (1966). The lunar observatories of Megalithic Man. *Nature* **212**, 1527-8.

77. Thom, A. (1966). Time-keeping with standing stones. *New Scientist*, **32**, 719-21.

78. Thom, A. (1966). Megaliths and mathematics. *Ant.* **40**, 121-8.

79. Thom, A. (1967). Contribution to 'Hoyle on Stonehenge: some comments'. *Ant.* **41**, 95-6.

80. Thom, A. (1967). *Megalithic sites in Britain*. Oxford: Oxford University Press. 174 pp.

81. Thom, A. (1968). Prehistoric observatories. *New Scientist* **38**, 32-5.

82. Thom, A. (1968). The metrology and geometry of cup and ring marks. *Systematics* **6**, 173-89.

83. Thom, A. (1969). Glastonbury as a possible megalithic observatory. In M. Williams (ed.), *Glastonbury: A study in patterns*, p. 5-7. London: RILKO.

84. Thom, A. (1969). The lunar observatories of Megalithic Man. *VA* **11**, 1-29.

85. Thom, A. (1969). The geometry of cup-and-ring marks. *Transactions of the Ancient Monuments Society* **16** (new series), 77-87.

86. Thom, A. (1970). Observing the moon in megalithic times. *JBAA* **80**, 93-9.

87. Thom. A. (1971). *Megalithic lunar observatories*. Oxford: Oxford University Press. 127 pp.

88. Thom. A. & Thom, A.S. (1971). The astronomical significance of the large Carnac menhirs. *JHA* **2**, 147-60.

89. Thom, A. & Thom, A.S. (1972). The Carnac alignments. *JHA* **3**, 11-26.

90. Thom, A. (1972). The uses of the alignments at Le Ménec, Carnac. *JHA* **3**, 151-64.

91. Thom, A. & Thom, A.S. (1973). A megalithic lunar observatory in Orkney: the Ring of Brogar and its cairns. *JHA* **4**, 111-23.

92. Thom, A. & Thom, A.S. (1973). The Kerlescan cromlechs. *JHA* **4**, 168-73.

93. Thom, A., Thom, A.S., Merritt, R.L. & Merritt, A.L. (1973). The astronomical significance of the Crucuno stone rectangle. *CA* **14**, 450-4.

94. Thom, A. & Thom, A.S. (1974). The Kermario alignments. *JHA* **5**, 30-47.

95. Thom, A. (1974). A megalithic lunar observatory in Islay. *JHA* **5**, 50-1.

96. Thom, A., Thom, A.S. & Thom, Alexander Strang (1974). Stonehenge. *JHA* **5**, 71-90.

97. Thom, A. (1974). Astronomical significance of prehistoric monuments in Western Europe. *PTRS* **A276**, 149-56.

98. Thom, A. (1974). Megalithic Astronomy. *Rendiconti del Seminario della Facoltà di Scienze della Università di Cagliari (Bologna)* **44** (supp.), 1-22.

99. Thom, A. (1974). Reply to 'On the interpretation of the Carnac Menhirs' by J.D. Patrick & C.J. Butler. *Irish Archaeological Research Forum* **1**(2), 40-4.

100. Thom, A., Thom, A.S. & Thom, Alexander Strang (1975). Stonehenge as a possible lunar observatory. *JHA* **6**, 19-30.

101. Thom, A. & Thom, A.S. (1975). Further work on the Brogar lunar observatory. *JHA* **6**, 100-14.

102. Thom, A. (1975). Carnac, un observatoire préhistorique. *Sciences et Avenir* **338**, 378-85.

103. Thom, A., Thom, A.S. & Gorrie, J.M. (1976). The two megalithic lunar observatories at Carnac. *JHA* **7**, 11-26.

104. Thom, A., Thom, A.S. & Foord, T.R. (1976). Avebury (1): A new assessment of the geometry and metrology of the ring. *JHA* **7**, 183-92.

105. Thom, A. & Thom, Alexander Strang (1976). Avebury (2): The West Kennet Avenue. *JHA* **7**, 193-7.

106. Thom, A. & Thom, A.S. (1977). Megalithic astronomy (Duke of Edinburgh lecture). *Journal of Navigation* **30**, 1-14.

107. Thom, A. & Thom, A.S. (1977). A fourth lunar foresight for the Brogar Ring. *JHA* **8**, 54-6.

108. Thom, A. (1977). Reply to 'Thom's survey of the Avebury Ring' by P.R. Freeman. *JHA* **8**, 135-6.

109. Thom, A. & Merritt, R.L. (1977). Some megalithic sites in Shetland. *JHA* **9**, 54-60.

110. Thom, A. & Thom, A.S. (1977). Rings and menhirs: geometry and astronomy in the Neolithic age. In E.C. Krupp (ed.), *In Search of Ancient Astronomies*, p. 39-80. New York: Doubleday.

111. Thom, A., Thom, A.S. & Thom, Alexander Strang (1977). Stonehenge. In K. Critchlow & G. Challifour (eds.), *Earth mysteries: A study in patterns*, p. 11-19. London: RILKO.

112. Thom. A. & Thom, A.S. (1977). Geometrie des Alignments de Carnac. In *Metrologie et Astronomie Préhistorique* (Université de Rennes), 1-15.

113. Thom, A. & Thom, A.S. (1977). The megalithic yard. *Journal of the Institute of Measurement and Control* **10**, 488-92.

114. Thom, A. & Thom, A.S. (1978). *Megalithic remains in Britain and Brittany*. Oxford: Oxford University Press. 192 pp.

115. Thom, A. (1978). The distances between stones in stone rows. *JRSS* **A141**, 253-7.

116. Thom, A. & Thom, A.S. (1978). A reconsideration of the lunar sites in Britain. *JHA* **9**, 170-9.

117. Thom, A. & Thom, A.S. (1979). The standing stones in Argyllshire. *GAJ* **6**, 5-10.

118. Thom, A. & Thom, A.S. (1979). Another lunar site in Kintyre. *AA* **1**, S97-8.

119. Thom, A. & Thom, A.S. (1980). A new study of all megalithic lunar lines. *AA* **2**, S78-89.

120. Thom, A. & Thom, A.S. (1980). The astronomical foresights used by Megalithic Man. *AA* **2**, S90-4.

121. Thom, A., Thom, A.S. & Burl, H.A.W. (1980). *Megalithic rings*. Oxford: British Archaeological Reports (BAR 81). 405 pp.

122. Thom, A. & Thom, A.S. (1981). A lunar site in Sutherland. *AA* **3**, S71-3.

123. Thom, A., Thom, A.S., Merritt, R.L. & Merritt, A.L. (1981). La Signification Astronomique du Rectangle de Crucuno. *Kadath* **42**, 19-21.

124. Thom, A. & Thom, A.S. (1982). Statistical and philosophical arguments for the astronomical significance of standing stones with a section on the solar calendar. In D.C. Heggie (ed.), *Archaeoastronomy in the Old World*, p. 53-82. Cambridge: Cambridge University Press.

125. Thom, A. & Thom, A.S. (1983). Observation of the moon in megalithic times. *AA* **5**, S57-66.

126. Thom, A. & Thom, A.S. (1984). The two major megalithic observatories in Scotland. *AA* **7**, S129-44.

127. Thom, A. (1984). Moving and erecting the menhirs. *PPS* **50**, 382-4.

128. Thom, A. & Thom, A.S. (1988). The metrology and geometry of Megalithic Man. In C.L.N. Ruggles (ed.), *Records in stone: Papers in memory of Alexander Thom*, p. 132-151. Cambridge: Cambridge University Press.

129. Thom, A., Thom, A.S. & Burl, H.A.W. (1988). *Stone Rows and Standing Stones*. Oxford: British Archaeological Reports, in press.

My father did not write the following ten papers. He was, however, deeply involved in the background work for them all; he inspected the scale plans produced, applied his geometrical constructions in the deductions, did weeks of work on the statistical analysis where necessary and helped us edit the manuscripts.

1a. Thom, A.S. (1953). Design of a right-angled bend with constant velocities at the walls. *ARCR* no. 14623 (Current Paper no. 135).

2a. Thom, A.S. & Foord, T.R. (1977). The Island of Eday. *JHA* **8**, 198-9.

3a. Freer, R. & Quinio, J-L. (1977). The Kerlescan alignments. *JHA* **8**, 52-4.

4a. Thom, A.S. (1980). The stone rings of Beaghmore: Geometry and astronomy. *Ulster Journal of Archaeology* **43**, 15-19.

5a. Merritt, R.L. & Thom, A.S. (1980). Le Grand Menhir Brisé. *Arch. J.* **137**, 27-39.

6a. Thom, A.S. (1980). A solstitial site near Peterborough? *AA* **2**, S95.

7a. Thom, A.S. (1981). Megalithic lunar observatories : an assessment of 42 lunar alignments. In C.L.N. Ruggles & A.W.R. Whittle (eds.), Astronomy and society in Britain during the period 4000 - 1500 B.C., p. 13-61. Oxford: British Archaeological Reports (BAR 88).

8a.   Thom, A.S. (1981). Review of 'Megalithic science: Ancient mathematics and astronomy in northern Europe' by D.C. Heggie. *AAB* **5**, 24-7.

9a.   Thom, A.S. & Merritt, R.L. (1983). Some Stone Rings in Scandinavia. *Arch J.* **140**, 109-19.

10a. Thom, A.S. (1984). The solar and lunar observatories of the megalithic astronomers. In E.C. Krupp (ed.), *Archaeoastronomy and the roots of science* (A.A.A.S. Selected Symposium 71), p. 83-168.

Acknowledgements

My sincere thanks are due to Mrs H.M. Gustin for her invaluable assistance in helping A.T. and keeping his papers in such excellent order over the last decade or so. The above list was begun by her several years ago, with my father's guidance, and my task was made much more easy by having it.

I wish also to thank Mrs H. Forrest for her patience and attention to detail while painstakingly typing draft after draft of most of the work published over the last ten years.

4

A catalogue of the Alexander Thom Archive held in the National Monuments Record of Scotland

LESLEY FERGUSON

Introduction

Professor Alexander Thom visited several hundred sites during the course of his work on prehistoric stone circles and related alignments, many more than once, and his notebooks record some 800 visits, of which over 600 were to Scottish sites. In his 40 years of research and fieldwork on archaeoastronomy Thom produced over 600 drawings and index cards, filled over 100 notebooks and published some 50 articles and five books. Few other individuals in recent times have generated such a wealth of material and information in pursuing an interest outwith their chosen profession.

The earliest papers referring to Thom's surveys relate to that of the stone circle at Torhousekie, Wigtownshire, which was carried out on 17 August 1938. The notes, which are neatly written on loose-leaf paper, consist of theodolite readings, measurements and sketches of individual stones, and lengthy calculations of declinations, azimuths and altitudes from the circle. At some later date, after a further visit to the site, a detailed plan of the circle was produced, complete with pencil annotations and sightlines to other stones in the vicinity as well as to the setting and rising positions of the sun and the star Capella.

This appears to have been the first of about 20 surveys carried out by Thom in south-west Scotland during August 1938 and May 1939, and it provided a model for the kind of detail which he recorded on his field visits.

In July 1939 Thom and his family spent a holiday in Kintyre, Knapdale and Lorn, on the west coast of Scotland, during which at least 35 sites were visited and surveyed in just 24 days. This, the first of many holidays combining relaxation with fieldwork, was nevertheless quite leisurely when compared with some later expeditions.

During the war, while based at the Royal Aircraft Establishment at Farnborough, Thom had little opportunity to carry on with his fieldwork, but he did manage to do a little preliminary work at Stonehenge and Avebury in Wiltshire in September 1943, and investigated a site at Winchfield, Hampshire, in April 1944. However, after his appointment as Professor of Engineering Science at Oxford in 1945 he resumed his research in earnest, and in two weeks in August 1946 he surveyed 11 sites in the Outer Hebrides and northern Argyll. From this time on Thom spent most of his summer and Easter vacations surveying and researching prehistoric sites.

From the notebooks it is possible to follow Thom's travels around the country. To take 1955 as an example: in early April Thom was in Perthshire and Angus, having travelled north from Oxford on 28 March (Easter Monday). On his return journey he visited eight sites in the Lake District on the weekend of 15-17 April. In mid-July we find him in Devon and Cornwall, and then in Aberdeenshire, Perthshire and Inverness in August, where at least 19 sites, scattered from Perth to Culloden, were visited in the space of just six days. On Wednesday 13 September he was at Stainton Dale and Fylingdales in Yorkshire, and on the weekend of 24-25 September he visted three sites in Derbyshire. In total, notes were taken at 60 sites; no mean feat when it is remembered that this was the age before motorway travel.

In his notebooks Thom records details of some of his car journeys; on one journey from Oxford, he left at 7 a.m. and reached Dunlop, Kilmarnock, at 8.30 p.m; $354\frac{1}{2}$ miles at an average speed of 30 m.p.h. The car appears to have achieved only 15 miles per gallon and the cost of the petrol for the journey would have been £5/10/-, a considerable sum of money for the 1950s.

Thom was usually accompanied on his travels by members of his family or friends. Many of the latter were also engineers, some having been taught by Thom himself. William MacGregor, a Glasgow engineering lecturer, helped with some of the first surveys in the south-west of Scotland in 1939. Continuing his career in the University, MacGregor eventually taught Thom's son, Archie. Another of Thom's students was Dr James Orr, with whom he co-wrote an engineering paper in 1931. A plan of the stones at Cnoc Fada, Mull, was drawn by Orr in the mid 1930s and it would seem that Thom was already encouraging others in the subject at this time. Another assistant in the field was Archibald Black, at that time a Reader in Engineering Science at Oxford and later Professor of Mechanical Engineering at Southampton. Black shared a love of yachting with Thom and sailed on the *Torridon* to the Scottish islands on a fieldwork holiday in 1948.

The first paper on megalithic research which Thom published appeared in 1954 in the *Journal of the British Astronomical Association*. Although another paper was published the following year, it was not until after his retirement in 1961 that Thom began to write articles on a regular basis. From 1964 onwards he had at least one article published every year up to 1984.

In the years immediately following his retirement Thom's fieldwork seems to have been limited compared with earlier years, but in 1966 he began a more intensive field programme concentrating almost entirely on Scottish monuments and taking in Shetland for the first time.

In 1970, Thom took up the invitation from Glyn Daniel, the editor of *Antiquity*, to survey the extensive and complex alignments and monuments at Carnac in Brittany. He returned there each spring until 1973 and again in 1976. Throughout this period he was also engaged in fieldwork throughout Britain from Orkney to Stonehenge and Avebury. However, by this time he was in his eighties and was unable to maintain the intensive travelling of earlier years. His fieldwork stopped in 1978, although his research and publications continued until his death in 1985.

As an engineer Thom was constantly aware of the importance of accuracy; his surveys were generally carried out to a very high standard. There is no doubt that his plans and notebooks form a unique and valuable collection of surveys of prehistoric stone monuments.

Acknowledgments

I should like to express my thanks to Dr A.S. Thom for his help while preparing this paper, to Miss C.H. Cruft, Mr G.S. Maxwell, Mr J.B. Stevenson, Mrs R. Nichol, Mr I. Fleming, Mrs H. Malaws (NMRW), and Mr J. Lanigan and Mr D. Ball (both NMRE) for assistance with the catalogue and to Dr A. Burl who kindly allowed me to examine a collection of drawings by Professor Thom temporarily in his possession. I am especially grateful to Dr J.N.G. Ritchie, Mrs D.M. Reynolds and Mr K. Ferguson for their advice and encouragement.

The Catalogue

In 1985, Professor Thom deposited his drawings, notebooks and card index (MS/430/1-107) in the National Monuments Record of Scotland (NMRS) in Edinburgh so that they should be available for research by scholars.

This catalogue comprises the drawings of stone circles, standing stones, alignments and other monuments produced by Professor Alexander Thom between 1938 and 1978.

The drawings have been separated into countries (England, France, Scotland and Wales) and within these the sites are arranged by the old counties and then alphabetically. The format of the entries is as follows:

1   Name of site
2   Classification
3   NMRS site number
4   National Grid Reference
5   NMRS reference of the individual drawing within the Drawings Collection of the National Monuments Record of Scotland. The suffixes used indicate that a drawing is a copy (/c) or that the original is not in NMRS (/co, copy only).
6   Professor Thom's catalogue number
7   Inscription on drawing
8   Details of drawing
9   Scale (as defined by Thom)
10 Medium
11 Paper size (height × width in mm)
12 Paper type if *not* drawn on cartridge paper
13 Date if written on the drawing

ENGLAND

Cambridgeshire

ROBIN HOOD AND LITTLE JOHN
Stones
TL19NW 16   TL13959838
DC4863
D3/1
Insc. *Horizon profile from Robin Hood and Little John, Peterborough*
Profile of the horizon from the stones.
No scale
Black ink
306 mm × 220 mm
Plastic film

Cornwall

ALTARNUN
Stone Circle
SX27NW 29   SX23617815
DC4605
S1/2
Insc. *Nine Stones*
Annotated plan of stone circle.
1½ in : 10 ft
Black ink and pencil
557 mm × 377 mm

BOSCAWEN-UN
Stone Circle
SW42NW 28   SW41222736
DC4617
S1/13
Insc. *Boscawen-Un*
Annotated plan of stone circle.
1 in : 8 ft
Black ink and pencil
556 mm × 340 mm

BOSKEDNAN
Stone Circle
SW43NW 40   SW43423513
DC4615
S1/11
Insc. *Nine Maidens, Ding Dong*
Plan of stone circle.
1 in : 8 ft
Black ink and pencil
556 mm × 379 mm

DULOE
Stone Circle
SX25NW 3   SX23575830
DC4605
S1/3
Insc. *Duloo*
Plan of stone circle.
1 in : 4 ft
Black ink and pencil
557 mm × 377 mm

FERNACRE
Stone Circle
SX17NW 42   SX14487998
DC4609
S1/7
Insc. *Rough Tor*
Plan of stone circle.
1 in : 10 ft
Black ink and pencil
557 mm × 396 mm

GOODAVER
Stone Circle
SX27NW 6   SX20907513
DC4620/co
S1/17
Insc. *Trezibbet*
Plan of stone circle.
1 in : 10 ft

Annotated dyeline
573 mm × 385 mm

GOODAVER
(see above)
DC4766/co
S1/17
Insc. *Trezibbet*
Plan of stone circle. Annotated 'Survey
    by Major A.F. Prain, R.E.'.
1 in : 10 ft
Dyeline
407 mm × 350 mm

HURLERS (THE), CENTRE
Stone Circle
SX27SE 18   SX25827139
DC4604
S1/1
Insc. *The Hurlers, Central Circle*
Annotated plan of stone circle.
1 in : 10 ft
Black ink and pencil
558 mm × 377 mm
1955

HURLERS (THE), NE
Stone Circle
SX27SE 18   SX25847146
DC4603
S1/1
Insc. *The Hurlers, North Circle*
Plan of stone circle.
1 in : 10 ft
Black ink and pencil
557 mm × 350 mm

HURLERS (THE), SW
Stone Circle
SX27SE 18   SX25817133
DC4602
S1/1
Insc. *The Hurlers, South Circle*
Annotated plan of stone circle.
1 in : 10 ft

Black ink and pencil
558 mm × 382 mm

***HURLERS (THE), SW, NE and
CENTRE***
Stone Circles
SX27SE 18   centre SX258 714
DC4777
S1/1
Insc. *The Hurlers*
Plan of the stone circles. Annotated
    'Survey by Major A.F.Prain 1965'.
1 in : 20 ft
Black ink
765 mm × 510 mm
Plastic film
1965

***HURLERS (THE), SW, NE and
CENTRE***
(see above)
DC4793
S1/1
Insc. *Hurlers*
Annotated plan of the stone circles.
1 in : 22 ft
Black ink and pencil
758 mm × 558 mm

LEAZE
Stone Circle
SX17NW 26   SX13667728
DC4608
S1/6
Insc. *Leaze*
Plan of stone circle.
$\frac{1}{8}$ in : 1 ft
Black ink and pencil
555 mm × 351 mm

MERRY MAIDENS
Stone Circle
SW42SW 1   SW43272450
DC4618
S1/14

Insc. *Merry Maidens*
Plan of stone circle.
1½ in : 10 ft
Black ink and pencil
556 mm × 381 mm

NINE MAIDENS
Stone Row
SW96NW 1　　　　　　SW93636754
　　　　　　　　　　　-SW93686763
DC4611
S1/9
Insc. *Merry Maidens*
Plan of stone row.
1 in : 20 ft
Pencil
558 mm × 377 mm

NINE MAIDENS
(see above)
DC4612
S1/9
Plan of the stone row with elevations of
　the stones.
¾ in : 5 ft
Pencil
559 mm × 380 mm

NINE MAIDENS
(see above)
DC4613
S1/9
Insc. *Maidens*
Plan of stone row.
1 in : 10 ft & 1 in : 20 ft
Pencil
558 mm × 380 mm

NINE MAIDENS
(see above)
DC5349
S1/9
Plan and elevations of the stones with a
　horizon drawing and a base plan.
⅝ in : 20 ft & ½ in : 5 ft

Black ink
219 mm × 297 mm

PORTHMEOR
Enclosure
SW43NW 52　　　　　　SW44473666
DC4616
S1/12
Insc. *Porthmeor*
Plan of enclosure.
1 in : 10 ft
Black ink and pencil
557 mm × 336 mm

STANNON
Stone Circle
SX18SW 2　　　　　　SX12628001
DC4610
S1/8
Insc. *Dinnever Hill*
Plan of stone circle.
1 in : 10 ft
Black ink and pencil
556 mm × 391 mm

STRIPPLE STONES
Henge Monument
SX17NW 25　　　　　　SX14377521
DC4606
S1/4
Insc. *Stripple Stones*
Plan of henge monument.
¾ in : 10 ft
Black ink and pencil
556 mm × 378 mm

TREGESEAL
Stone Circle
SW33SE 60　　　　　　SW38653236
DC4619
S1/16
Insc. *Botallack*
Plan of stone circle.
1½ in : 10 ft

Black ink and pencil
556 mm × 378 mm

TRIPPET STONES
Stone Circle
SX17NW 8   SX13127501
DC4607
S1/5
Insc. *Treswigger*
Plan of stone circle.
¹₈ in : 1 ft
Black ink and pencil
556 mm × 376 mm

WENDRON, SOUTH
Stone Circle
SW63NE 7   SW683 365
DC4614
S1/10
Insc. *Nine Maidens, Cambourne*
Plan of stone circle.
1¹₂ in : 10 ft
Black ink and pencil
558 mm × 383 mm

Cumberland

BLAKELEY RAISE
Stone Circle
NY01SE 6   NY06011403
DC4531
L1/16
Insc. *Blakeley Moss*
Annotated plan of stone circle with
    outline sketch of Screel Hill,
    Kirkcudbright.
1 in : 8 ft
Blue ink and pencil
555 mm × 385 mm

BLAKELEY RAISE
(see above)

DC4772
L1/16
Insc. *Blakeley Moss*
Plan of stone circle with sketch of Screel
    Hill, Kirkcudbright.
¹₄ in : 2 ft
Black ink
369 mm × 288 mm
Plastic film

BLAKELEY RAISE
(see above)
DC4895
L1/16
Insc. *Blakeley Moss*
Horizon   profile   of   Screel   Hill,
    Kirkcudbright.
No scale
Black ink
254 mm × 202 mm
Plastic film

BRATS HILL
Stone Circle
NY10SE 1   NY173 023
DC4521
L1/6
Plan of stone circle.
¹₈ in : 1 ft
Pencil
555 mm × 364 mm
Tracing paper

BRATS HILL
(see above)
DC4522
L1/6
Insc. *Burnmoor (E)*
Plan of stone circle.
¹₈ in : 1 ft
Black ink and pencil
558 mm × 381 mm
1955

BRATS HILL
(see above)
DC4523
L1/6
Insc. *Burnmoor Circle E*
Annotated plan of stone circle.
⅛ in : 1 ft
Blue ink and pencil
558 mm × 380 mm
1959

BRATS HILL
(see above)
DC4515
L1/6
Insc. *Burnmoor Circles*
Plan showing the positions of the stone circles to each other with chart showing circle centres as measured in 1966.
¾ in : 100 ft
Pencil
557 mm × 317 mm

BRATS HILL
(see above)
DC4520
L1/6
Insc. *Burnmoor, 1st survey*
Annotated plan showing position of stone circle in relation to circles C and D.
1 in : 100 ft
Pencil
559 mm × 326 mm

CASTLE RIGG
Stone Circle
NY22SE 1   NY29142363
DC4505
L1/1
Insc. *Castle Rigg*
Annotated plan of stone circle.
⅛ in : 1 ft
Black ink and pencil
559 mm × 385 mm

CASTLE RIGG
(see above)
DC4506
L1/1
Plan of stone circle showing the flattened circle shape.
No scale
Red and green ink
553 mm × 372 mm
Tracing paper

CASTLE RIGG
(see above)
DC4507
L1/1
Plan of stone circle. Annotated '1st survey. Drawn too large by 1 ft on diam[eter]'.
No scale
Black ink and pencil
458 mm × 372 mm
Linen backed tracing paper

CASTLE RIGG
(see above)
DC4508
L1/1
Insc. *Castle Rigg, 1959 survey*
Plan of stone circle.
⅛ in : 1 ft
Blue ink and pencil
558 mm × 381 mm

CASTLE RIGG
(see above)
DC4776
L1/1
Insc. *Castle Rigg*
Drawing illustrating the stones and horizon as seen from the centre of the circle and showing the rising and setting positions of various celestial bodies.
No scale
Black, red and green ink
845 mm × 155 mm

ELVA PLAIN
Stone Circle
NY13SE 4　　　　　　　　　NY177 317
DC4509
L1/2
Insc. *Elver Plain, Setmurthy*
Plan of stone circle.
$\frac{1}{8}$ in : 1 ft
Black ink and pencil
562 mm × 382 mm

GIANT'S GRAVE
Standing Stones
SD18SW 12　　　　　　　　SD136 803
DC4528
L1/11
Insc. *Giants Graves*
Plan of standing stones.
$\frac{1}{4}$ in : 1 ft
Black ink and pencil
557 mm × 295 mm

GIANT'S GRAVE
(see above)
DC4855/co
L1/11
Insc. *Giants Graves*
Plan of standing stones with a hillside
　sketch.
$\frac{1}{5}$ in : 1 ft
Dyeline copy
203 mm × 117 mm

GLASSONBY
Cairn Circle
NY53NE 2　　　　　　　　NY57283935
DC4526
L1/9
Insc. *Glassonby*
Plan of stone circle.
$\frac{1}{8}$ in : 1 ft
Black ink and pencil
560 mm × 370 mm

GREY CROFT
Stone Circle
NY00SW 3　　　　　　　　NY03340238
DC4527
L1/10
Insc. *Seascale*
Plan of stone circle.
$\frac{1}{8}$ in : 1 ft
Black ink and pencil
559 mm × 383 mm

LACRA A
Stone Circle
SD18SE 2　　　　　　　　SD15128125
DC4529
L1/12
Insc. *Lacra E*
Plan of stone circle.
$\frac{1}{8}$ in : 1 ft
Black ink and pencil
558 mm × 318 mm

LACRA B
Stone Circle
SD18SW 3　　　　　　　　SD15008096
DC4529
L1/13
Insc. *Lacra S*
Plan of stone circle.
$\frac{1}{8}$ in : 1 ft
Black ink and pencil
558 mm × 318 mm

LITTLE MEG
Cairn
NY53NE 14　　　　　　　NY57703748
DC4525
L1/8
Insc. *Little Meg*
Plan of cairn.
$\frac{1}{4}$ in : 1 ft
Black ink and pencil
559 mm × 374 mm

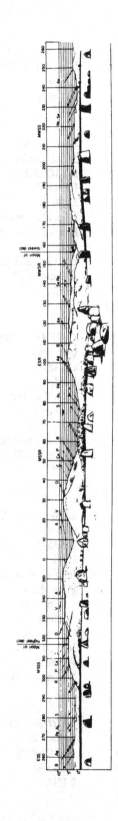

Fig. 4.1. Castle Rigg, Cumberland: Extended elevation drawing illustrating the horizon as seen from the centre of the circle and showing the rising and setting positions of various velestial bodies (A. Thom; DC4776).

LONG MEG AND HER DAUGHTERS
Stone Circle
NY53NE 5   NY57113721
DC4524
L1/7
Insc. *Long Meg and Her Daughters*
Annotated plan of stone circle.
$^3_8$ in : 10ft
Black ink and pencil
553mm × 388mm

LOW LONGRIGG NE
Stone Circle
NY10SE 2   NY172 027
DC4511
L1/4
Insc. *Burnmoor A, 1952 Rough Survey*
Plan of stone circle with tracing from the
    6-in. OS map showing the positions of
    circles A,B,C, D and E.
$^1_{10}$ in : 1 ft
Black ink and pencil
558 mm × 381 mm

LOW LONGRIGG NE
(see above)
DC4512
L1/4
Insc. *Burnmoor A, Circle A 1959 Survey*
Annotated plan of stone circle.
$^1_{10}$ in : 1 ft
Blue ink and pencil
558 mm × 384 mm
1959

LOW LONGRIGG NE
(see above)
DC4515
L1/6
Insc. *Burnmoor Circles*
Plan showing the position of the stone
    circle in relation to the others with
    chart showing circle centres as
    measured in 1966.

$^3_4$ in : 100 ft
Pencil
557 mm × 317 mm

LOW LONGRIGG SW
Stone Circle
NY10SE 2   NY172 027
DC4513
L1/4
Insc. *Burnmoor B*
Plan of stone circle.
$^1_{10}$ in : 1 ft
Blue ink and pencil
559 mm × 375 mm
1952

LOW LONGRIGG SW
(see above)
DC4514
L1/4
Insc. *Burnmoor, Circle B, 1959 Survey*
Plan of stone circle.
$^1_{10}$ in : 1 ft
Blue ink and pencil
557 mm × 381 mm
1959

LOW LONGRIGG SW
(see above)
DC4515
L1/4
Insc. *Burnmoor Circles*
Plan showing the position of the stone
    circle in relation to the others with
    chart showing circle centres as
    measured in 1966.
$^3_4$ in : 100 ft
Pencil
557 mm × 317 mm

STUDFOLD
Stone Circle
NY02SW 6   NY040 223
DC4530
L1/14

Insc. *Dean Moor*
Annotated plan of stone circle.
1 in : 10 ft
Pencil
558 mm × 380 mm

SUNKENKIRK, SWINSIDE
Stone Circle
SD18NE 5   SD17178818
DC4510
L1/3
Insc. *Swinside, Sunkenkirk*
Plan of stone circle.
$\frac{1}{8}$ in: 1 ft
Black ink and pencil
560 mm × 382 mm

WHITEMOSS NE
Stone Circle
NY10SE 1   NY172 024
DC4518
L1/5
Insc. *Burnmoor Circle D, 1959 Survey*
Plan of stone circle.
$\frac{1}{10}$ in : 1 ft
Blue ink and pencil
558 mm × 380 mm
1959

WHITEMOSS NE
(see above)
DC4519
L1/5
Insc. *Burnmoor Circle D*
Plan of stone circle.
$\frac{1}{10}$ in : 1 ft
Blue ink and pencil
559 mm × 381 mm
1952

WHITEMOSS NE
(see above)
DC4515
L1/5
Insc. *Burnmoor Circles*

Plan showing the position of the stone
circle in relation to the others with a
chart showing circle centres as
measured in 1966
$\frac{3}{4}$ in : 100 ft
Pencil
557 mm × 317 mm

WHITEMOSS NE
(see above)
DC4520
L1/5
Insc., *Burnmoor, 1st Survey*
Annotated plan showing the position of
stone circle in relation to circles D and
E.
1 in : 100 ft
Pencil
559 mm × 326 mm

WHITEMOSS SW
(see above)
DC4516
L1/5
Insc. *Burnmoor Circle C, 1959 Survey*
Plan of stone circle with chart showing
adjusted distances between theodolite
stations.
$\frac{1}{10}$ in : 1 ft
Blue ink and pencil
557 mm × 379 mm
1959

WHITEMOSS SW
(see above)
DC4517
L1/5
Insc. *Burnmoor, Circle C*
Plan of stone circle.
$\frac{1}{10}$ in : 1 ft
Blue ink and pencil
559 mm × 382 mm
1952

WHITEMOSS SW
(see above)
DC4520
L1/5
Insc. *Burnmoor, 1st Survey*
Annotated plan showing position of stone
    circle in relation to circles C and E.
1 in : 100 ft
Pencil
559 mm × 326 mm

WHITEMOSS SW
(see above)
DC4515
L1/5
Insc. *Burnmoor Circles*
Plan showing the position of the stone
    circle in relation to the others with
    chart showing circle centres as
    measured in 1966.
¾ in : 100 ft
Pencil
557 mm × 317 mm

Derbyshire

BARBROOK
Stone Circle
SK27NE 13   SK27857558
DC4446
D1/7
Insc. *Barbrook*
Annotated plan of stone circle.
¼ in : 1 ft
Black ink and pencil
558 mm × 379 mm

BIG MOOR
Cairn (possible)
SK27NE 53   SK27117600
DC4449
D1/13
Insc. *Big Moor*

Plan of cairn.
1½ in : 10 ft
Black ink and pencil
558 mm × 379 mm

FROGGAT EDGE
Stone Circle
SK27NW 38   SK24957679
DC4445
D1/5
Insc. *Froggat Edge*
Plan of stone circle.
1½ in : 10 ft
Black ink and pencil
558 mm × 372 mm

MOSCAR MOOR
Stone Circle
SK28NW 2   SK21528685
DC4448
D1/9
Insc. *Moscar Moor*
Plan of stone circle.
1½ in : 10 ft
Blue ink and pencil
558 mm × 382 mm

NINE LADIES
Stone Circle
SK26SW 15   SK24916349
DC4444
D1/3
Insc. *Nine Ladies, Stanton Moor*
Annotated plan of stone circle.
¼ in : 1 ft
Black ink and pencil
556 mm × 380 mm

OWLER BAR
Stone Circle
SK27NE 72   SK283 772
DC4447
D1/8
Insc. *Owler Bar*
Plan of stone circle.

$^1_8$ in : 1 ft
Blue ink and pencil
558 mm × 380 mm

Devon

BRISWORTHY
Stone Circle
SX56NE 64   SX56466549
DC4622
S2/3
Insc. *Brisworthy*
Plan of stone circle.
$^1_8$ in : 1 ft
Blue ink and pencil
565 mm × 377 mm

CHOLWICHTOWN
Stone Circle
SX56SE 33   SX584 622
DC4623
S2/7
Insc. *Lee Moor*
Plan of stone circle.
$^1_8$ in : 1 ft
Black ink and pencil
560 mm × 378 mm

GREY WETHERS, N & S
Stone Circles
SX68SW 1   SX638 831
DC4794
S2/1
Insc. *Grey Wethers*
Annotated plan of stone circles. Annotated 'Shaded stones are those standing before "reconstruction". (See survey by Rev. W.C. Lukis F.S.A.1879)'.
$^1_8$ in : 1 ft
Blue ink and pencil
763 mm × 557 mm
1954

MERRIVALE
Stone Circle
SX57SE 8   SX556 743
DC4621
S2/2
Insc. *Merrivale*
Plan of stone circle.
$^1_8$ in : 1 ft
Blue ink and pencil
564 mm × 360 mm

POSTBRIDGE
Cairn Circle
SX67NE 6   SX67517869
DC4614 (reverse)
S2/8
Insc. *Near Postbridge*
Plan of cairn circle.
$^1_4$ in : 1 ft
Black ink and pencil
557 mm × 383 mm

POSTBRIDGE
(see above)
DC4624
S2/8
Insc. *near Postbridge*
Plan of cairn circle.
$^1_4$ in : 1 ft
Black ink
555 mm × 379 mm
Linen backed tracing paper

RINGMOOR
Cairn Circle
SX56NE 60   SX563 658
DC4623
S2/4
Insc. *Ringmoor*
Annotated plan of cairn circle.
$^1_8$ in : 1 ft
Blue ink and pencil
560 mm × 378 mm

RINGMOOR
(see above)
DC4627
S2/4
Insc. *Ringmoor*
Plan of cairn circle.
¹₈ in : 1 ft
Black ink and pencil
560 mm × 355 mm

TROWLESWORTHY
Stone Circle
SX56SE 38   SX576 639
DC4623
S2/5
Insc. *Trowlesworthy*
Plan of stone circle.
¹₈ in : 1 ft
Black ink and pencil
560 mm × 378 mm

Dorset

HAMPTON DOWN
Stone Circle
SY58NE 53   SY596 865
DC4627
S4/3
Insc. *Hampton Down*
Plan of stone circle.
¹₈ in : 1 ft
Blue ink and pencil
560 mm × 355 mm

KINGSTON RUSSELL
Stone Circle
SY58NE 6   SY577 878
DC4626
S4/2
Insc. *Kingston Russell*
Plan of stone circle.
¹₈ in : 1 ft

Blue ink and pencil
561 mm × 376 mm

NINE STONES
Stone Circle
SY69SW 17   SY610 904
DC4625
S4/1
Insc. *Ninestones, Winterbourne Abbas*
Annotated plan of stone circle.
¹₄ in : 1 ft
Blue ink and pencil
560 mm × 330 mm

Lancashire

DRUIDS' TEMPLE
Stone Circle
SD27SE 23   SD292 739
DC4540
L5/1
Insc. *Birkrigg Common*
Plan of stone circle.
¹₈ in : 1 ft
Black ink and pencil
558 mm × 382 mm

THREE BROTHERS, ROCKING STONES
Stones
SD47SE   SD49457346
DC4541
L5/2
Insc. *Three Brothers*
Plan of stones. Annotated 'Poor survey'.
¹₈ in : 1 ft
Pencil
558 mm × 380 mm

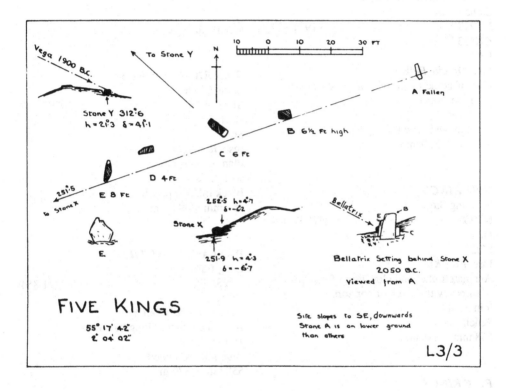

Fig. 4.2. Five Kings, Northumberland: Plan and elevation of the standing stones (A. Thom; DC4894).

Northumberland

DUDDO FOUR STONES
Stone Circle
NT94SW 8   NT931 437
DC4538
L3/1
Insc. *Duddo, Felkington*
Plan of stone circle with a sketch of one
    of the stones.
$\frac{1}{4}$ in : 1 ft
Blue ink and pencil
559 mm × 380 mm

FIVE KINGS
Standing Stones
NT90SE 6   NT957 000
DC4539
L3/3
Insc. *Five Kings*
Annotated plan and elevation of standing
    stones with sketch of horizon.
$\frac{1}{8}$ in : 1 ft
Pencil
558 mm × 378 mm

FIVE KINGS
(see above)
DC4894
L3/3
Insc. *Five Kings*
Annotated plan and elevations of standing
    stones with horizon sketch.
$\frac{1}{2}$ in : 8 ft
Black ink
250 mm × 205 mm

LILBURN
Stone Circle
NT92SE 44   NT971 205
DC4457
L3/4

Insc. *Lilburn*
Plan of stone circle.
No scale
Pencil
556 mm × 383 mm

Oxfordshire

ROLLRIGHT STONES
Stone Circle
SP23SE 14   SP295 308
DC4630
S6/1
Insc. *Rollright*
Plan of stone circle.
$\frac{1}{8}$ in : 1 ft
Black ink and pencil
558 mm × 394 mm

WAYLAND'S SMITHY
Long Barrow
SU28NE 4   SU281 854
DC4666
Insc. *Weyland's Smithy*
Annotated plan of long barrow.
$1\frac{1}{2}$ in : 10 ft
Blue ink and pencil
558 mm × 379 mm

Shropshire

BLACK MARSH
Stone Circle
SO39NW 4   SO324 999
DC4451
D2/2
Insc. *Black Marsh*
Plan of stone circle with sketch of
    hillside.
$1\frac{1}{2}$ in : 10 ft
Blue ink and pencil
558 mm × 372 mm

BLACK MARSH
(see above)
DC4765/co
D2/2
Insc. *Black Marsh*
Plan of stone circle.
1½ in : 10 ft
Dyeline
399 mm × 365 mm

MITCHELL'S FOLD
Stone Circle
SO39NW 2　　　　　SO304 983
DC4450
D2/1
Insc. *Mitchell's Fold*
Plan of stone circle.
⅛ in : 1 ft
Blue ink and pencil
558 mm × 378 mm

PEN-Y-CWM
Cairn
SO37NW 3　　　　　SO313 788
DC4452
D2/3
Insc. *Pen-Y-Cwm*
Plan of cairn.
⅛ in : 1 ft
Pencil
558 mm × 374 mm

Somerset

DEVIL'S BED AND BOLSTER
Long Barrow
ST85SW 1　　　　　ST815 533
DC4370
Insc. *Devil's Bed and Bolster*
　(incomplete)
Plan of long barrow.
No scale

Pencil
558 mm × 328 mm

STANTON DREW, CENTRE
Stone Circle
ST66SW 2　　　　　ST600 633
DC4662
S3/1
Insc. *Stanton Drew*
Plan of stone circle.
½ in : 10 ft
Pencil
253mm × 232mm

STANTON DREW, CENTRE
(see above)
DC4785
S3/1
Insc. *Stanton Drew*
Annotated plan of stone circles.
　Annotated 'Survey by A. Thom'.
3 in : 100 ft
Black ink and pencil
760 mm × 557 mm
1957

Westmorland

CASTLEHOWE SCAR
Stone Circle
NY51NE 6　　　　　NY587 154
DC4533
L2/11
Insc. *Castlehowe Scar*
Plan of stone circle.
¼ in : 1 ft
Black ink and pencil
561 mm × 377 mm

COCKPIT (THE)
Stone Circle
NY42SE 11　　　　　NY482 222

DC4532
L2/2
Insc. *Tarnmoor, Ullswater*
Plan of stone circle.
⅛ in : 1 ft
Blue ink and pencil
560 mm × 382 mm

GAMELANDS
Stone Circle
NY60NW 8   NY640 081
DC4537
L2/14
Insc. *Orton*
Plan of stone circle.
1 in : 20 ft
Black ink and pencil
562 mm × 384 mm

GAMELANDS
(see above)
DC4784
L2/14
Insc. *Composite Survey of Four Circles.*
   Each reduced to the same diameter
   and orientated on major axis.
Plan.
No scale
Black ink
363 mm × 284 mm
Tracing paper

GUNNERKELD
Stone Circle
NY51NE 4   NY568 177
DC4775
L2/10
Insc. *Gunnerwell*
Plan of stone circle.
Annotated 'Note by AST. I re-surveyed
   this   and   the   farm   is   called
   Gunnerswell, pronounced Gooners-
   well. How did Aubrey Burl get
   ...Keld?'.
⅛ in : 1 ft

Black ink and pencil
559 mm × 384 mm

HIGH STREET
Stone Circle
NY41NE 2   NY45711925
DC4541
L2/3
Insc. *High Street*
Plan of stone circle.
⅛ in : 1 ft
Pencil
558 mm × 380 mm

HIGH STREET
(see above)
DC4896
L2/3
Insc. *High Street*
Plan of stone circle.
⅝ in : 5 ft
Black ink
308 mm × 271 mm
Plastic film

IRON HILL
Stone Circle
NY51SE 7   NY596 147
DC4534
L2/12
Insc. *Harberwain*
Plan of stone circle.
⅛ in : 1 ft
Blue ink and pencil
560 mm × 381 mm

IRON HILL
(see above)
DC4535
L2/12
Insc. *Harberwain*
Plan of stone circle.
⅛ in : 1 ft
Pencil
560 mm × 336 mm

ODDENDALE
Stone Circle
NY51SE 8   NY59211291
DC4536
L2/13
Insc. *Oddendale*
Annotated plan of stone circle.
¹₈ in : 1 ft
Black ink and pencil
561 mm × 382 mm

Wiltshire

AVEBURY
Stone Circle and Henge Monument
SU16NW 22.1   SU103 699
DC4728
S5/3
Insc. *The Avebury Arcs*
Plan illustrating the positions of the
    stones relative to the arcs.
³₈ in : 100ft
Black ink
680 mm × 416 mm
Plastic film

AVEBURY
(see above)
DC4749
S5/3
Insc. *Avebury*
Plan of stone circles.
1¹₂ in : 100 ft
Black ink and pencil
1204 mm × 409 mm
1964

AVEBURY
Stone Circle and Henge Monument
SU16NW 22.1   SU103 699
DC4802
S5/3
Insc. *Avebury*

Annotated plan of the stone arcs.
1 mm : 1 ft
Pencil
758 mm × 562 mm

AVENUE (THE)
Stone Avenue
SU14SW 14   SU12704258
DC4725   -SU14164155
S5/1
Insc. *'Moon Walk', Stonehenge*
Annotated plan of avenue.
1:2500 and 1 in : 1000 ft
Pencil
762 mm × 555 mm

AVENUE (THE)
(see above)
DC4726
S5/1
Insc. *Moon Walk Parallel Banks near
    Stonehenge*
Annotated plan of the avenue with section
    along line of banks and diagrams
    showing moon setting positions on the
    walk at the minor standstill. Annotated
    'AT's artwork. A.S.T. 13/4/86'.
1:100 and 1 in : 1000 ft
Black ink
430 mm × 267 mm
Tracing paper

COATE
Stone Circle
SU18SE 32   SU181 824
DC4629
S5/6
Insc. *Day House Lane, near Swindon*
Plan of stone circle. Annotated
    'orientation unknown'.
1 in : 20 ft
Black ink and pencil
558 mm × 389 mm
1958

COATE
(see above)
DC4767/co
S5/6
Insc. *Dayhouse Lane near Swindon*
Plan of stone circle.
½ in : 10 ft
Dyeline
347 mm × 285 mm

FIGSBURY RING
Fort
SU13SE 27   SU188 338
DC4753
Insc. *Figsbury Ring*
Annotated plan showing the location of
   the fort.
½ in : 100 ft
Pencil
381 mm × 364 mm

SANCTUARY (THE)
Stone and Timber Circles
SU16NW 22.3   SU11846802
DC4702
S5/2
Insc. *The Sanctuary, Survey of Concrete
   Posts*
Plan of stone and timber circles.
1 in : 8 ft
Black ink and pencil
762 mm × 555 mm

SANCTUARY (THE)
(see above)
DC4703
S5/2
Insc. *The Sanctuary*
Annotated plan of stone circle.
Annotated 'Rough survey (prismatic).
   Note by A.S.T. 13/4/86. This did not
   satisfy A.T. and I know he re-did this
   site later, presumably 1960 or 61. I do
   not know the date of this one. A.S.T.'.
1 in : 16 ft

Black ink and pencil
465 mm × 372 mm

SANCTUARY (THE)
(see above)
DC4808
S5/2
Insc. *The Sanctuary, Survey of Concrete
   Posts*
Plan of stone circle.
1 in : 8 ft
Black ink
742 mm × 499 mm
Plastic film

STONEHENGE
Henge Monument and Stone Circles
SU14SW 4   SU12244218
DC4431
S5/1
Insc. *Stonehenge. Pantograph to 1/250
   from Atkinson's nominal 1/96 Z.Y.
   holes*
Annotated plan of stone circles.
Annotated '18 Jan. '85. Always im-
   patient, A.T., not finding a clean sheet,
   pantographed this on to the back of a
   Clava plan made years before.
   R.J.C.A. had sent us a "nominal" 1/96
   tracing of his plan, and we used this
   device to get the Z.Y. holes on to our
   1/250 plan. We of course could not
   survey below the grass. This tracing is
   really useless as it is slightly smaller
   than it should be, see A.T.'s pencil
   notes. Our new plans were supposed to
   be, and were, up to date and accurate.
   A.S. Thom'.
1 : 250
Black ink and pencil
556 mm × 380 mm

STONEHENGE
(see above)
DC4704
S5/1

Insc. *Stonehenge*
Original survey drawing of the stones.
1 : 84
Pencil
720 mm × 528 mm
1973

STONEHENGE
(see above)
DC4705
S5/1
Original survey drawing of the henge and
  stones.
1 : 250
Black ink and pencil
720 mm × 528 mm

STONEHENGE
(see above)
DC4706
S5/1
Insc. *Pantograph reduction of large scale
  plan from Atkinson*
Plan of the stone circle.
No scale
Black ink
760 mm × 558 mm

STONEHENGE
(see above)
DC4706 (reverse)
S5/1
Plan of parts of the stone circle.
Annotated 'Same as other side but done
  in parts'.
No scale
Black ink
760 mm × 558 mm

STONEHENGE
(see above)
DC4707
S5/1
Annotated elevation drawings of the
  standing stones.

1 : 84
Pencil
761mm × 556mm

STONEHENGE
(see above)
DC4708
S5/1
Insc. *Stonehenge*
Printed copy of plan of the henge and
  stone circle. Annotated 'An early trial
  to scale print, as the tracing grew'.
1 : 250
Black ink and pencil
748 mm × 547 mm
1973

STONEHENGE
(see above)
DC4708/c
S5/1
Plan of henge monument and stone circle.
Annotated 'Returned from R.J.C.A. on 5
  Aug 73 with red marks'. and 'April 84.
  For Kadath:- I wonder why I surveyed
  this. Evidently R.J.C.A. did not red
  ring it and I assume his red rings are
  what he reckons are "proddable"
  Aubrey Holes. A.S.T. 18/4/84'.
1 : 250
Dyeline
733 mm × 534 mm

STONEHENGE
(see above)
DC4709
S5/1
Insc. *Stonehenge*
Contour survey of the henge and stones.
  Annotated 'These men did the
  levelling shown here, for A.T.
  Surveyed and drawn by:
WO 2 (AIG) D. Bowden
WO 2 (AIG) G.F. Laverty
GNR C.E. Tressidder
GNR S. Hope

GNR T. Kelley
By kind permission of the commandant,
The Royal School of Artillery. Aug.
1973'.
1 in : 250 ft
Black ink
768 mm × 586 mm
Plastic film

STONEHENGE
(see above)
DC4710
S5/1
Annotated plan of the stone circle with
added ellipses.
1 in : 7 ft
Black ink and pencil
576 mm × 461 mm

STONEHENGE
(see above)
DC4711
S5/1
Insc. *Stonehenge*
Elliptical plan.
No scale
Pencil
592 mm × 398 mm
Tracing paper

STONEHENGE
(see above)
DC4712
S5/1
Insc. *Stonehenge*
Elliptical plan.
1 : 84
Black ink
492 mm × 237 mm

STONEHENGE
(see above)
DC4713
S5/1
Insc. *Stonehenge*

Ellipses.
1 : 84
Pencil
358 mm × 323 mm

STONEHENGE
(see above)
DC4714
S5/1
Plan of stones.
No scale
Blue ink
535mm × 497mm
Linen backed tracing paper

STONEHENGE
(see above)
DC4715/co
S5/1
Plan of stones.
No scale
Dyeline
540 mm × 514 mm

STONEHENGE
(see above)
DC4716/co
S5/1
Insc. *Stonehenge*
Plan of circle and henge monument.
1 : 250
Dyeline
743 mm × 553 mm

STONEHENGE
(see above)
DC4717/co 2 copies
S5/1
Insc. *Stonehenge*
Annotated plan of stones.
1/84
Dyeline
763 mm × 581 mm & 742 mm × 553
    mm

STONEHENGE
(see above)
DC4718
S5/1
Tracing of several of the central stones but not in relation to each other.
No scale
Pencil
378 mm × 252 mm
Tracing paper

STONEHENGE
(see above)
DC4719
S5/1
Insc. *Stonehenge Spirals*
Plan of the stone circle.
No scale
Black ink and pencil
330 mm × 239 mm
Tracing paper

STONEHENGE
(see above)
DC4720/co
S5/1
Plan of the Y and Z holes.
No scale
Photocopy and pencil
352 mm × 296 mm

STONEHENGE
(see above)
DC4721
S5/1
Overlay for DC4720.
No scale
Pencil
401 mm × 314 mm
Tracing paper

STONEHENGE
(see above)
DC4727
S5/1

Insc. *Stonehenge Post Holes*
Plan showing a section of the rings and the postholes. Annotated 'AT's artwork'.
¼ in : 5 ft
Black ink
211 mm × 188 mm
Linen backed tracing paper

STONEHENGE
(see above)
DC4756
S5/1
Insc. *Post Holes*
Annotated 'Re-plot of Nov. 1922 sketches, pages 1-7 and key plan. A.S. Thom 25/7/73'.
1 : 125
Pencil
452 mm × 404 mm

STONEHENGE
(see above)
DC4803
S5/1
Plan of the spirals.
10 mm : 8 ft
Black ink
359 mm × 311 mm
Linen backed tracing paper

WEST KENNET AVENUE
Stone Avenue
SU16NW 22.2   SU10326979
DC4729   -SU11846802
S5/3
Insc. *Avebury: West Kennet Avenue*
Plan of avenue. Annotated 'Fieldworkers Alasdair and Avril. Drawn by A.T. 13/4/86'.
1 : 500
Black ink and pencil
2000 mm × 323 mm

WEST KENNET AVENUE
(see above)
DC4729 (reverse)
S5/3
Annotated plan of avenue.
1 : 1200
Black ink and pencil
2000 mm × 323 mm

WINTERBOURNE BASSETT
Stone Circle
SU07NE 5   SU093 755
DC4628
S5/5
Insc. *Winterbourne Bassett*
Plan of stone circle.
¾ in : 10 ft
Blue ink and pencil
558 mm × 382 mm

WOODHENGE
Henge Monument and Timber Circles
SU14SE 6   SU15064338
DC4795
S5/4
Insc. *Woodhenge*
Annotated plan of timber circles.
⅛ in : 1 ft
Black ink and pencil
764 mm × 559 mm
1957

WOODHENGE
(see above)
DC4796
S5/4
Insc. *For Woodhenge*
Plan of the ellipses. Annotated 'Survey
    was made with a tape having a stretch
    of 0.6% (to 0.5 at 50 ft). Hence the
    megalithic fathom would have meas-
    ured 5.44 / 1.006 or 5.41 ft. Hence 5.41
    ft is the unit used in setting out this
    diagram. It is thus applicable to the
    survey plotted with no stretch
    correction'.

No scale
Black ink
556 mm × 475 mm
Tracing paper

Yorkshire

FYLINGDALES
Standing Stones
NZ90SW 21   NZ920 037
DC4542
L6/2
Insc. *Fylingdales*
Plan of standing stones.
1 in : 4 ft
Black ink and pencil
558 mm × 380 mm

STAINTON DALE
Cairn Circle
SE99NE 5   SE98249698
DC4543
L6/3
Insc. *Stainton Dale*
Plan of stone circle. Annotated
    'Somewhat rough and incomplete
    survey of a cairn circle. The level
    inside is still high. Filled with stones.
    Remainder probably built into wall
    which passes close [by]'.
1 in : 4 ft
Black ink and pencil
558 mm × 380 mm

FRANCE

'Morbihan'

CARNAC
DC4697
Insc. *Sites near the Carnac Alignments*
Plan showing site locations in the Carnac area including Kerlescan, Petit-Ménec, Crucuny, Champ de Menhirs and Kervilor.
1 cm : 75 m
Black ink
612 mm × 485 mm
Plastic film

CARNAC
DC4697/c
Annotated dyeline copy of DC4697.
No scale
Dyeline
690 mm × 512 mm

CARNAC
DC4698
Sketch contour plan of the Carnac area.
Annotated '20/1/85 Perhaps A.T. made this from the French maps using pantograph?. It's the Carnac area. A.T. I still have the air-photo'.
No scale
Pencil
761 mm × 557 mm

CHAMP DE LA CROIX, CRUCUNY
Megalithic Enclosure
DC4750
Insc. *Crucuny, Champ de la Croix*
Annotated plan of the enclosure and the menhir at Le Manio.
³₈ in : 10 ft
Black ink
360 mm × 313 mm

CHAMP DE MENHIR, CARNAC
Standing Stones
DC4687
Insc. *Champ de Menhir*
Annotated plan of standing stones.
Annotated 'Surveyed by A.S.T. Key put on later by Dr T.R. Foord, an "independent assessor of the look of the stones"'.
¾ in : 10 ft
Pencil
558 mm × 378 mm
1971

CHAMP DE MENHIRS, CARNAC
Standing Stones
DC4743
Insc. *Champ de Menhirs*
Plan showing the positions of the stones. Annotated 'Some of the stones shown may be outcrops but those shown [by] hatches are considered to be either upright or fallen menhirs. The site is on a low flat topped hill, the highest point is near *A*.'
¾ in : 10 ft
Black ink
403 mm × 301 mm
Plastic film

CRUCONO
Megalithic Enclosure
DC4699/co
Insc. *Cruconno, Plouharnel*
Xerox copy of Dryden's plan of the stones. Annotated 'Ground plan of square of stones near Courconna-Morbihan with exact positions of the stones.' and 'A.S.T. surveyed this the

first trip.   This is Dryden's (see bottom) A.S.T. 20/1/85'.
No scale
Photocopy
363 mm × 248 mm

CRUCUNY, PLOUHARNEL
Megalithic Enclosure and Standing Stone
DC4696
Insc. *Crucuny*
Annotated plan of the enclosure and standing stone. Annotated 'The discrepancy in Az[imuth] of Le Maneo stone evidently arises because this is on a different sheet of the 2½" maps. The az[imuth] from the photo map seems reliable but we do not know the kind of error from e.g. the difference in altitude (level)'.
¾ in : 10 ft
Pencil
764 mm × 557 mm

KERLESCAN, CARNAC
Megalithic   Enclosure   and   Stone Alignments
DC4695
Insc. *Kerlescan/Petit Menec*
Plan of the stone alignments and enclosure. Annotated 'Pantograph used from the 1:500's to make this. A.S.T. 20/1/85'.
1 in : 170 ft
Black ink and pencil
672 mm × 407 mm

KERLESCAN, CARNAC
Megalithic   Enclosure   and   Stone Alignments
DC4735
Insc. *Kerlescan & Petit Menec*
Annotated plan of the enclosure and alignments.
No scale
1207 mm × 323 mm
Plastic film

KERLESCAN, CARNAC
Megalithic Enclosure
DC4736
Insc. *Kerlescan West Cromlech*
Plan of the enclosure. Annotated 'barrel-shaped cromlech, Kerlescan'.
1 cm : 5 m
Black ink
312 mm × 240 mm
Linen backed tracing paper

KERLESCAN, CARNAC
Megalithic Enclosure
DC4738/c
Plan of enclosure.
1 cm : 10 m
Annotated dyeline
437 mm × 387 mm

KERLESCAN, CARNAC
Megalithic Enclosure
DC4739
Plan of enclosure.
No scale
Black ink
463 mm × 364 mm
Polythene

KERLESCAN, CARNAC
Megalithic Enclosure
DC4740
Plan of enclosure.
1 cm : 10 m
Black and orange ink
557 mm × 408 mm

KERLESCAN, CARNAC
Megalithic Enclosure
DC4744
Insc. *Kerlescan cromlech*
Plan of the enclosure.
1 in : 50 ft
Black ink
478 mm × 371 mm
Linen backed tracing paper

KERLESCAN, CARNAC
Megalithic Enclosure
DC4690
Annotated plan of the enclosure
Annotated 'Essential dimensions are given in the figure in rods etc.'
No scale
Black ink
332 mm × 246 mm
Linen backed tracing paper

KERLESCAN, CARNAC
Megalithic Enclosure
DC4737
Plan showing the geometry of the enclosure.
No scale
Black ink
298 mm × 198 mm
Linen backed tracing paper

KERMARIO, CARNAC
Stone Alignmnents
DC4682
Annotated plan of alignments.
Annotated 'Note 20/1/85. Kermario 1/500 plan. The windmill is above. A.S. Thom'.
1 : 500
Black ink and pencil
3730 mm × 434 mm

KERMARIO, CARNAC
Stone Alignments
DC4688
Insc. *Kermario. Key plan showing sections for analysis*
Plan of the alignments.
No scale
Dyeline
514 mm × 157 mm

KERMARIO, CARNAC
Stone Alignments
DC4734

Insc. *Kermario Alignments*
Annotated plan of the alignments.
Annotated 'By pantograph for Giot, Rennes'.
1 : 1000
Black ink
1377 mm × 258 mm
Plastic film

KERMARIO, CARNAC
Stone Alignments
DC4741
Insc. *The triangles which produce the knees*
Diagram showing a geometrical interpretation of the alignments.
No scale
Black ink
307 mm × 220 mm
Plastic film

KERVILOR, CARNAC
Standing Stones
DC4685
Insc. *Kervilor*
Annotated plan of standing stones.
1 in : 100 ft
Blue ink and pencil
559 mm × 378 mm
1971

KERVILOR, CARNAC
Megalithic Enclosures, Standing Stones and Stone Alignments
DC4685 (reverse)
Insc. *Alignments, Cromlechs and Associated Menhirs near Carnac*
Sketch plan showing the location of the monuments.
No scale
Pencil
559 mm × 378 mm

KERVILOR, CARNAC
Stones

DC4686
Insc. *Kervilor, Lower Site (Bulldozed)*
Annotated plan of site. Annotated
'20/1/85. We were taken to this
bulldozed site which was in a man's
garden. He was going to build a house
on it. We fixed the boulders on the
small scale plan. A.S.T.'.
$1\frac{1}{2}$ in : 100 ft
Pencil
557 mm × 357 mm

LE MANIO, CARNAC
Standing Stone
DC4696
Insc. *Le Manio menhir*
Elevation drawing and sketch of standing
stone.
No scale
Pencil
764 mm × 557 mm

LE MANIO, CARNAC
Standing Stone and Enclosure
DC4742
Plan of standing stone *M* at Le Manio and
the enclosure with an inset showing the
ground plan of menhir *L*.
1 cm : $2\frac{1}{2}$ m
Black ink
421 mm × 337 mm
Plastic film

LE MÉNEC, CARNAC
Megalithic   Enclosure   and   Stone
Alignments
DC4681
Insc. *Le Ménec-West End*
Plan of the stone alignments and
enclosure.
1 : 500
Black ink and pencil
1848 mm × 458 mm
1970

LE MÉNEC, CARNAC
Megalithic   Enclosure   and   Stone
Alignments
DC4694
Insc. *East Cromlech, Le Ménec*
Annotated plan of stone alignments and
enclosure. Annotated 'Replotted after
1971 remeasurement'.
$\frac{3}{8}$ in : 10 ft
Black ink and pencil
758 mm × 559 mm

LE MÉNEC, CARNAC
Megalithic   Enclosures   and   Stone
Alignments
DC4730
Insc. *Le Ménec Alignments*
Annotated plan of the alignments and
enclosures at the E and W ends.
1 : 1000
Black ink
1200 mm × 261 mm
Plastic film

LE MÉNEC, CARNAC
Megalithic   Enclosure   and   Stone
Alignments
DC4731
Insc. *Le Ménec West End*
Annotated plan of the alignments and the
W end enclosure.
$\frac{1}{2}$ in : 20 ft
Black ink
458 mm × 338 mm
Linen backed tracing paper

LE MÉNEC, CARNAC
Megalithic Enclosure
DC4732/co
Insc. *West Cromlech Le Ménec*
Annotated plan of stone alignments and
enclosure.
1 cm : 5 m
Dyeline
497 mm × 314 mm

LE MÉNEC, CARNAC
Megalithic Enclosures and Stone
 Alignments
DC4733/co
Insc. *Le Ménec*
Annotated plan of the alignments and the
 enclosures at the E and W ends.
½ in : 100 ft
Annotated dyeline
639 mm × 231 mm

LE MÉNEC, CARNAC
Megalithic Enclosures and Stone
 Alignments
DC4749
Insc. *Le Ménec*
Plan of the stone alignments and
 enclosures.
No scale
Black ink and pencil
1204 mm × 409 mm

LE MÉNEC, CARNAC
Megalithic Enclosure and Stone
 Alignments
DC4692
Insc. *East Cromlech, Le Ménec*
Annotated plan of the enclosure and stone
 alignments.
⅜ in : 10 ft
Black ink and pencil
770 mm × 517 mm
Linen backed tracing paper

LE MÉNEC, CARNAC
Megalithic Enclosure and Stone
 Alignments
DC4693
Insc. *Le Ménec (East) 1st survey 1970*
Annotated plan of the stone alignments
 and enclosure. Annotated 'We revised
 it next year, 1971. A.S.T. 20/1/85'.
1 : 500
Pencil
557 mm × 392 mm

LE MÉNEC, CARNAC
Megalithic Enclosure
DC4751
Insc. *Cromlech at Menec by A.T.*
Plan of the west enclosure.
½ in : 10 ft
Black ink and pencil
758 mm × 560 mm

PETIT-MÉNEC, CARNAC
Stone Alignments
DC4684
Insc. *Little (Petit) Ménec*
Annotated plan of stone alignments.
1 : 500
Pencil
757 mm × 560 mm

PETIT-MÉNEC, CARNAC
Stone Alignment
DC4683
Insc. *Western end of Petit Ménec*
Annotated plan of alignments with sketch
 on back. Annotated 'Link up thro'
 Lane to main road of the small scale
 plan and coords of these stones. A.S.T.
 21/1/85'.
1 : 500
Pencil
559 mm × 378 mm

ST-PIERRE-QUIBERON
Megalithic Enclosure
DC4691
Insc. *St. Pierre Quiberon*
Annotated plan of enclosure.
Annotated 'Menhirs as they now stand
 with a Type 1 egg superimposed'.
No scale
Black ink
423 mm × 341 mm
Linen backed tracing paper

ST-PIERRE-QUIBERON
Megalithic Enclosure
DC4745
Insc. *AT' s plot of St. Pierre*
Plan of the enclosure.
³₈ in : 10 ft
Pencil
560 mm × 380 mm

ST-PIERRE-QUIBERON
Megalithic Enclosure
DC4746
Insc. *Cromlech at St. Pierre, Quiberon*
Plans of the enclosure and its geometry.
³₈ in : 10 ft
Dyeline and black ink
387 mm × 280 mm

ST-PIERRE-QUIBERON
Megalithic Enclosure
DC4747
Plan of the enclosure.
No scale
Black ink
305 mm × 280 mm
Linen backed tracing paper

ST-PIERRE-QUIBERON
Stone Alignments
DC4748
Insc. *St. Pierre alignments*
Plan of the alignments.
1 : 500
Pencil
758 mm × 558 mm

ST-PIERRE-QUIBERON
Stones
DC4689
Insc. *Beach at St. Pierre*
Plan of stones.
1¹₂ in : 100 ft
Black ink
394 mm × 259 mm
Linen backed tracing paper

SCOTLAND

Aberdeenshire

ARDLAIR
Recumbent Stone Circle
NJ52NE 4   NJ55272794
DC4398
B1/18
Insc. *Holywell*
Plan of recumbent stone circle and
    outlying stones *A* and *B*.
Annotated 'Tape stretch: 0.65% in full
    length'.
1¹₂ in : 10 ft
Blue ink and pencil
558 mm × 380 mm

ARDLAIR
(see above)
DC4761/co
B1/18
Insc. *Holywell or Ardlair*
Plan of recumbent stone circle with inset
    showing positions of outlying stones *A*
    and *B*.
1¹₂ in : 10 ft
Dyeline copy
298 mm × 258 mm

AUCHNAGORTH (UPPER)
Stone Circle
NJ85NW 2   NJ83905627
DC4386
B1/5
Insc. *Upper Auchnagorth*
Plan of stone circle. Annotated 'snow
    shower during survey'.
1¹₂ in : 10 ft
Pencil
556 mm × 380 mm

BALQUHAIN
Recumbent Stone Circle
NJ72SW 2   NJ73502408
DC4393
B1/11
Insc. *Balquhain, Chapel Garroch*
Annotated plan of recumbent stone circle.
⅛ in : 1 ft
Blue ink and pencil
558 mm × 369 mm

BROOMEND OF CRICHIE
Stone Circle and Henge Monument
NJ71NE 6   NJ77921967
DC4413
B2/12
Insc. *Broomend of Crichie*
Plan of stone circle and henge monument.
⅛ in : 1 ft
Pencil
558 mm × 380 mm

CASTLE FRASER
Recumbent Stone Circle
NJ71SW 3   NJ71501253
DC4407
B2/3
Insc. *West Mains, Castle Fraser*
Plan of recumbent stone circle and outliers.
1 in : 8 ft
Black ink
557 mm × 374 mm

CHAPEL O' SINK (THE)
Cairn
NJ71NW 4   NJ70601895
DC4397
B1/16
Insc. *Westerton*
Plan of cairn.
⅛ in : 1 ft
Pencil
558 mm × 379 mm

CULLERLIE
Stone Circle
NJ70SE 2   NJ78500428
DC4410
B2/7
Insc. *Cullerlie (Echt)*
Plan of stone circle with a circle and ellipse superimposed. Annotated '8 rings of stones inside circle, of which the individual stones were surveyed in one case only'.
1 in : 4 ft
Black ink and pencil
558 mm × 324 mm

DEERPARK
Stone Circle
NJ61NE 1   NJ68331564
DC4412
B2/10
Insc. *Monymusk*
Plan of stone circle.
1 in : 4 ft
Black ink and pencil
557 mm × 386 mm

EASTER AQUORTHIES
Recumbent Stone Circle
NJ72SW 12   NJ73232079
DC4387
B1/6
Insc. *Aquhorthies, Manar (Newbiggin)*
Plan of recumbent stone circle.
⅛ in : 1 ft
Blue ink and pencil
556 mm × 380 mm

EASTER AQUORTHIES
(see above)
DC4763/co
B1/6
Insc. *Aquorthies Manar*
Plan of recumbent stone circle.
¼ in : 2 ft
Dyeline copy
290 mm × 277 mm

INSCHFIELD
Recumbent Stone Circle
NJ62NW 6   NJ62332934
DC4395
B1/14
Insc. *Inchfield*
Plan of recumbent stone circle.
$\frac{1}{8}$ in : 1 ft
Blue ink and pencil
558 mm × 324 mm

KIRKTON OF BOURTIE
Recumbent Stone Circle
NJ82SW 2   NJ80082487
DC4388
B1/7
Insc. *Kirktown of Bourtie*
Plan of recumbent stone circle.
$\frac{1}{8}$ in : 1 ft
Blue ink and pencil
556 mm × 380 mm

LOANHEAD OF DAVIOT
Recumbent Stone Circle
NJ72NW 1   NJ74772885
DC4402
B1/26
Insc. *Loanhead of Daviot*
Annotated plan of the recumbent stone
    circle   and   enclosed   cremation
    cemetery.
$1\frac{1}{2}$ in : 10 ft
Blue ink and pencil
558 mm × 380 mm

LOANHEAD OF DAVIOT
(see above)
DC4403
B1/26
Insc. *Daviot*
Plan of the recumbent stone circle and
    enclosed   cremation   cemetery.
    Annotated '(Rough Survey 1962). See
    1963 survey but also see back of
    sheet'.
$1\frac{1}{2}$ in : 10 ft

Pencil
558 mm × 380 mm
1962

LOANHEAD OF DAVIOT
(see above)
DC4403 (reverse)
B1/26
Insc. *Loanhead of Daviot*
Plan of recumbent stone circle.
$1\frac{1}{2}$ in : 10 ft
Pencil
558 mm × 380 mm

MAINS OF HATTON
Recumbent Stone Circle
NJ64SE 6   NJ69934254
DC4401
B1/25
Insc. *Charlesfield*
Plan of recumbent stone circle.
$\frac{1}{8}$ in : 1 ft
Blue ink and pencil
558 mm × 380 mm

MAINS OF HATTON
(see above)
DC4762/co
B1/25
Insc. *Charlesfield*
Plan of recumbent stone circle.
$\frac{1}{4}$ in : 2 ft
Dyeline copy
377 mm × 339 mm

MELGUM
Stone Circle
NJ40NE 1   NJ47140524
DC4411
B2/8
Insc. *Tarland*
Plan of stone circle.
1 in : 10 ft
Black ink and pencil
556 mm × 372 mm

MIDMAR KIRK
Recumbent Stone Circle
NJ60NE 3   NJ69940649
DC4417
B2/17
Insc. *Midmar Church*
Plan of recumbent stone circle with elevations of the stones. Annotated 'azimuth to be calculated from field book'.
$1\frac{1}{2}$ in : 10 ft
Blue ink and pencil
559 mm × 379 mm

OLD RAYNE
Recumbent Stone Circle
NJ62NE 1   NJ67982799
DC4394
B1/13
Insc. *Old Rayne*
Plan of recumbent stone circle.
$\frac{1}{8}$ in : 1 ft
Blue ink and pencil
558 mm × 381 mm

RINGING STONE (THE)
Recumbent Stone Circle
NJ54NW 12   NJ53164564
DC4396
B1/15
Insc. *Rothiemay*
Plan of recumbent stone circle.
$\frac{1}{10}$ in : 1 ft
Pencil
558 mm × 379 mm

SANDS OF FORVIE
Ring Cairn
NK02NW 3   NK01082628
DC4404
B1/27
Insc. *Sands of Forvie*
Annotated plans of ring cairn.
1 in : 10 ft & $1\frac{1}{2}$ in : 10 ft
Pencil
558 mm × 380 mm

SHELDON
Stone Circle
NJ82SW 1   NJ82292493
DC4389
B1/8
Insc. *Sheldon of Bourtree*
Plan of stone circle.
Annotated 'incomplete survey, see repeat'.
$\frac{1}{8}$ in : 1 ft
Pencil
556 mm × 380 mm

SHELDON
(see above)
DC4390
B1/8
Insc. *Sheldon*
Plan of stone circle with elevation of a stone.
$\frac{1}{8}$ in : 1 ft
Blue ink and pencil
556 mm × 376 mm

SHELDON
(see above)
DC4758/co
B1/8
Insc. *Sheldon of Bourtie*
Annotated plan of stone circle.
$\frac{1}{4}$ in : 2 ft
Dyeline copy
466 mm × 324 mm

SHETHIN
Cairn
NJ83SE 6   NJ88153280
DC4392
B1/10
Insc. *Fountain Hill, Tarves*
Annotated plan of cairn.
$\frac{1}{4}$ in : 1 ft
Pencil
558 mm × 380 mm

SOUTH LEY LODGE
Recumbent Stone Circle
NJ71SE 3   NJ76671325
DC4415
B2/14
Insc. *Leylodge*
Annotated plan of recumbent stone circle.
$\frac{1}{8}$ in : 1 ft
Blue ink and pencil
558 mm × 382 mm

SOUTH LEY LODGE
(see above)
DC4764/co
B2/14
Insc. *Leylodge*
Plan of recumbent stone circle.
$\frac{1}{4}$ in : 2 ft
Photocopy
252 mm × 250 mm

SOUTH YTHSIE
Stone Circle
NJ83SE 12   NJ88493039
DC4391
B1/9
Insc. *South Ythsie*
Plan of stone circle.
$\frac{1}{4}$ in : 1 ft
Blue ink
558 mm × 381 mm

STRICHEN HOUSE
Recumbent Stone Circle
NJ95SW 2   NJ93675447
DC4384
B1/1
Insc. *Strichen*
Plan of recumbent stone circle. Annotated
   'said to be re-erected'.
$1\frac{1}{2}$ in : 10 ft
Pencil
556 mm × 380 mm

SUNHONEY
Recumbent Stone Circle
NJ70NW 9   NJ71590569
DC4406
B2/2
Insc. *Sunhoney*
Plan of recumbent stone circle.
$1\frac{1}{2}$ in : 10 ft
Black ink
558 mm × 380 mm

TOMNAGORN
Recumbent Stone Circle
NJ60NE 1   NJ65130774
DC4416
B2/16
Insc. *Tomnagorn*
Annotated plan of recumbent stone circle.
$1\frac{1}{2}$ in : 10 ft
Blue ink and pencil
559 mm × 378 mm

TOMNAGORN
(see above)
DC4771
B2/16
Insc. *Tomnagorn*
Plan of recumbent stone circle.
$1\frac{1}{2}$ in : 10 ft
Black ink
465 mm × 341 mm
Plastic film

TOMNAVERIE
Recumbent Stone Circle
NJ40SE 1   NJ48650349
DC4411
B2/9
Insc. *Tomnaverie*
Plan of recumbent stone circle.
1 in : 8 ft
Black ink and pencil
556 mm × 372 mm

TYREBAGGER
Recumbent Stone Circle
NJ81SE 11   NJ85951321
DC4405
B2/1
Insc. *Standing Stone Farm, Dyce*
Annotated plan of recumbent stone circle.
1½ in : 10 ft
Black ink and pencil
556 mm × 380 mm

WANTONWELLS
Recumbent Stone Circle
NJ62NW 2   NJ61872729
DC4600
B1/12
Insc. *Wantonwells*
Plan of recumbent stone circle.
⅛ in : 1 ft
Pencil
558 mm × 379 mm

WESTER ECHT
Recumbent Stone Circle
NJ70NW 2   NJ73850834
DC4414
B2/13
Insc. *New Wester Echt*
Plan of recumbent stone circle. Annotated
    'This is the number (B2/13) given on
    list for A. Burl'.
⅛ in : 1 ft
Pencil
558 mm × 366 mm

WHITE COW WOOD
Ring Cairn (possible)
NJ95SW 5   NJ94725192
DC4385
B1/3
Insc. *White Cow Wood*
Plan of cairn. Annotated 'North point
    uncertain' and '(Ring of stones on
    inside of bank)'.
1½ in : 10 ft

Pencil
556 mm × 380 mm

WHITEHILL WOOD
Recumbent Stone Circle
NJ61SW 3   NJ64321350
DC4418
B2/18
Insc. *Tillyfourie Hill*
Plan of recumbent stone circle.
⅛ in : 1 ft
Pencil
559 mm × 379 mm

WHITEHILL WOOD
(see above)
DC4774/co
B2/18
Insc. *Tillyfourie Hill*
Plan of recumbent stone circle.
⅛ in : 1 ft
Dyeline copy
306 mm × 267 mm

YONDER BOGNIE
Recumbent Stone Circle
NJ64NW 15   NJ60064577
DC4399
B1/23
Insc. *Yonder Bognie*
Annotated plan of recumbent stone circle.
⅛ in: 1 ft
Blue ink and pencil
558 mm × 380 mm

YONDER BOGNIE
(see above)
DC4760/co
B1/23
Insc. *Yonder Bognie*
Annotated plan of recumbent stone circle.
¼ in : 2 ft
Dyeline copy
375 mm × 307 mm

Angus

BLACKGATE
Stone Circle
NO45SE 8   NO48445286
DC4593
P3/2
Insc. *Blackgate*
Annotated plan of stone circle.
$\frac{1}{10}$ in : 1 ft
Black ink and pencil
513 mm × 383 mm

COROGLE BURN
Standing Stones
NO36SW 2   NO34886017
                           & NO34876014
DC4595
P3/1
Insc. *Glen Prosen*
Plan of standing stones.
$\frac{1}{8}$ in : 1 ft
Black ink and pencil
558 mm × 385 mm

COROGLE BURN
(see above)
DC5348
P3/1
Insc. *Glen Prosen*
Plan of standing stones.
$\frac{3}{8}$ in : 10 ft
Black ink
255 mm × 204 mm

Argyll

ACHACHA
Cairn
NM94SW 8   NM94364076
DC4551
M8/1

Insc. *Loch Creran*
Annotated plan of cairn.
1 cm : 2 ft
Blue ink and pencil
557 mm × 382 mm

ACHARRA
Standing Stone
NM95SE 3   NM98665455
DC5329
M7/1
Insc. *Acharra Duror, Appin*
Plan and elevations of standing stone.
$\frac{3}{10}$ in : 6 ft
Black ink
68 mm × 63 mm

ARDNACROSS, MULL
Cairns and Standing Stones
NM54NW 3   NM541 491
DC5325
M1/9
Insc. *Sketch of site at Ardnacross*
Annotated plan of cairns and standing
   stones with elevation of the upright
   stone and hillside sketches. Annotated
   'There is now only one upright stone.
   Another outlier (fallen) lies about
   30 yds. north of circle'.
$\frac{3}{8}$ in : 10 ft
Black ink
203 mm × 91 mm
1947

ARDNAVE, ISLAY
Enclosure
NR27SE 10   NR27347327
DC4380
A7/22
Insc. *Ardnave, Islay*
Annotated plan of enclosure.
1 : 50
Black ink and pencil
317 mm × 240 mm

ARNICLE
Standing Stone
NR73NW 7   NR73493506
DC4376
A4/9
Insc. *Stones near Ben Turc distances by
    pacing only*
Plan of stones.
Annotated 'This is evidently McKay's
    Cross, and is A4/9. AST 18/1/85'.
1 in : 20 yds
Black ink and pencil
560 mm × 378 mm

ARNICLE
(see above)
DC4377
A4/9
Insc. *Cross Mhic-Aoida, Ben Turc*
Plan of stones.
1 : 500
Black ink and pencil
574 mm × 386 mm

ARNICLE
(see above)
DC4824
A4/9
Plan of stones with profiles showing the
    position of the moon at Beinn
    Bheigeir, Islay and Cnoc Moy.
⅛ in : 10 ft
Black ink
346 mm × 246 mm
Plastic film

AUCHNAHA
Chambered Cairn
NR98SW 4   NR93298170
DC4382
A10/3
Insc. *Ballimore near Kilfinnan*
Annotated plan of chambered cairn with
    sight lines to outliers and hills.
No scale
Black and blue ink

560 mm × 380 mm
1951

AUCHNAHA
(see above)
DC4382
A10/3
Plan of stones.
⅜ in : 10 ft
Pencil
560 mm × 380 mm

AUCHNAHA
(see above)
DC4840
A10/3
Insc. *Site at Ballimore near Kilfinnan*
Annotated plan of chambered cairn.
¹⁄₁₀ in : 2 ft
Black ink
203 mm × 168 mm
1951

AVINAGILLAN
Standing Stone
NR86NW 1   NR83906745
DC4819
A3/8
Insc. *Stone at W[est] L[och] Tarbert,
    Knapdale*
Plan and elevation of standing stone.
No scale
Black ink
65 mm × 57 mm
1939

BALLINABY, ISLAY
Standing Stone
NR26NW 13   NR21996720
DC4837
A7/5
Plan of standing stone with profile
    showing position of the setting moon
    behind the horizon.
No scale

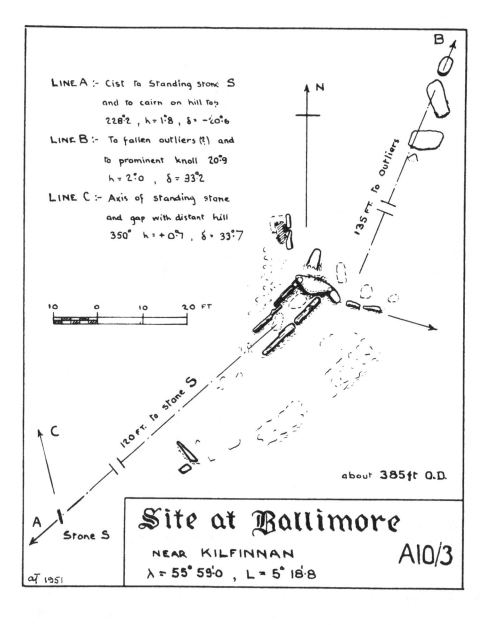

LINE A :- Cist to Standing stone S
and to cairn on hill top
228°2 , h = 1°8 , δ = -20°6

LINE B :- To fallen outliers (?) and
to prominent knoll 20°9
h = 2°0 , δ = 33°2

LINE C :- Axis of standing stone
and gap with distant hill
350° h = +0°7 , δ = 33°7

N

B

135 ft. to outliers

10   0   10   20 FT

120 FT. To stone S

about **385ft** O.D.

C

A   Stone S

Site at Ballimore

NEAR KILFINNAN
λ = 55° 59·0 , L = 5° 18·8

A10/3

aT 1951

Fig. 4.3. Auchnaha, Argyll: Plan of the chambered cairn with sight lines to outliers and hills
(A. Thom; DC4840).

Black ink
312 mm × 216 mm
Plastic film

BALLISCATE, MULL
Standing Stones
NM45SE 1   NM49965413
DC5324
M1/8
Insc. *Sketch of site near Tobermory*
Annotated plan and elevation of standing
   stones.
$\frac{1}{2}$ in : 10 ft
Black ink
103 mm × 80 mm
1946

BALLOCHROY
Standing Stones
NR75SW 3   NR73095242
DC4533
A4/4
Insc. *Ballochroy, Kintyre*
Annotated plan of standing stones.
$\frac{1}{8}$ in : 1 ft
Pencil
561 mm × 377 mm

BALLOCHROY
(see above)
DC4821
A4/4
Insc. *Ballochroy, Kintyre*
Plan of standing stones with drawings of
   the midsummer sunset over Corra
   Bheinn, Jura, and midwinter sunset
   over Cara Island
No scale
Black ink
157 mm × 110 mm
1939

BALLOCHROY
(see above)
DC4822

A4/4
Plan of the standing stones with horizon
   profiles showing the sun setting over
   Cara Island and Corra Bheinn, Jura.
$\frac{1}{16}$ in : 1 ft
Black ink
204 mm × 155 mm
Plastic film

BALLYMEANOCH
Standing Stones, Henge Monument and
   Kerb Cairn
NR89NW 14   NR83379641
NR89NW 18   NR83319627
NR89NW 40   NR83399642
DC4800
A2/12
Insc. *Duncracaig, Argyllshire*
Annotated plan of the standing stones,
   henge monument and kerb cairn.
$\frac{3}{8}$ in : 10 ft
Black ink and pencil
761 mm × 558 mm

BALLYMEANOCH
Standing Stones and Kerb Cairn
NR89NW 14   NR83379641
NR89NW 40   NR83399642
DC4811
A2/12
Plan of the standing stones and kerb cairn
   with horizon profiles showing the
   setting position of the moon.
No scale
Black ink
240 mm × 225 mm
Plastic film

BARBRECK
Standing Stones
NM80NW 19   NM83150641
DC4360
A2/3
Insc. *Barbreck House*
Survey drawings of the standing stones.
$\frac{1}{8}$ in : 1 ft & $\frac{1}{20}$ in : 1 ft

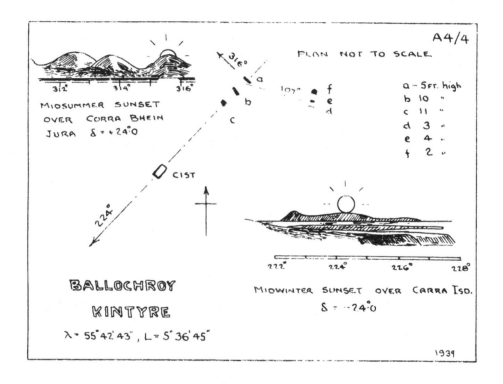

Fig. 4.4. Ballochroy, Argyll: Plan of the standing stones with drawings of the midsummer sunset over Corra Bheinn, Jura and midwinter sunset over Cara Island (A. Thom; DC4821)

Pencil
560 mm × 380 mm

BARBRECK
(see above)
DC4845
A2/3
Insc. *Barbreck House*
Plan of standing stones. Two of the stones
  are annotated 'Suggested circle'.
¹₂ in : 4 ft
Black ink
392 mm × 255 mm

BARNASHAIG
Standing Stone with Cup-marks and
  Standing Stones
NR78NW 1   NR72988640
NR78NW 6   NR72808613
NR78NW 7   NR72698594
DC4373
A3/4
Insc. *Tayvallich*
Annotated plan of the standing stones
  with elevation of Stone *A* and sketch
  showing view from Circle *B* towards
  Cnoc Reamhar.
1¹₂ in : 100 ft
Pencil
558 mm × 379 mm

BARNASHAIG
Standing Stone
NR78NW 1   NR72988640
DC4669
A3/4
Insc. *Tayvallich, Profile from North Stone*
  (1970)
Sketch plan.
No scale
Pencil
302 mm × 239 mm
Graph paper
1970

BARNLUASGAN
Enclosure
NR79SE 19   NR78289070
DC4372
A3/3
Insc. *Bellanoch*
Annotated plan of enclosure.
¹₈ in : 1 ft
Blue ink and pencil
558 mm × 384 mm

BARR LEATHAN, MULL
Stone Setting
NM73SW 4   NM72753425
DC4546
M2/2
Insc. *Loch Don, Mull*
Plan of stone setting and sketch of stones
  on moor. Annotated 'Some notes in the
  card index might clear this DUART
  site up'.
¹₈ in : 1 ft
Pencil
555 mm × 378 mm

BEACHARR
Standing Stone
NR64SE 2   NR69264331
DC4823
A4/5
Horizon profile showing the moon setting
  behind Beinn an Oir, Jura, and Beinn
  Shiantaidh, Jura.
No scale
Black ink
224 mm × 191 mm
Plastic film

CAMAS AN STACA, JURA
Standing Stone
NR46SE 1   NR46416479
DC4830
A6/1
Insc. *Stone at Camus An Stacca*

Annotated plan and elevation of standing
  stone with sketch of Sgorr nam
  Faoileann, Islay.
No scale
Black ink
115 mm × 95 mm
1949

CAMAS AN STACA, JURA
(see above)
DC4831
A6/1
Insc. *Camus an Stacca, Jura*
Chart showing the moon setting behind
  Beinn Bheigeir, Islay
No scale
Black ink
198 mm × 124 mm
Plastic film

CARNASSERIE
Standing Stones
NM80SW 22   NM83450078
                           & NM83450079

DC4809
A2/6
Insc. *Carnasserie*
Sketch of standing stones.
No scale
Black ink
198 mm × 70 mm
Plastic film

CARRAGH AN TARBERT, GIGHA
Standing Stone
NR65SE 22   NR65555228
DC4826
A4/17
Horizon profiles showing the moon
  setting behind Beinn Shiantaidh, Jura,
  and rising behind Meall Reamhar.
No scale
Black ink
210 mm × 208 mm
Plastic film

CARSE
Standing Stones
NR76SW 1   NR74256163
                           & NR74146166

DC4374
A3/6
Insc. *Loch Stornoway*
Plan and elevations of standing stones.
No scale
Pencil
550 mm × 374 mm

CARSE
(see above)
DC4375
A3/6
Insc. *Loch Stornoway*
Annotated plan of standing stones.
3 in : 100 ft
Pencil
565 mm × 376 mm

CARSE
(see above)
DC4817
A3/6
Insc. *Loch Stornoway*
Annotated plan and elevations of standing
  stones.
½ in : 10 ft
Black ink
151 mm × 92 mm
1939

CNOC A'CHARRAGH, COLONSAY
Standing Stone
NR49NW 3   NR42669940
DC4828
A5/2
Insc. *Colonsay, near North End*
Annotated plan and elevation of standing
  stone with sketch of horizon to the
  west.
No scale
Black ink
95 mm × 88 mm
1948

CNOC FADA, DERVAIG, MULL
Standing Stones
NM45SW 5   NM43595305
DC4544
M1/5
Insc. *Standing stones N. of Dervaig (about ¹⁄₂m N of group near Tober. Road)*
Annotated plan of standing stones. Annotated 'Dr James Orr did this in the mid 1930's AST 19.1.85.'.
¹⁄₄ in : 1 ft
Pencil
556 mm × 379 mm

CNOC FADA, DERVAIG, MULL
Standing stones
NM45SW 4   NM43905203
DC4544
M1/5
Insc. *Standing stones above Dervaig, near Tobermory Road*
Annotated plan of standing stones with skyline elevations.
¹⁄₈ in : 1 ft
Pencil
556 mm × 379 mm

CNOC FADA, DERVAIG, MULL
(see above)
DC4899/co
M1/5
Insc. *Alignment near Dervaig, Mull*
Plan of stones.
¹⁄₂ in : 10 ft
Photocopy
206 mm × 94 mm

CRETSHENGAN
Standing Stone
NR76NW 3   NR70726688
DC4818
A3/5
Profile showing position of setting sun on Dubh Bheinn, Jura, from standing stone.

No scale
Black ink
311 mm × 205 mm

CULTOON, ISLAY
Stone Circle
NR15NE 1   NR19565698
DC4378
A7/15
Insc. *Cooltoon, Islay*
Plan of stone circle.
1 : 250
Black ink and pencil
557 mm × 378 mm

CULTOON, ISLAY
(see above)
DC4379
A7/15
Insc. *Cooltoon, Islay*
Plan of stone circle.
1 : 250
Black ink
516 mm × 334 mm
Linen backed tracing paper
1973

DUACHY
Standing Stones
NM82SW 1   NM80142052
DC4358
A1/4
Insc. *Loch Seil*
Survey drawing of the standing stones.
1 in : 20 ft
Pencil
560 mm × 380 mm
1949

DUACHY
(see above)
DC4844
A1/4
Insc. *Stones Near Loch Seil*
Plan of standing stones.
1 in : 30 ft

Black ink
183 mm × 80 mm
1947

DUNAMUCK
Standing Stones
NR89SW 27   NR84839248
DC4369
A2/14
Insc. *Dunamuck II*
Plan and elevation of the stones with a
    sketch of a hillside.
$^1_8$ in : 1 ft
Pencil
560 mm × 379 mm

DUNAMUCK
(see above)
DC4812
A2/14
Insc. *Dunamuck II*
Annotated plan and elevations of standing
    stones with hillside sketches.
1 cm : 5 ft
Black ink
203 mm × 131 mm
1951

DUNAMUCK
Standing Stones
NR89SW 28   NR84709290
DC4369
A2/21
Insc. *Dunamuck I*
Plan and elevation of the stones with a
    sketch of a hillside.
$^1_8$ in : 1 ft
Pencil
560 mm × 379 mm

DUNAMUCK
(see above)
DC4813
A2/21
Insc. *Dunamuck I*

Annotated plan of standing stones with
    elevations of stones *A* and *C* and a
    hillside sketch.
$^7_8$ in : 10 ft
Black ink
199 mm × 122 mm
1951

ESCART
Standing Stones
NR86NW 2   NR84606674
DC4375
A4/1
Insc. *Escart*
Annotated plan of standing stones.
$^1_8$ in : 1 ft
Black ink and pencil
565 mm × 376 mm

ESCART
(see above)
DC4820
A4/1
Plan of standing stones with horizon
    profile showing the moon setting
    behind Sheirdrim Hill.
$^3_8$ in : 5 ft
Black ink
200 mm × 180 mm
Plastic film

GLAC MHOR, DERVAIG, MULL
Standing Stones
NM45SW 7   NM43855164
DC4900
M1/6
Insc. *Sketch of stones nr. Dervaig, Mull*
Annotated plan of stones with hillside
    sketch.
No scale
Black ink
100 mm × 78 mm

GLENGORM, MULL
Standing Stones

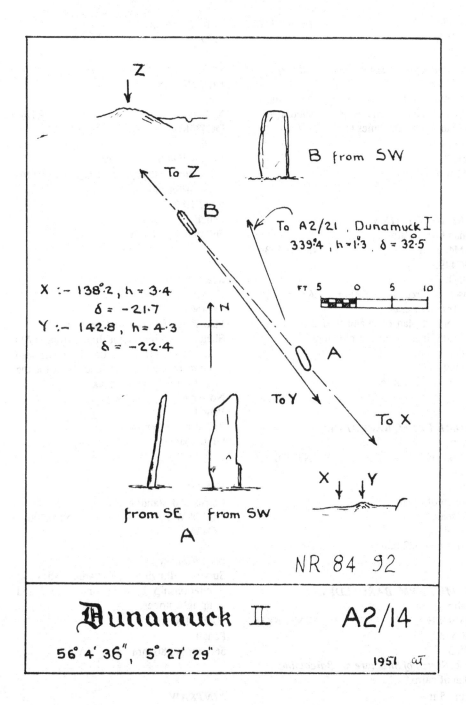

Fig. 4.5. Dunamuck, Argyll: Plan and elevation of the standing stones with hillside sketches
(A. Thom; DC4812).

NM45NW 2   NM43475713
DC4545
M1/7
Insc. *Glengorm Estate about ¼ mile from
    Castle*
Annotated plan of standing stones and
    surrounding enclosure. Annotated
    'Survey by Dr James Orr'.
³₈ in : 1 ft
Pencil
557 mm × 380 mm

GREADAL FHINN
Chambered Cairn
NM46SE 1   NM47656397
DC4495
M5/1
Insc. *Greadal Fhinn, Kilchoan,
    Ardnamurchan*
Annotated plan of chambered cairn with
    sight lines to various hill sides.
1 cm : 10 ft
Pencil
558 mm × 377 mm

HOME FARM, BARCALDINE
Cairn
NM94SE 1   NM956 416
DC4551
M8/3
Sketch of cairn.
No scale
Pencil
557 mm × 382 mm

HOME FARM, BARCALDINE
Cairn
NM94SE 1   NM956 416
DC5330
M8/3
Insc. *Sketch of Structure nr. Barcaldine*
Plan of cairn.
¼ in : 5 ft
Black ink

81 mm × 55 mm
1939

KILLUNDINE
Cairns
NM54NE 2   NM58674968
DC4358
M6/1
Insc. *Killundine, Morven*
Annotated survey drawing of cairns
    showing sight lines.
¼ in : 10 ft
Pencil
560 mm × 380 mm

KILMARTIN
Cairn
DC4668
Insc. *Kilmartin Linear Cemetery*
Sketch showing sight lines from 1700
    B.C. to 1900 B.C. from the cairn with
    chart of declinations and dates for the
    star Capella on the back.
No scale
Pencil
239 mm × 170 mm
Graph paper

KINTRAW
Cairns and Standing Stones
NM80SW 1   NM83060497
DC4361
A2/5
Insc. *Kintraw*
Survey drawing of area showing
    relationship between the cairns and
    upright stones.
1½ in : 100 ft
Pencil
560 mm × 380 mm

KINTRAW
(see above)
DC4362

A2/5
Insc. *Kintraw, Argyll*
Annotated plan of cairns and upright stones with hillside sketch.
¹₈ in : 40 ft
Black ink and pencil
560 mm × 380 mm

KINTRAW
(see above)
DC4363
A2/5
Insc. *Kintraw*
Plan of the cairns showing position in relation to ledges and upright stones.
1 in : 40 ft
Black ink and pencil
435 mm × 210 mm

KINTRAW
(see above)
DC4847
A2/5
Plan of the cairns showing position in relation to ledges and upright stones.
⁵₁₆ in : 30 ft
Black ink
205 mm × 204 mm
Plastic film

KNOCKROME, JURA
Standing Stone
NR57SW 3   NR54847144
DC4833
A6/4
Insc. *Alignment at Knockrome, Jura*
Annotated plan and elevation of standing stones with hillside sketches. Annotated '[Drawn by] A.T. and A.N. Black'.
No scale
Black ink
188 mm × 77 mm

KNOCKROME, JURA
(see above)
DC4834
A6/4
Insc. *Knockrome*
Annotated chart showing the moon setting over Crackaig Hill, Jura.
No scale
Black ink
195 mm × 138 mm
Plastic film

KNOCKSTAPPLE
Standing Stone
NR71SW 10   NR70261241
DC4827
A4/19
Profile showing the declination of the moon behind the horizon.
No scale
Black ink
196 mm × 192 mm
Plastic film

LARACH NA H-IOBAIRTE
Enclosure
NM91SE 8   NM96571423
DC4359
A1/10
Insc. *Loch Avich*
Plan of enclosure.
¹₈ in : 1 ft
Black ink and pencil
560 mm × 380 mm

LOCHBUIE, MULL
Cairn
NM62NW 4   NM61552525
DC4548
M2/14
Insc. *Loch Buie, 2nd circle*
Plan of cairn with sketches on back.
¹₈ in : 1 ft
Blue ink and pencil
561 mm × 380 mm

LOCHBUIE, MULL
(see above)
DC4549
M2/14
Insc. *Loch Buie*
Plot showing the theodolite angles.
No scale
Pencil
560 mm × 381 mm

LOCHBUIE, MULL
Stone Circle and Standing Stone
NM62NW 1   NM61782511
DC4550
M2/14
Insc. *Loch Buie*
Annotated plan and elevations of the
    stones.
^{1}₈ in : 1 ft
Blue ink and pencil
557 mm × 380 mm

LOW STILLAIG
Standing Stones
NR96NW 5   NR93166835
NR96NW 8   NR93506779
DC4373
A10/5 & A10/6
Insc. *Stillaig*
Annotated plan and elevations of standing
    stones.
No scale
Pencil
558 mm × 379 mm

LOW STILLAIG
(see above)
DC4841
A10/5 and A10/6
Profile of horizon showing the moon
    setting behind Cruach Breacain
No scale
Black ink
199 mm × 190 mm
Plastic film

MACHRIHANISH
Cairn
NR62SW 2   NR64462065
DC4825
A4/15
Insc. *Sketch of site at Machriehanish*
Annotated sketch plan of cairn.
^{3}₅ in : 40 yds
Black ink
110 mm × 92 mm
1939

MAOL MOR, DERVAIG, MULL
Standing Stones
NM45SW 5   NM43595305
DC4898
M1/4
Insc. *Alignment nr Dervaig Mull*
Plan of stone alignment and elevation of
    stones from south.
^{1}₂ in : 10 ft
Black ink
102 mm × 74 mm

MINGARY, MULL
Standing Stones
NM45NW 5   NM41355524
DC4897
M1/3
Plan of stones and profile of Carn Mhor
    summit showing the position of the
    setting moon.
No scale
Black ink
202 mm × 104 mm
Plastic film

MOINE MHOR
Standing Stones and Cairn
NR89SW 5   NR80839409
NR89SW 6   NR80909408
DC4370
A2/25
Insc. *Small Stones on Moine Mhor
    (Crinan Moss)*

Drawing of the standing stones with a
plan showing the relationship between
the two. Annotated 'Note: This seems
to be Corryvreckan-I see Hilda wrote
"Plan AT 1981" while preparing it for
B.A.R. book. AST 18/1/85. See
another sheet'.
$\frac{1}{8}$ in : 1 ft & 1 : 2500
Black ink and pencil
558 mm × 328 mm

MOINE MHOR
(see above)
DC4371
A2/25
Insc. *Small Stones at Moine Mhor,
Crinan Moss*
Plans of standing stones and cairn.
No scale
Black ink and pencil
558 mm × 385 mm

MOINE MHOR
(see above)
DC4814
A2/25
Insc. *'Small Stones', Moine Mhor Crinan
Moss*
Plan of standing stones and cairn.
$\frac{1}{2}$ in : 4 ft
Black ink
353 mm × 238 mm
Plastic film

SALACHARY
Standing Stones
NM80SW 16   NM84050403
DC4815
A2/26
Insc. *Salachary*
Plan of standing stones with horizon
sketches.
$\frac{1}{2}$ in : 4 ft
302 mm × 213 mm
Plastic film

SANNAIG, JURA
Standing Stones
NR56SW 4   NR51846480
DC4832
A6/3
Insc. *Sannaig, Jura*
Annotated plan and elevation of standing
stone with a hillside sketch.
$\frac{3}{10}$ in : 6 ft
Black ink
94 mm × 92 mm
1949

SCALASAIG, COLONSAY
Standing Stones
NR39SE 18   NR38669376
DC4829
A5/8
Insc. *Near Scalasaig*
Annotated plan and elevation sketch
looking along the standing stones to
horizon.
$\frac{3}{4}$ in : 10 ft
Black ink
104 mm × 81 mm

SKIPNESS
Stone
NR95NW 1   NR90635876
DC4862
A4/21
Insc. *Moon-rise behind Ben Tarsuinn
from Skipness stone*
Diagram of horizon showing position of
the rising moon.
No scale
Black ink
291 mm × 200 mm
Plastic film

STRONTOILLER
Stone Circle
NM92NW 8   NM90672914
DC4356
A1/2
Insc. *Loch Nell*

Plan of stone circle.
$^1_{10}$ in : 1 ft
Black ink and pencil
376mm × 276mm
1939

STRONTOILLER
(see above)
DC4357
A1/2
Annotated plan of area showing relationship of the stone circle, standing stone and cairn.
No scale
Black ink
624 mm × 369 mm
Heavy tissue paper

SUIE, MULL
Barrow and Standing Stones
NM32SE 7   NM37062185
DC4547
M2/8
Insc. *Bunessan B*
Plan of barrow and standing stone. Annotated 'See notes re. Arinish on back. AST' and 'Line of 16 stones probably not wall'.
1 in : 10 ft
Blue ink and pencil
559 mm × 362 mm

SUIE, MULL
(see above)
DC5327
M2/7
Insc. *Dail Na Carraigh*
Annotated plan and elevations of standing stones.
$^1_2$ in : 14 ft
Black ink
122 mm × 118 mm
1949

TAOSLIN, MULL
Standing Stone
NM32SE 1   NM39732238
DC5328
M2/8
Insc. *Ross of Mull*
Annotated plan and elevation of standing stone with horizon sketch.
$^3_{10}$ in : 6 ft
Black ink
71 mm × 58 mm

TARBERT, JURA
Standing Stones
NR68SW 1   NR60628229
NR68SW 2   NR60898220
DC4835/co
A6/5
Insc. *Stones at East Loch Tarbet, Jura*
Annotated plan and elevation of standing stones with sketches of views from *A* and *B*.
$^1_4$ in : 5 ft
Dyeline
115 mm × 103 mm

TARBERT, JURA
(see above)
DC4836
A6/5
Diagram showing the sun rising on Beinn Tarsuin, Arran, as viewed from the standing stones.
No scale
Black ink
256 mm × 152 mm
Plastic film

TEMPLE WOOD
Stone Circle and Standing Stones
NR89NW 3   NR82829760
NR89NW 6   NR82639782
DC4364
A2/8
Insc. *Temple Wood, Slockavullin*

Annotated drawing of stone circle indicating relationship with standing stones. Annotated 'Surveyed 1939 by A. Thom and Wm MacGregor'.
³₈ in : 10 ft
Black ink and pencil
560 mm × 380 mm
1939

TEMPLE WOOD
(see above)
DC4365
A2/8 & A2/12
Insc. *Temple Wood*
Survey drawing showing position of stone circle and standing stones.
1¹₂ in : 100 ft
Black ink and pencil
558 mm × 365 mm

TEMPLE WOOD
(see above)
DC4366
A2/8
Insc. *Temple Wood (1939)*
Plan of stone circle.
¹₈ in : 1 ft
Black ink and pencil
558 mm × 365 mm
1939

TEMPLE WOOD
(see above)
DC4367/co
A2/8
Insc. *Alignments near Temple Wood*
Photocopy of plans showing positions for observing the moon c.1770 B.C.
³₈ in : 10 ft
Photocopy
380 mm × 330 mm

TEMPLE WOOD
(see above)

DC4790
A2/8
Insc. *Menhirs Near Temple Wood*
Annotated plan showing positions for observing moon c.1770 B.C. Insets show the circle plan and the position of the setting moon.
¹₄ in : 10 ft
Black ink and pencil
387 mm × 322 mm

TEMPLE WOOD
(see above)
DC4791
A2/8
Insc. *Slockavullin*
Annotated plan illustrating the lunar declinations given by notch behind stone circle.
No scale
Black ink and pencil
378 mm × 280 mm

TEMPLE WOOD
(see above)
DC4810
A2/8
Annotated plans of the stone circle with insets showing the moon setting behind Bellanoch Hill. Final publication version of DC4790.
¹₂ in : 18ft
Black ink
282 mm × 206 mm
Plastic film

TIRGHOIL, MULL
Standing Stone
NM32SE 6   NM35322242
DC5326
M2/6
Insc. *Ross of Mull*
Plan and elevation of standing stone with hillside sketch.
⁵₁₆ in : 6 ft

Black ink
63 mm × 59 mm

Ayrshire

GARLEFFIN
Standing Stones
NX08SE 1   NX08738172
DC4453
G1/4
Insc. *Ballantrae*
Plan of standing stones.
3 in : 100 ft
Pencil
558 mm × 382 mm

GARLEFFIN
(see above)
DC4865
G1/4
Insc. *Ballantrae*
Plan of standing stones.
⅝ in : 50 ft
Black ink
285 mm × 210 mm
Plastic film

HAGGSTONE MOOR
Cairns and Standing Stone
NX07SE 20   NX06147251
NX07SE 27   NX065 726
NX07SE 29   NX067 728
DC4867
G3/2
Insc. *Haggstone Moor*
Plan of standing stone and cairn showing
 relationship with other monuments.
 Diagrams of the moon setting behind
 the Mull of Kintyre from 'Long Tom'
 and the standing stone.
No scale
Black ink

207 mm × 200 mm
Plastic film

OLD PARK OF THE GLEICK
Standing Stone
NX07SE 28   NX06027159
DC4867
G3/2
Insc. *Haggstone Moor*
Plan of standing stone showing
 relationship with other monuments.
 Diagrams of the moon setting behind
 the Mull of Kintyre from 'Long Tom'
 and Haggstone Moor standing stone.
No scale
Black ink
207 mm × 200 mm
Plastic film

Banffshire

BLACKHILL OF DRACHLAW
Stone Circle
NJ64NE 6   NJ67294633
DC4400
B1/24
Insc. *Blackhill of Drachlaw*
Annotated plan of stone circle.
³₁₆ in : 1 ft
Blue ink and pencil
558 mm × 380 mm

NORTH BURRELDALES
Stone Circle
NJ65SE 1   NJ67585491
DC4426
B4/2
Insc. *Burreldales*
Plan of stone circle.
1½ in : 1 ft
Blue ink and pencil
556 mm × 379 mm

ROTHIEMAY
Recumbent Stone Circle
NJ54NE 6   NJ55084872
DC4427
B4/4
Insc. *Milltown*
Annotated plan of recumbent stone circle.
¹₈ in : 1 ft
Blue ink and pencil
558 mm × 381 mm

WHITEHILL WOOD
Stone Circle and Cairn
NJ65SE 12   NJ67825051
DC4425
B4/1
Insc. *Carnoussie House*
Plan of stone circle and cairn.
1 in : 10 ft
Blue ink and pencil
556 mm × 379 mm

Berwickshire

BORROWSTON RIG
Stone Circle
NT55SE 5   NT55765231
DC4678
G9/10
Insc. *Borrowston Rig*
Tracing of the stone circle from the
    RCAMS Inventory for Berwickwshire.
No scale
Pencil
558 mm × 376 mm
Tracing paper

BORROWSTON RIG
(see above)
DC4788
G9/10
Insc. *Borrowston Rig*

Annotated plan of stone circle with
    elevation of the cup-marked stone.
¾ in : 10 ft
Blue and red ink and pencil
762 mm × 558 mm

BORROWSTON RIG
(see above)
DC4798
G9/10
Insc. *Borrowston Rig. Stones on moor
    above G9/10, Earnsheugh Water*
Plan of stones.
1¹₂ in : 100 ft
Black ink and pencil
766 mm × 560 mm
1958

BORROWSTON RIG
(see above)
DC4798
G9/10
Plan of stone circle. Annotated 'See later
    survey'.
¾ in: 10 ft
Black ink and pencil
766 mm × 560 mm

Buteshire

LARGIZEAN, BUTE
Standing Stones
NS05NE 7   NS08465535
DC4839/co
A9/7
Insc. *Nr Stravanan Bay, Bute*
Annotated plan and elevation of standing
    stones with hillside sketch.
¹₈ in : 10 ft
Photocopy
207 mm × 82 mm

MACHRIE MOOR 2, ARRAN
Stone Circle
NR93SW 1.3   NR91133242
DC4381
A8/6
Insc. *Machrie Moor*
Plan of stone circle.
1 in : 8 ft
Black ink and pencil
558 mm × 382 mm

MID SANNOX
Standing Stone
NS04NW 3   NS01444578
DC4838
A8/1
Insc. *Cioch na h'-Oighe from Stone at Mid Sannox, Arran*
Annotated chart showing the horizon and the position where the sun would reappear after setting.
No scale
Black ink
111 mm × 108 mm
Graph paper

Caithness

BATTLE MOSS, LOCH OF YARROWS
Stone Rows
ND34SW 22   ND31284402
DC4555
N1/8
Insc. *Loch of Yarrows*
Plan of stone rows.
$^1_8$ in: 1 ft
Black ink and pencil
555 mm × 397 mm

BATTLE MOSS, LOCH OF YARROWS
(see above)
DC5334
N1/7

Plan of the stone rows.
$^7_{16}$ in : 10 ft
Black ink
203 mm × 124 mm

BORGUE
Standing Stone
ND12NW 13   ND12632665
DC5342
N1/21
Plan and elevation of standing stone with horizon profile showing the moon rising.
No scale
Black ink
210 mm × 150 mm

CAMSTER
Stone Rows
ND24SE 3   ND26024377
DC4559
N1/14
Insc. *Camster*
Plan of stone rows.
$^3_4$ in : 10 ft
Black and red ink
555 mm × 383 mm
1969

CAMSTER
(see above)
DC4560
N1/14
Insc. *Stone Rows near Grey Cairns of Camster*
Plan of stone rows. Annotated 'First survey'.
1 in : 8 ft
Black ink and pencil
559 mm × 380 mm

CAMSTER
(see above)
DC5336

N1/14
Plan of stone rows.
¹₈ in : 2 ft
Black ink
211 mm × 172 mm

CNOC NA MARANAICH
Standing Stone
ND13SW 11   ND13203315
DC5341
N1/20
Plan and elevation of stone with horizon
    profile showing the position of the
    midsummer setting sun.
No scale
Black ink
205 mm × 140 mm

CREAG BHREAC MHOR
Stone Rows
ND06NW 8   ND01176595
DC4552
N1/3
Insc. *Dunreay*
Plan of stone rows.
¹₁₀ in : 1 ft
Red and blue ink and pencil
558 mm × 378 mm

CREAG BHREAC MHOR
(see above)
DC5333
N1/3
Insc. *Dunreay*
Plan of stone rows.
⁹₁₆ in : 10 ft
Red and black ink
254 mm × 203 mm

DIRLOT
Stone Rows
ND14NW 6   ND12284856
DC5337
N1/17

Plan of stone rows.
³₈ in : 10 ft
Black ink
306 mm × 256 mm

DIRLOT
(see above)
DC5338
N1/17
Horizon profile showing position of
    setting moon.
No scale
Black ink
222 mm × 205 mm

DIRLOT
Stones
ND14NW   ND127 489
DC5340
N1/24
Plan of stones.
1 in : 100 ft
Black ink
251 mm × 180 mm

FORSE
Stone Circle
ND23NW 10   ND20773630
DC4553
N1/5
Insc. *Forse, Latheron*
Plan of stone circle.
³₄ in : 10 ft
Pencil
557 mm × 384 mm

GARRYWHIN
Cairn and Stone Rows
ND34SW 18   ND31404131
DC4557
N1/9
Insc. *Garrywhin, Wattenan (Groots Loch)*
Plan of stone rows. Annotated 'Taking
    mean of all stones (bar large one). Grid

lines should be moved -.08 ft i.e. practically zero'.

1 in : 10 ft

Black ink and pencil

559 mm × 380 mm

GARRYWHIN

(see above)

DC4667

N1/9

Insc. *Watenan*

Plan showing positions of the stone rows and the cairn. Annotated 'From the S end of the ridge that lies bet[ween] Groats L[och] and Broughwhin is a cairn 18' diameter with kist stones radiate NE to SW. On the top of a low ridge, 80y E of the N end of Broughwhin Loch is a cairn. Running to S has been a setting of stones only 3 upstanding in 1910'.

No scale

Blue ink and pencil

210 mm × 184 mm

Tracing paper

GARRYWHIN

(see above)

DC5335

N1/9

Insc. *Wattenan, Ulbster*

Plan of stone rows.

$^7_{16}$ in : 10 ft

Black and red ink

252 mm × 203 mm

GUIDEBEST

Stone Circle

ND13NE 3   ND18023510

DC4558

N1/13

Insc. *Latheron Wheel Burn*

Annotated plan of stone circle.

1 in : 20 ft

Black ink and pencil

559 mm × 380 mm

HILL O'MANY STANES, MID CLYTH

Stone Rows

ND23NE 6   ND29513840

DC4792

N1/1

Insc. *Mid Clyth*

Annotated plan of stone rows. Annotated 'Tape stretch 6'(?). Very blowy. Stretch probably increased from 5" to 7" '.

$^1_{10}$ in : 1 ft

Blue ink and pencil

763 mm × 559 mm

1957

HILL O'MANY STANES, MID CLYTH

(see above)

DC5331

N1/1

Plan of stones showing stances for various declinations with horizon profiles.

$^3_4$ in : 50 ft

Black ink

215 mm × 204 mm

HILL O'MANY STANES, MID CLYTH

(see above)

DC5332

N1/1

Horizon profile showing position of the rising moon from stone rows.

No scale

Black ink

304 mm × 245 mm

LOCH OF YARROWS

Standing Stones

ND34SW 39   ND31634311

DC4554

N1/8

Insc. *Yarrow Loch*

Plan of standing stones.

$^1_4$ in : 1 ft

Blue ink and pencil
558 mm × 377 mm

LOCH OF YARROWS
(see above)
DC4556
N1/8
Insc. *Near 2 stones at Loch of Yarrows*
Annotated plan of standing stones and
    others in area.
3 in : 100 ft
Pencil
558 mm × 382 mm

LOWER CAMSTER
Standing Stones
ND24NE 7   ND25984585
DC4561
N1/18
Insc. *Lower Camster*
Annotated plan of standing stones.
³₄ in : 10 ft
Pencil
558 mm × 377 mm

LOWER CAMSTER
(see above)
DC5339
N1/18
Insc. *Lower Camster*
Plan of standing stones.
³₄ in : 10 ft
Black ink
299 mm × 272 mm

WATENAN FARM
Stone Rows
ND34SW 23   ND31524118
DC4667
N1/9
Insc. *Watenan*
Plan showing positions of the stone rows
    and the cairn. Annotated 'From the S
    end of the ridge that lies between

Groats Loch and Broughwhin is a cairn
18' diameter with kist stone radiate NE
to SW. On the top of a low ridge, 80y
E of the N end of Broughwhin Loch is
a cairn. Running to S has been a setting
of stones only 3 upstanding in 1910'.
No scale
Blue ink and pencil
210 mm × 184 mm
Tracing paper

Dumfriesshire

AULDGIRTH
Stone Circle
NX98NW 11   NX91848522
DC4473
G6/2
Insc. *Auldgirth*
Annotated plan of stone circle.
1 in : 10 ft
Black ink and pencil
560 mm × 377 mm

GIRDLE STANES
Stone Circle
NY29NE 13   NY25359615
DC4480
G7/5
Insc. *Girdle Stanes, White Esk*
Annotated plan of stone circle.
1 in : 20 ft
Blue ink
560 mm × 379 mm

KIRKHILL
Stone Circle
NY19NW 12   NY13969592
DC4477
G7/3
Insc. *Wamphray Glen*
Plan of stone circle.
¹₈ in : 1 ft

Blue ink
560 mm × 380 mm

LITTLE HARTFELL
Stone Circle
NY28NW 4   NY22408806
DC4452
G7/6
Insc. *Whitcastles*
Incomplete plan of stone circle.
$^3_4$ in : 10 ft
Pencil
558 mm × 374 mm

LITTLE HARTFELL
(see above)
DC4787
G7/6
Insc. *Whitcastles*
Annotated plan of stone circle. Annotated
    'Tape stretch:- add 0.4%'.
1 in : 10 ft
Blue ink and pencil
763 mm × 559 mm

LITTLE HARTFELL
(see above)
DC4789
G7/6
Insc. *Whitcastles*
Plan of stone circle.
No scale
Black ink and pencil
407 mm × 313 mm

LOUPIN' STANES
Stone Circle
NY29NE 11   NY25709663
DC4448 (reverse)
G7/4
Insc. *Loupin' Stanes*
Plan of outlying stones.
No scale
Blue ink and pencil
558 mm × 382 mm

LOUPIN' STANES
(see above)
DC4478
G7/4
Insc. *Loupin' Stanes*
Plan of outlying stones.
$^3_8$ in : 10 ft & $^1_8$ in : 3 ft
Blue ink and pencil
561 mm × 381 mm

LOUPIN' STANES
(see above)
DC4479
G7/4
Insc. *Loupin' Stanes*
Plan of stone circle with elevations of
    stones *A* and *B*.
$^1_8$ in : 1 ft
Blue ink and pencil
560 mm × 379 mm

MONIAIVE
Stones
NX79SE   NX77 90
DC4474
G6/5
Insc. *Moniave*
Annotated plan of stones.
$^3_4$ in : 10 ft
Pencil
560 mm × 379 mm

SEVEN BRETHREN
Stone Circle
NY28SW 3   NY21718269
DC4475
G7/2
Insc. *Seven Brethren*
Annotated plan of stone circle. Annotated
    'See new survey'.
$^1_{10}$ in : 1 ft
Black ink and pencil
557 mm × 374 mm
1939

SEVEN BRETHREN
(see above)
DC4476
G7/2
Insc. *Seven Brethren*
Plan of stone circle.
¹₈ in : 1 ft
Blue ink and pencil
558 mm × 370 mm
1957

TWELVE APOSTLES (THE)
Stone Circle
NX97NW 19   NX94707940
DC4471
G6/1
Insc. *Stone Circle 'Twelve Apostles'*
Plan of stone circle.
³₈ in : 10 ft
Black ink
556 mm × 370 mm
Linen backed tracing paper

TWELVE APOSTLES (THE)
(see above)
DC4472
G6/1
Insc. *Twelve Apostles*
Annotated plan of stone circle. Annotated
   'A very careful survey 1939 A Thom
   and J Orr'.
1 mm : 1 ft
Black ink and pencil
559 mm × 381 mm
1939

East Lothian

KINGSIDE HILL
Stone Circle
NT66NW 13   NT62636503
DC4679
Insc. *E. Lothian 240*

Tracing of the stone circle from the
   RCAMS Inventory of East Lothian.
   Annotated 'Is this "Crow Stones"?
   Sheet 75, 619 653. No on visit to Crow
   Stones.'
1 in : 27.4 ft
Pencil
558 mm × 378 mm
Tracing paper

KINGSIDE SCHOOL
Stones
NT66SW 2   NT64336421
DC4487
G9/13
Insc. *Kell Burn*
Plan of stones and alignment.
1 in : 20 ft
Pencil
559 mm × 379 mm

KINGSIDE SCHOOL
(see above)
DC4856/co
G9/13
Insc. *Alignment at Kingside*
Annotated plan of stones.
No scale
Dyeline
202 mm × 79 mm

NINE STONES (THE)
Stone Circle
NT66NW 14   NT62546549
DC4486
G9/11
Insc. *Nine Stone Rig*
Plan of stone circle with sketch of Crow
   Stones.
¹₄ in : 1 ft
Blue ink and pencil
559 mm × 379 mm

YADLEE
Stone Circle

NT66NE 3   NT65406732
DC4679
Insc. *Zadlee*
Tracing of the stone circle from the
    RCAMS Inventory for East Lothian.
1 in : 23 ft
Pencil
558 mm × 378 mm
Tracing paper

Fife

LUNDIN LINKS
Standing stones
NO40SW 1   NO40480272
DC4849
P4/1
Insc. *Standing Stones of Lundin*
Sketch plan of standing stones with
    diagrams illustrating the moon rising
    over the Bass Rock and setting behind
    Cormie Hill.
$^5_{16}$ in : 20 ft
Black ink
263 mm × 224 mm
Plastic film

Inverness-shire

AN CARRA, BEINN A'CHARRA, NORTH UIST
Standing Stone
NF76NE 1   NF78646909
DC4494
H3/9
Insc. *Beinn a' Charra*
Plan and elevation of standing stone.
    Annotated on back 'Surprise! These
    eng[ineer']s drawings were Alan
    Watson Thom's: done at Cambridge.
    He was killed in the car crash when he
    was 21. AST 19/1/85'.

1 in : 10 ft
Pencil
558 mm × 372 mm

AN CARRA, BEINN A'CHARRA, NORTH UIST
(see above)
DC4883
H3/9
Insc. *Stone on Ben A Charra, North Uist*
Plan and elevation of standing stone.
$^1_2$ in : 10 ft
Black ink
119 mm × 97 mm

AN CARRA, SOUTH UIST
Standing Stone
NF73SE 1   NF77033211
DC4889/co
H5/1
Insc. *An Carra, South Uist*
Elevation of standing stone.
$^5_{16}$ in : 6 ft
Dyeline
62 mm × 40 mm

AVIEMORE
Stone Circle and Chambered Cairn
NH81SE 1   NH89701347
DC4437
B7/12
Insc. *Aviemore*
Annotated plan of stone circle and
    chambered cairn.
1^1_2 in : 10 ft
Blue ink and pencil
558 mm × 385 mm

BALNUARAN OF CLAVA, NE
Stone Circle and Ring Cairn
NH74SE 1   NH75764447
DC4431
B7/1
Insc. *Clava, North Circle*

Plan of stone circle and round cairn.
1 in : 10 ft
Black ink and pencil
556 mm × 380 mm

BALNUARAN OF CLAVA, NE, SW and CENTRE
Chambered Cairns and Stone Circles
NH74SE 1   NH75764447
NH74SE 3   NH75684438
NH74SE 4   NH75684438
DC4797
B7/1
Insc. *Clava*
Annotated plan of the stone circles and
    chambered cairns.
1 in : 22 ft
Black ink and pencil
761 mm × 558 mm

BALNUARAN OF CLAVA, NE, SW AND CENTRE
(see above)
DC4864
B7/1
Insc. *Clava Cairns*
Plan of chambered cairns and stone
    circles.
½ in : 10 ft
Black ink
253 mm × 203 mm

BARPA LANGASS, NORTH UIST
Chambered Cairn
NF86NW 6   NF83816571
DC4679
Insc. *Barpa Langass*
Tracing from the RCAMS Inventory of
    the Outer Hebrides.
1 in : 14.34 ft
Pencil
558 mm × 378 mm
Tracing paper

BELLADRUM
Chambered Cairn
NH54SW 8   NH51604158
DC4554
B7/14
Insc. *Belladrum*
Plan of chambered cairn.
¼ in : 1 ft
Blue ink and pencil
558 mm × 377 mm

BERNERAY, BARRA
Stone Circle
NL58SE   NL564 803
DC4501
H6/5
Insc. *Berneray, Barra*
Annotated plan of stone circle. Annotated
    'Somewhat hurried survey (1962)
    ("Molita" in poor anchorage)'.
No scale
Blue ink and pencil
558 mm × 378 mm

BORVE, BARRA
Standing Stones
NF60SE 10   NF65270144
DC4892/co
H6/1
Insc. *Site at Borve, Barra*
Annotated plan of standing stones with
    sketch of hillside.
½ in : 50 ft
Dyeline
112 mm × 80 mm
1949

CLACHAN ERISCO, BORVE, SKYE
Stone Circle
NG44NE 1   NG45194801
DC4502
H7/5
Insc. *Clacha Erisco*
Plan of stone circle. Annotated 'poor
    survey'.
⅛ in : 1 ft

Pencil
560 mm × 380 mm

CLACH AN T'SAGAIRT, NORTH UIST
Stone
NF87NE 14   NF87857605
DC4858
H3/2
Insc. *Clach an 't Saigairt, North Uist*
Elevation drawings of stone.
$^3{}_{10}$ in : 6 ft
Black ink
64 mm × 47 mm

CLACH AN T'SAGAIRT, NORTH UIST
(see above)
DC4859
H3/2
Insc. *Boreray as seen from Clach an Sagairt, 48.5 miles.*
Diagram showing horizon and setting sun.
No scale
Black ink
180 mm × 151 mm
Plastic film

CLACH MHIC LEOID, HARRIS
Standing Stone
NG09NW 4   NG04099720
DC4879
H2/2
Profile of Boreray showing the setting sun viewed from the standing stone.
No scale
Black ink
143 mm × 116 mm
Plastic film

CLACH MHOR A' CHE', WEST FORD, NORTH UIST
Standing Stone
NF76NE 3   NF77006621

DC4373
H3/12
Insc. *Clach Mhor a' Che' and Dun Na Carnaigh*
Annotated plan of standing stone and cairn with elevation of stone and sketch showing view to Craig Hasten.
No scale
Pencil
558 mm × 379 mm

CLACH MHOR A' CHE', WEST FORD, NORTH UIST
(see above)
DC4884
H3/12
Insc. *Clach Mhor a Che and Dun Na Carnaigh*
Annotated plan of chambered cairn and standing stone.
$^3{}_4$ in : 10 ft
Black ink
203 mm × 131 mm
1948

CLADDACH, ILLERAY, NORTH UIST
Stone Circle (possible)
NF86SW 34   NF80466431
DC4886
H3/15
Insc. *Sketch not to scale of Suspected Circle at Claddach-Illeray*
Plan of stone circle.
No scale
Black ink
134 mm × 94 mm

CLADH MAOLRITHE, BERNERAY
Standing Stone
NF98SW 7   NF91228068
DC4880
H3/1
Insc. *Cladh Maolrithe Bernera*
Plan and elevation of standing stone with sketch of Spuir islet.
No scale

Black ink
157 mm × 93 mm
1948

CRINGRAVAL, CLACHAN, NORTH UIST
Cairn
NF86SW 21   NF81526451
DC4495
H3/14
Insc. *Cringraval*
Annotated sketch of cairn and standing
   stones.
1 cm : 10 ft
Pencil
558 mm × 377 mm

CRINGRAVAL, CLACHAN, NORTH UIST
(see above)
DC4885
H3/14
Insc. *Site on Cringraval. Not to scale*
Plan of cairn and sketch of hillside.
^{3}₈ in : 10 ft
Black ink
135 mm × 110 mm

CROFTCROY
Stone Circle and Chambered Cairn
NH63SE 2   NH68353318
DC4441
B7/17
Insc. *Farr*
Plan of stone circle and chambered cairn.
   Annotated 'I helped survey this in the
   back garden of P.O. Buildings have
   now been put up on the site. AST
   18/1/85'.
1½ in : 10 ft
Blue ink and pencil
560 mm × 373 mm

CROIS CHNOCA BREACA, SOUTH UIST
Standing Stone
NF73SW 3   NF73403366
DC4890/co
H5/2
Insc. *Crois Chnoca Breaca, South Uist*
Elevation drawings of stone.
^{3}₁₀ in : 6 ft
Dyeline
64 mm × 43 mm

CROIS MHIC JAMAIN, NORTH UIST
Standing Stones
NF87NE 8   NF89377819
DC4888/co
H3/22
Insc. *Crois Mhic Jamain, North Uist*
Annotated plan of standing stones.
½ in : 10 ft
Dyeline
96 mm × 91 mm
1949

CULDOICH
Stone Circle and Chambered Cairn
NH74SE 2   NH75114378
DC4432
B7/2
Insc. *Milltown of Clava*
Plan of stone circle and chambered cairn.
1½ in : 10 ft
Black ink and pencil
558 mm × 379 mm

DALCROSS MAINS
Stone Circle and Chambered Cairn
NH74NE 15   NH77984846
DC4435
B7/6
Insc. *Castle Dalcross*
Plan of stone circle and chambered cairn.
1½ in : 10 ft
Blue ink and pencil
556 mm × 380 mm

DAVIOT
Stone Circle and Chambered Cairn
NH74SW 5   NH72744119
DC4434
B7/5
Insc. *Daviot*
Annotated plan of stone circle.
1½ in : 10 ft
Blue and red ink and pencil
556 mm × 380 mm

DELFOUR
Stone Circle and Chambered Cairn
NH80NW 1   NH84420858
DC4436
B7/10
Insc. *Easter Delfour, Alvie*
Plan of stone circle and chambered cairn
   with elevation of outlying stone.
1½ in : 10 ft
Blue ink and pencil
558 mm × 377 mm

DRUID TEMPLE
Stone Circle and Chambered Cairn
NH64SE 23   NH68514201
DC4442
B7/18
Insc. *Druid Temple, Inverness*
Annotated plan of stone circle and
   chambered cairn.
1½ in : 10 ft
Blue ink and pencil
558 mm × 380 mm

DRUIM A'CHARRA, BREIVIG,
BARRA
Standing Stones
NL69NE 1   NL68909903
DC4500
H6/3
Insc. *Brevig, Barra*
Plan and elevation of the stones with a
   hillside sketch.
3 in : 100 ft & ⅛ in : 1 ft

Pencil
558 mm × 384 mm

DRUIM A' CHARRA, BREIVIG,
BARRA
(see above)
DC4893
H6/3
Insc. *Brevig, Barra*
Annotated plan and elevation of standing
   stone with hillside sketches.
¾ in : 50 ft and ⅝ in : 10 ft
Black ink
252 mm × 202 mm

DUN NA CARNAICH, WEST FORD,
NORTH UIST
Chambered Cairn
NF76NE 4   NF76996617
DC4373
H3/12
Insc. *Clach Mhor a' Che' and Dun Na*
   Carnaigh
Annotated plan of standing stone and
   cairn with elevation of stone and
   sketch showing view to Craig Hasten.
No scale
Pencil
558 mm × 379 mm

DUN NA CARNAICH, WEST FORD,
NORTH UIST
(see above)
DC4884
H3/12
Insc. *Clach Mhor a Che and Dun Na*
   Carnaigh
Annotated plan of chambered cairn and
   standing stone.
¾ in : 10 ft
Black ink
203 mm × 131 mm
1948

FIR BHREIGE, NORTH UIST
Standing Stones
NF77SE 12   NF77007029
                             & NF77037027
DC4494
H3/5
Insc. *Fir Bhreige*
Annotated plan and elevations of the
   standing stones.
1 in : 10 ft
Pencil
558 mm × 372 mm

FIR BHREIGE, NORTH UIST
(see above)
DC4882
H3/5
Insc. *Firbhreige (W)*
Plans and elevations of standing stones.
½ in : 10 ft
Black ink
118 mm × 106 mm

GASK
Stone Circle and Chambered Cairn
NH63NE 10   NH67943585
DC4439
B7/15
Insc. *Mains of Gask*
Plan of stone circle and chambered cairn
   with elevation of outlying stone.
$\frac{1}{10}$ in : 1 ft
Blue ink and pencil
558 mm × 380 mm

GLEN BORERAIG
Hut-circle
NG51NE 2   NG59211743
DC4503
H7/6
Insc. *Glen Borreraig*
Plan of hut-circle. Annotated 'On 19/1/85
   I see this is a pantographed copy,
   unfinished . Where is the original?
   A.S.T.'.
$\frac{1}{8}$ in : 1 ft

Blue ink and pencil
560 mm × 380 mm

GRAMISDALE, BENBECULA
Stone Circle and Chambered Cairn
NF85NW 2   NF82505614
DC4498
H4/1
Insc. *Circle at N. Ford*
Plan of stone circle.
1 in : 10 ft
Pencil
558 mm × 375 mm

GRENISH
Stone Circle and Chambered Cairn
NH91NW 5   NH90781550
DC4438
B7/13
Insc. *Loch Nan Carraigean, near
   Aviemore*
Annotated plan of stone circle and
   chambered cairn.
$\frac{1}{8}$ in : 1 ft
Blue ink and pencil
558 mm × 380 mm

KINCHYLE OF DORES
Stone Circle and Chambered Cairn
NH63NW 5   NH62153896
DC4443
B7/19
Insc. *River Ness*
Plan of stone circle and chambered cairn.
$\frac{1}{8}$ in : 1 ft
Blue ink and pencil
558 mm × 379 mm

NA CLACHAN BHREIGE, SKYE
Stone Circle
NG51NW 1   NG54321768
DC4504
H7/9
Insc. *Strathaird*

Plan of stone circle.
^{1}₈ in : 1 ft
Pencil
560 mm × 380 mm

POBULL FHINN, BEN LANGASS, NORTH UIST
Stone Circle
NF86NW 7   NF84276502
DC4496
H3/17
Insc. *Pobull Fhinn*
Annotated plan of stone circle.
1 in : 10 ft
Pencil
556 mm × 383 mm

RU ARDVULE, SOUTH UIST
Standing Stone
NF72NW 3   NF72732860
DC4891/co
H5/3
Insc. *Ru Ardvule, South Uist*
Plan and elevation of standing stone.
^{1}₂ in : 10 ft
Dyeline
100 mm × 96 mm
1949

SORNACH COIR FHINN, LOCH A'PHOBUILL, NORTH UIST
Stone Circle
NF86SW 28   NF82876305
DC4495
H3/18
Insc. *Sornach Coir Fhinn, Loch Eport*
Annotated plan of stone circle.
1 cm : 10 ft
Pencil
558 mm × 377 mm

SORNACH COIR FHINN, LOCH A' PHOBUILL, NORTH UIST
(see above)

DC4497
H3/18
Insc. *Sornach Coir Fhinn*
Plan of stone circle.
^{3}₄ in : 10 ft
Blue ink and pencil
558 mm × 381 mm

SORNACH COIR FHINN, LOCH A' PHOBUILL, NORTH UIST
(see above)
DC4887
H3/18
Insc. *Stones at Leacach an Tigh Chloiche from Circle Sornach Coir Fhinn*
Profile plan showing the setting sun.
No scale
Black ink
101 mm × 85 mm
Graph paper

SUIDHEACHADH SEALG, BENBECULA
Stone Circle and Chambered Cairn
NF85NW 3   NF82475522
DC4499
H4/2
Insc. *Suidheachadh Sealg*
Plan of stone circle.
Annotated 'Note: measurements made to nearest foot'.
^{1}₁₀ in : 1 ft
Blue ink and pencil
556 mm × 350 mm

TORDARROCH
Stone Circle and Chambered Cairn
NH63SE 3   NH68013350
DC4440
B7/16
Insc. *West Farr*
Annotated plan of stone circle and chambered cairn. Annotated 'I helped here. Later I took rubbings of the cup markings on the largest recumbent

stone and sent the rubbings to Ron.
Morris. A.S.T. 18/1/85'.
$\frac{1}{10}$ in : 1 ft
Blue ink and pencil
558 mm × 378 mm

TULLOCHGORM
Ring Cairn
NH92SE 3   NH96482130
DC4412
B7/4
Insc. *Nethybridge*
Plan of ring cairn.
$1\frac{1}{2}$ in : 10 ft
Black ink and pencil
557 mm × 386 mm

UNIVAL, LEACACH AN TIGH CHLOICHE, NORTH UIST
Chambered Cairn and Standing Stone
NF86NW 4   NF80036685
DC4854
H3/11
Insc. *Haskeir from Leacach an Tigh Chloiche*
Elevation sketch of island viewed from chambered cairn.
No scale
Black ink
103 mm × 57 mm
Graph paper

Kincardineshire

AQUHORTHIES
Recumbent Stone Circle
NO99NW 1   NO90189634
DC4419
B3/1
Insc. *Aquhorthies N*
Plan of recumbent stone circle.
$1\frac{1}{2}$ in : 10 ft

Black ink and pencil
558 mm × 380 mm

CAIRNFAULD
Stone Circle
NO79SE 1   NO75359406
DC4421
B2/11
Insc. *Cairnfauld*
Plan of stone circle.
1 in : 8 ft
Black ink and pencil
558 mm × 379 mm

CAMPSTONE HILL, RAEDYKES SE
Ring Cairn
NO89SW 9   NO83309060
DC4421
B3/3
Insc. *Raedykes S*
Annotated plan of ring cairn.
$1\frac{1}{2}$ in : 10 ft
Black ink and pencil
558 mm × 379 mm

CAMPSTONE HILL, RAEDYKES 4
Ring Cairn
NO89SW 6   NO83229066
DC4422
B3/4
Insc. *Raedykes N*
Annotated plan of ring cairn.
$1\frac{1}{2}$ in : 10 ft
Black ink and pencil
558 mm × 377 mm

CLUNE WOOD
Recumbent Stone Circle and Cairn
NO79SE 2   NO79469495
DC4424
B3/7
Insc. *Clune Wood*
Annotated plan of recumbent stone circle
    and cairn.

1 in : 10 ft
Blue ink and pencil
558 mm × 379 mm

CLUNE WOOD
(see above)
DC4759/co
B3/7
Insc. *Clune Wood*
Annotated plan of recumbent stone circle
    and cairn.
1 in : 10 ft
Dyeline copy
270 mm × 248 mm

ESLIE THE GREATER
Recumbent Stone Circle
NO79SW 2   NO71719159
DC4408
B2/4
Insc. *Esslie the Greater*
Annotated plan of the recumbent stone
    circle. Annotated 'Tape stretch =
    0.43%'.
1½ in : 10 ft
Black ink and pencil
558 mm × 390 mm
1955

ESLIE THE LESSER
Recumbent Stone Circle
NO79SW 1   NO72259215
DC4409
B2/5
Insc. *Esslie the Less*
Plan of recumbent stone circle and ring
    cairn.
1½ in: 10 ft
Black ink and pencil
557 mm × 344 mm

GLASSEL
Stone Circle
NO69NW 2   NO64889966

DC4423
B3/6
Insc. *Kynoch Plantation, Glassel near
    Torphins*
Plan of stone circle.
1 cm : 1 ft
Black ink and pencil
556 mm × 380 mm

NINE STONES (THE), GARROL
Recumbent Stone Circle
NO79SW 8   NO72339122
DC4409
B2/6
Insc. *Garrol Wood*
Plan of recumbent stone circle and ring
    cairn.
1½ in : 10 ft
Black ink and pencil
557 mm × 344 mm

OLD BOURTREEBUSH
Recumbent Stone Circle
NO99NW 2   NO90359608
DC4420
B3/2
Insc. *Old Bourtree Bush, Aquhorthies S*
Plan of recumbent stone circle.
1½ in : 10 ft
Black ink and pencil
558 mm × 377 mm

OLD BOURTREEBUSH
(see above)
DC4757/co
B3/2
Insc. *Aquhorthies S (Old Bourtreebush)*
Plan of recumbent stone circle.
1½ in : 10 ft
Dyeline copy
435 mm × 325 mm

Kirkcudbrightshire

BAGBIE
Standing Stone
NX45NE 4   NX49775620
DC4464
G4/13
Insc. *Kirkmabreck, Bagbie*
Plan of stone.
1 in : 100 ft
Pencil
561 mm × 377 mm

BAGBIE
Standing Stone and Cairn
NX45NE 4   NX49775620
NX45NE 5   NX49795640
DC4464
G4/13
Insc. *Kirkmabreck, Bagbie*
Plan of stone in relation to cairn.
¹₄ in : 1 ft
Pencil
561 mm × 377 mm

BAGBIE
Cairn
NX45NE 5   NX49795640
DC4816
G4/13
Insc. *Kirkmabreck*
Annotated plan of cairn.
¹₄ in : 1 ft
Black ink
435 mm × 303 mm
Tracing paper

BARHOLM
Stones
NX55SW 45   NX521 529
DC4468
G5/4
Insc. *near Ravenshall Point*
Plan of stones.

No scale
Pencil
558 mm × 360 mm

CAULDSIDE BURN
Stone Circle
NX55NW 8   NX52955711
DC4465
G4/14
Insc. *Cambret East*
Annotated plan of stone circle.
¹₈ in : 1 ft
Black ink and pencil
557 mm × 379 mm

CAULDSIDE BURN
(see above)
DC4466
G4/14
Annotated astronomical plan showing the rising and setting positions of various stars from the centre of the stone circle.
No scale
Black ink and pencil
510 mm × 380 mm

CAULDSIDE BURN
(see above)
DC4467
G4/14
Insc. *Stone Circle, SE of Cambret Hill*
Plan showing azimuthal positions of stones in circle in relation to various stars.
¹₄ in : 2 ft
Black ink and pencil
557 mm × 394 mm
Linen backed tracing paper

COMMUNION STONES
Stone Rows
NX87NE 1   NX85917905
DC4470

G5/10
Insc. *Communion Stones*
Plan of stone rows.
$\frac{1}{4}$ in : 1 ft
Black ink and pencil
558 mm × 380 mm

COMMUNION STONES
(see above)
DC4870
G5/10
Insc. *Communion Stones*
Plan of stone rows.
$\frac{5}{8}$ in : 5 ft
Black ink
252 mm × 202 mm

DRUMFERN
Stone Circle
NX37SE 10   NX39997099
DC4461
G4/3
Insc. *Drumfern Cairn, Drannandow*
Plan of stone circle with sightlines to
    White Cairn.
1 in : 10 ft
Black ink and pencil
558 mm × 378 mm

EASTHILL
Stone Circle
NX97SW 1   NX91937388
DC4469
G5/9
Insc. *Maxwellton*
Plan of stone circle.
$\frac{1}{8}$ in : 1 ft
Black ink and pencil
557 mm × 380 mm

GLENQUICKEN
Stone Circle
NX55NW 5   NX50965821
DC4463

G4/12
Insc. *Cambret Moor*
Plan of stone circle and two unidentified
    nearby sites.
No scale
Black ink
560 mm × 394 mm
Linen backed tracing paper

GLENQUICKEN
(see above)
DC4784
G4/12
Insc. *Composite Survey of Four Circles.*
    Each reduced to same diameter and
    orientated on major axis
Plan.
No scale
Black ink
363 mm × 284 mm
Tracing paper

HOLM OF DALTALLOCHAN
Stone Circle (possible)
NX59SE 4   NX55289422
DC4459
G4/1
Insc. *Carsphairn*
Plan of stone circle.
$\frac{1}{8}$ in : 1 ft
Black and red ink and pencil
556 mm × 380 mm

LAIRDMANNOCH
Stone Circle
NX66SE 2   NX66296143
DC4462
G4/9
Insc. *Loch Mannoch*
Plan of stone circle.
$\frac{3}{8}$ in : 1 ft
Black ink and pencil
561 mm × 365 mm

THIEVES (THE)
Standing Stones
NX47SW 2   NX40447159
DC4460
G4/2
Insc. *The Thieves*
Annotated plan of standing stones.
¹₄ in : 1 ft
Black ink and pencil
556 mm × 380 mm

URQUHART
Stone Circle
NJ26SE 7   NJ28956407
DC4428
B5/1
Insc. *Urquhart*
Plan of stone circle.
1 in : 10 ft
Blue ink and pencil
556 mm × 381 mm

Midlothian

Nairn

NEWBRIDGE
Cairn and Standing Stones
NT17SW 8   NT12347260
DC4488
G9/14
Insc. *Newbridge*
Plan of cairn and standing stones.
Annotated on back 'AST re-surveyed
this with Az's to Arthur's Seat. Hope
to get it to BAR Alignment Book.
20/1/85'.
³₈ in : 10 ft
Blue ink and pencil
558 mm × 376 mm

LITTLE URCHANY
Stone Circle and Chambered Cairn
NH84NE 1   NH86654857
DC4429
B6/1
Insc. *Little Urchany*
Annotated plan of stone circle and
chambered cairn.
¹₈ in : 1 ft
Blue ink and pencil
559 mm × 380 mm

MOYNESS
Stone Circle and Cairn
NH95SE 7   NH95275366
DC4430
B6/2
Insc. *Moyness*
Plan of stone circle and cairn. Annotated
'many stones lying about not
surveyed'.
1 in : 8 ft
Pencil
558 mm × 380 mm

Morayshire

BOAT OF BALLIEFURTH
Standing Stones
NJ02SW 4   NJ01172468
NJ02SW 5   NJ01092464
NJ02SW 6   NJ01022456
DC4433
B7/3
Insc. *Dulnanbridge*
Plan of standing stones.
1 in : 8 ft
Pencil
557 mm × 380 mm

Orkney

COMET STONE
Standing Stone
HY21SE 13   HY29631331
DC4571
O1/1
Insc. *Brogar Comet Stone*
Annotated plan of standing stone.
1 : 500
Black and blue ink
317 mm × 314 mm
Linen backed tracing paper

COMET STONE
(see above)
DC4572
O1/1
Insc. *Brogar Comet Stone*
Annotated plan of standing stone.
1 : 500
Black ink and pencil
557 mm × 376 mm

COMET STONE
(see above)
DC5358
O1/1
Profile of horizon showing the moon
    setting near Ravie Hill, as viewed from
    the standing stone.
No scale
Black ink
230 mm × 189 mm

HALL OF CLESTRAN
Standing Stone
HY20NE 4   HY29360729
DC4663
Insc. *Hall of Clestran*
Annotated tracing from 1:10000 map to
    show the location of the stone.
    Annotated 'Approx. position of large
    stone pulled to side of field. It was
    "unsafe", he said. A.S.T. 20/1/85'.

1 : 10000
Black ink and pencil
489 mm × 173 mm
Plastic film

KAME OF CORRIGALL
Quarries
HY32SW   HY331 206
DC4573
O1/5
Insc. *Quarries on ridge at Kame of
    Corrigall*
Annotated plan of quarries.
1 : 250
Black ink
409 mm × 344 mm

KAME OF CORRIGALL
(see above)
DC4573/c
O1/5
Insc. *Quarries on ridge at Kame of
    Corrigall*
Annotated plan of quarries.
1 : 250
Dyeline
409 mm × 344 mm

KAME OF CORRIGALL
(see above)
DC4574
O1/5
Insc. *Quarries on Kame of Corrigall,
    Orkney*
Annotated plan of quarries.
1 : 250
Blue ink and pencil
557 mm × 363 mm

KAME OF CORRIGALL
(see above)
DC5357
O1/1
Profile of quarries.
No scale

Black ink
248 mm × 190 mm

RING OF BRODGAR
Henge Monument
HY21SE 1 HY29451335
DC4567
O1/1
Insc. *Ring of Brodgar*
Annotated plan and profile of henge
   monument.
1½ in : 50 ft
Black ink
489 mm × 383 mm
Linen backed tracing paper

RING OF BRODGAR
(see above)
DC4568
O1/1
Insc. *Orkney*
Annotated plan of henge monument and
   the surrounding area. Annotated on
   back 'This must have been done on
   one of the occasions I was not helping.
   He is trying to get relative locations of
   Stenness ring, Barnhouse stone, Watch
   stone, and several stones reputed to be
   standing stones, in the wall by the road
   between Watch stone and Brogar R.
   The 1 in : 20 ft plan is the blow up of
   the "2nd" set near the stumps, but I can
   not say what *F* and *A* are at the Brogar
   end. Probably the small *O* is Stenness?,
   with the *N* line through it. AST
   19/1/85'.
1 in : 400 ft & 1 in : 20 ft
Pencil
560 mm × 373 mm

RING OF BRODGAR
(see above)
DC4569
O1/1
Insc. *Brogar*

Plan of henge monument and small cairns
   in the area.
1 : 1250
Black ink and pencil
390 mm × 356 mm

RING OF BRODGAR
(see above)
DC4570
O1/1
Insc. *Brogar Cairns*
Plan of henge monument and cairns in the
   area.
1 : 1250
Black ink
427 mm × 397 mm
Plastic film

RING OF BRODGAR
(see above)
DC4679
O1/1
Insc. *Ring of Brodgar, Stenness*
Tracing of the henge monument from the
   RCAMS Inventory for Orkney.
1 in: 70.8 ft
Pencil
558 mm × 378 mm
Tracing paper

RING OF BRODGAR
(see above)
DC4754
O1/1
Insc. *Brogar Cairns*
Annotated plan of henge monument and
   cairns in the area. Annotated 'A.T.'s
   art work'.
1 : 1250
Black ink and pencil
407 mm × 387 mm
Linen backed tracing paper

RING OF BRODGAR
(see above)

DC4755
O1/1
Insc. *Brogar*
Contour survey of site.
3 in : 100 ft
Pencil
558 mm × 380 mm

RING OF BRODGAR
(see above)
DC4778
O1/1
Insc. *Ring of Brogar*
Plan and section of henge monument. Annotated 'The stones marked by Thomas as "prostrate" for which he does not also show a stump are marked *P*. Small stones inside ring are shown thus'.
1½ in : 50 ft
Black ink and pencil
441 mm × 412 mm
1971

RING OF BRODGAR
(see above)
DC4786
O1/1
Insc. *Brogar, 1972 survey*
Annotated plan of henge monument.
1 in : 20 ft
Black ink and pencil
757 mm × 558 mm
1972

RING OF BRODGAR
(see above)
DC5355
O1/1
Profile showing moon rising over Mid Hill at the minor standstill.
No scale
Black ink
240 mm × 180 mm

RING OF BRODGAR
(see above)
DC5356
O1/1
Profile showing moon setting on the cliffs at Hellia, Hoy, at the minor standstill.
No scale
Black ink
236 mm × 174 mm

ROSEVIEW
Mounds
HY22NW 10   HY24662537
DC4657
O1/1
Insc. *Ravie Hill, Orkney*
Location plan of mounds.
No scale
Pencil
500 mm × 380 mm
Tracing paper

STONES OF STENNESS
Henge Monument
HY31SW 2   HY30671252
DC5359
O1/2
Profile of horizon showing the moon setting on cairn, as viewed from the stone circle.
No scale
Black ink
240 mm × 169 mm

Peeblesshire

TWEEDSMUIR
Cairn and Standing Stones
NT02SE 10   NT09502400
NT02SE 12   NT095 239
DC4485
G9/4
Insc. *Tweedsmuir*

Plan of cairn and standing stones.
1 in : 10 ft
Pencil
559 mm × 380 mm

Perthshire

BLACKFAULDS
Stone Circle
NO13SW 15   NO14133167
DC4597
P2/9
Insc. *Guildtown*
Plan of stone circle.
¼ in : 1 ft
Blue and red ink and pencil
557 mm × 379 mm

BROUGHDEARG
Standing Stones
NO16NW 1   NO13746704
DC4599
P2/13
Insc. *Broughdarg, Glenshee*
Plan of standing stones.
¼ in : 1 ft
Blue ink and pencil
558 mm × 381 mm

CARSE
Stone Circle
NN84NW 2   NN80284846
DC4588
P1/5
Insc. *Weem*
Plan of stone circle.
⅛ in : 1 ft
Black ink
560 mm × 382 mm

CLACHAN AN DIRIDH
Stone Circle

NN95NW 5   NN92515574
DC4589
P1/18
Insc. *Clachan An Dirach, Pitlochry*
Plan of stone circle with sketch of hillside.
¼ in : 1 ft
Pencil
559 mm × 380 mm

CLACHAN AN DIRIDH
(see above)
DC4860
P1/18
Insc. *Clachan An Diridh*
Plan of stone circle.
⅛ in : 1 ft
Black ink
183mm × 140mm
Plastic film

CLACHAN AN DIRIDH
(see above)
DC5346
P1/18
Plan of standing stones.
¼ in : 1 ft
Black ink
224 mm × 142 mm

CLACH NA TIOM-PAN
Stone Circle
NN83SW 2   NN83013281
DC4588
P1/17
Insc. *Glen Almond*
Plan of standing stones.
¼ in : 1 ft
Black ink
560 mm × 382 mm

COLEN
Stone Circle
NO13SW 19   NO11063116

DC4575
P2/6
Insc. *Colen*
Plan of stone circle.
¼ in : 1 ft
Black ink and pencil
542 mm × 290 mm

COURTHILL, GLENBALLOCH
Stone Circle
NO14NE 12   NO18434807
DC4594
P2/4
Insc. *Courthill*
Plan of stone circle.
¼ in : 1 ft
Black ink and pencil
562 mm × 380 mm

CROFTMORAIG
Stone Circle
NN74NE 12   NN79754726
DC4590
P1/19
Insc. *Croftmoraig*
Annotated plan of stone circle.
⅛ in : 1 ft
Black ink and pencil
559 mm × 380 mm

DALCHIRIA
Standing Stones
NN81NW 3   NN82441588
DC4485
P1/2
Insc. *Muthill*
Plan and elevation of standing stones.
⅛ in : 1 ft
Pencil
559 mm × 380 mm

DALCHIRIA
(see above)
DC5344/co

P1/1
Insc. *Dalchirla, Muthill*
Plan and elevation of standing stone.
⅜ in : 4 ft
Dyeline
127 mm × 100 mm

DRUIDS SEAT WOOD
Stone Circle
NO13SW 20   NO12483132
DC4593
P2/3
Insc. *Blindwells*
Plan of stone circle.
¼ in : 1 ft
Black ink and pencil
513 mm × 383 mm

DULL
Stone Circle
NN84NW 4   NN80224873
DC4578
P1/4
Insc. *Weem W*
Plan of stone circle.
¼ in : 1 ft
Black ink and pencil
561 mm × 379 mm

DUNARDRY
Stone Circle
NN50SW 2   NN51810019
DC4383
A11/2
Insc. *Aberfoyle*
Annotated plan of stone circle.
⅛ in : 1 ft
Pencil
552 mm × 381 mm

DUNARDRY
Stone Circle
NN50SW 2   NN51810019
DC4843

A11/2
Insc. *Aberfoyle*
Plan of stone circle.
¹₂ in : 4 ft
Black ink
338 mm × 211 mm
Plastic film

EAST CULT, DUNKELD
Standing Stones
NO04SE 2   NO07254216
DC4595
P2/7
Insc. *East Cult*
Plan of standing stones.
¹₈ in : 1 ft
Black ink and pencil
558 mm × 385 mm

EAST CULT, DUNKELD
(see above)
DC4848
P2/7
Plans and elevations of the standing
    stones.
⁵₈ in : 10 ft
Black ink
202 mm × 123 mm

FORTINGALL CHURCH
Standing Stones
NN74NW 3   NN74524695
DC4576
P1/6
Insc. *Fortingall*
Plans of the standing stones.
No scale
Red ink and pencil
554 mm × 377 mm
Tracing paper

FORTINGALL CHURCH
(see above)
DC4577

P1/6
Insc. *Fortingall*
Plans of standing stones *A* and *B*.
No scale
Pencil
557 mm × 382 mm

FORTINGALL CHURCH
(see above)
DC4799
P1/6
Insc. *Fortingall*
Plan of standing stones.
³₄ in : 10 ft
Pencil
763 mm × 557 mm

FOWLIS WESTER
Stone Circles, Cairn and Standing Stone
NN92SW 1   NN92422492
DC4579
P1/10
Sketch plan of monuments.
No scale
Black ink and pencil
559 mm × 377 mm

FOWLIS WESTER
(see above)
DC4580
P1/10
Insc. *Fowlis Wester*
Annotated plan of stone circle, cairn and
    standing stone with elevation of
    standing stone to NE.
1 in : 10 ft
Blue and red ink and pencil
555 mm × 375 mm

FOWLIS WESTER
(see above)
DC5360
P1/10
Plan of stone circle, cairn and standing
    stone with horizon profiles.

No scale
Black ink
342 mm × 302 mm
Plastic film

GELLYBANKS FARM
Standing Stones
NO03SE 8   NO08213134
DC4581
P1/12
Insc. *Gellybanks nr Perth*
Annotated plan of standing stones.
$^3_8$ in : 1 ft
Pencil
556 mm × 380 mm

GELLYBANKS FARM
(see above)
DC4853/co
P1/12
Insc. *Gellybanks*
Plan of standing stones.
$^3_5$ in : 4 ft
Dyeline
94 mm × 89 mm

GLENHEAD
Standing Stones
NN70SE 3   NN75480045
DC4485
P1/1
Insc. *Doune*
Plan of standing stones.
$^1_{10}$ in : 1 ft
Pencil
559 mm × 380 mm

GLENHEAD
(see above)
DC4850/co
P1/2
Insc. *Doune*
Plan of standing stones.
$^3_4$ in : 10 ft

Dyeline copy
130 mm × 95 mm

KILLIN
Stone Circle
NN53SE 12   NN57703280
DC4575
P1/3
Insc. *Killin*
Plan of stone circle.
$^1_4$ in : 1 ft
Black ink and pencil
542 mm × 290 mm

LEYS OF MARLEE
Stone Circle
NO14SE 15   NO15994388
DC4592
P2/1
Insc. *Leys of Marlee*
Plan of stone circle.
$^1_8$ in : 1 ft
Black ink and pencil
560 mm × 380 mm

LITTLE FINDOWIE
Standing Stones
NN93NW 2   NN94483865
DC4585
P1/15
Insc. *Pulpit Stone, Little Findowie*
Annotated plan of standing stones.
1^1_2 in : 10 ft
Pencil
558 mm × 381 mm

LITTLE FINDOWIE
(see above)
DC5345
P1/15
Insc. *Pulpit Stone, Little Findowie*
Plan of standing stones.
$^5_8$ in : 10 ft

Black ink
312 mm × 205 mm

LUNDIN
Stone Circle
NN85SE 9   NN88065056
DC4578
P1/7
Insc. *Aberfeldy*
Plan of stone circle.
¼ in : 1 ft
Black ink and pencil
561 mm × 379 mm

MEIKLE FINDOWIE
Stone Circle
NN93NE 1   NN95903868
DC4586
P1/16
Insc. *Meikle Findowie*
Plan of stone circle.
¼ in : 1 ft
Blue and red ink and pencil
559 mm × 382 mm

MEIKLE FINDOWIE
(see above)
DC4587
P1/16
Insc. *Mickle Findowie*
Plan of stone circle.
¼ in : 1 ft
Black ink
554 mm × 380 mm
Linen backed tracing paper

MONCREIFFE HOUSE
Stone Circle
NO11NW 11   NO13281933
DC4591
P1/20
Insc. *Moncrieffe Ho.*
Plan of stone circle.
1 : 50

Black and blue ink
555 mm × 372 mm
1974

MONCREIFFE HOUSE
(see above)
DC4752
P1/20
Insc. *Moncrieffe House*
Annotated plan of stone circle.
1 : 50
Black ink
408 mm × 309 mm
Linen backed tracing paper

MONCREIFFE HOUSE
(see above)
DC4773
P1/20
Insc. *Moncrieffe House*
Plan of stone circle.
¼ in : 1 ft
Black ink
332 mm × 260 mm
Tracing paper

MONZIE CASTLE
Cairn
NN82SE 26   NN88162417
DC4582
P1/13
Insc. *Monzie*
Annotated plan of cairn with sketch of
    hillside and drawing of cup-marked
    stone.
1½ in : 10 ft
Blue ink and pencil
556 mm × 380 mm

MONZIE CASTLE
(see above)
DC4583
P1/13
Insc. *Monzie*

Sketch of cup and ring-marked stone.
No scale
Black ink and pencil
559 mm × 379 mm

MONZIE CASTLE
(see above)
DC4700
P1/13
Insc. *Monzie*
Plan of cairn.
No scale
Pencil
380 mm × 368 mm

MONZIE CASTLE
(see above)
DC4701
P1/13
Insc. *Monzie*
Tracing of cup and ring marks.
¾ in : 1 ft
Black ink
292 mm × 185 mm
Tracing paper

NEWTYLE
Standing Stones
NO04SW 7   NO04494107
DC4581
P2/12
Insc. *Dunkeld*
Annotated plan of standing stones.
¼ in : 1 ft
Pencil
556 mm × 380 mm

NEWTYLE
(see above)
DC5347/co
P2/12
Insc. Dunkeld
Plan of standing stones.
½ in : 5 ft

Dyeline
106 mm × 103 mm

SANDY ROAD, SCONE
Stone Circle
NO12NW 28   NO13272646
DC4598
P2/11
Insc. *New Scone Wood*
Plan of stone circle.
1½ in : 10 ft
Blue and red ink and pencil
558 mm × 379 mm

SHIANBANK
Stone Circles
NO12NE 7   NO15552730
DC4596
P2/8
Insc. *Shianbank*
Plans of stone circles.
1½ in: 10 ft
Blue ink and pencil
557 mm × 383 mm

SPITTAL OF GLENSHEE
Stone Circle
NO17SW 1   NO11717017
DC4600
P2/14
Insc. *Spittal of Glenshee*
Plan of stone circle.
⅛ in : 1 ft
Pencil
558 mm × 379 mm

TIGH-NA-RUAICH
Stone Circle
NN95SE 1   NN97625345
DC4592
P2/2
Insc. *Ballinluig*
Plan of stone circle.
¼ in : 1 ft

Black ink and pencil
560 mm × 380 mm

TULLYBEAGLES
Stone Circle and Cairn
NO03NW 7   NO01283620
DC4584
P1/14
Insc. *Tullybeagles*
Plans of stone circle and cairn.
1¹₂ in : 10 ft
Blue ink and pencil
556 mm × 381 mm

WOODSIDE
Stone Circle
NO15SE 9   NO18485005
DC4594
P2/5
Insc. *Hill of Drimmie*
Plan of stone circle.
¹₄ in : 1 ft
Black ink and pencil
562 mm × 380 mm

Ross & Cromarty

CALLANISH, LEWIS
Stone Circle and Alignment
NB23SW 1   NB21303300
DC4784
H1/1
Insc. *Composite Survey of Four Circles.
Each reduced to same diameter and
orientated on major axis*
Plan.
No scale
Black ink
363 mm × 284 mm
Tracing paper

CLACH AN TRUSHAL, LEWIS
Standing Stones
NB35SE 1   NB37555377
DC4875/co
H1/12
Insc. *Clach An Trushel, Lewis*
Elevation drawing of standing stones.
³₁₀ in : 6 ft
Dyeline
63 mm × 50 mm

CLACH STEIN, LEWIS
Standing Stone
NB53SW 5   NB51653174
DC4877/co
H1/14
Insc. *Clach Stein*
Annotated plan and elevation of standing
    stone with sketch of hills. Annotated
    'The main stone is fallen and broken.
    The small upright is orientated on
    Suilven (40 miles)'.
³₈ in : 5 ft
Dyeline
138 mm × 56 mm

CNOC NAN DURSAINEAN, GARRABOST, LEWIS
Chambered Cairn and Stone Circle
NB53SW 2   NB52383307
DC4493
H1/13
Insc. *Dursainean*
Plan of chambered cairn and stone circle.
1 in : 10 ft
Blue ink and pencil
560 mm × 384 mm

CNOC NAN DURSAINEAN, GARRABOST, LEWIS
(see above)
DC4876/co
H1/13
Insc. *Dursainean*
Annotated plan of the chambered cairn
    and stone circle with hillside sketches.

Annotated 'Intervisible with Clach Stein (1 ml.) and visible from H1/15 about 600 yd'.
½ in : 10 ft
Dyeline copy
137 mm × 100 mm
1949

DURSAINEAN, LEWIS
Standing Stone
NB53SW 7   NB52813340
DC4878/co
H1/15
Insc. *On Burn Side Below Dursainean*
Annotated plan of standing stone.
No scale.
Dyeline
137 mm × 37 mm

STEINACLEIT, SHADER, LEWIS
Stone Circle and Cairn
NB35SE 2   NB39625407
DC4492
H1/10
Insc. *Steinacleit*
Annotated plan of stone circle and cairn with sketch of bank '(paced)'.
¹⁄₁₀ in : 1 ft and 1 in : 20 yds
Blue and red ink and pencil
557 mm × 374 mm

TURSACHAN, AIRIDH NAM BIDEARAN, LEWIS
Stone Circle
NB22NW 1   NB23422989
DC4491
H1/5
Insc. *Callanish V*
Annotated plan of stone circle. Annotated 'Line of stones points to Mor Monach 10.2 mls'.
⁵⁄₁₆ in : 10 ft
Black ink
442 mm × 279 mm
Graph paper

TURSACHAN, AIRIDH NAM BIDEARAN, LEWIS
(see above)
DC4680
H1/5
Insc. *Callanish V*
Plan of stones.
No scale
Black ink and pencil
557 mm × 381 mm

TURSACHAN, AIRIDH NAM BIDEARAN, LEWIS
(see above)
DC4873
H1/5
Profile of hill from the north stone showing the setting moon.
No scale
Black ink
219 mm × 199 mm

TURSACHAN, BARRAGLOM, GREAT BERNERA, LEWIS
Standing Stones
NB13SE 2   NB16423424
DC4874
H1/8
Insc. *Great Berneray, Loch Roag*
Plan of standing stones with hillside sketches.
No scale
Black ink
146 mm × 79 mm
1947

TURSACHAN, CNOC FILLIBHIR BHEAG, LEWIS
Stone Circle
NB23SW 2   NB22503269
DC4489
H1/3
Insc. *Callanish III, Cnoc Fillibhir Bheag*
Plan of stone circle.
1½ in : 10 ft
Black ink and pencil

556 mm × 380 mm
1972

***TURSACHAN, CNOC FILLIBHIR
BHEAG, LEWIS***
(see above)
DC4490
H1/3
Ellipses.
No scale
Pencil
555 mm × 379 mm

Roxburghshire

BLACK KNOWE
Cairn
NT71NE 32   NT75061552
DC4482
G8/5
Insc. *Dere Street I*
Plan of cairn with sketch of Browndean
   Law. Annotated '[Stone] 36 in high.
   Flat topped providing a stand from
   which 6 stone can be seen over others.
   About 50 yds to east there are two
   stones oriented on hill at about 203.6
   h = 1.05'.
$\frac{1}{8}$ in : 1 ft
Blue ink and pencil
560 mm × 379 mm

BLACK KNOWE
(see above)
DC4482 (reverse)
G8/5
Plan of cairn.
No scale
Pencil
560 mm × 379 mm

BLACK KNOWE
(see above)
DC4871/co
G8/5
Insc. *Dere Street I*
Plan of stone circle with sketch of
   Browndean Law.
No scale
Photocopy
200mm × 128mm

BURGH HILL
Stone Circle
NT40NE 17   NT47010624
DC4486
G9/15
Insc. *Allan Water near Hawick*
Annotated plan of stone circle.
$\frac{1}{8}$ in : 1 ft
Black ink and pencil
559 mm × 379 mm

BURGH HILL
(see above)
DC4679
Insc. *Roxburgh 1011, Burgh Hill,
   Teviothead Parish*
Tracing of the stone circle from the
   RCAMS Inventory for Roxburghshire.
1 in : 18.6 ft
Pencil
558 mm × 378 mm
Tracing paper

FIVE STANES
Stone Circle
NT71NE 36   NT75261686
DC4483
G8/7
Insc. *Five Stanes (Dere Street III)*
Annotated plan of stone circle.
$1\frac{1}{2}$ in : 10 ft
Blue ink and pencil
560 mm × 379 mm

NINESTONE RIG
Stone Circle
NY59NW 6   NY51809731
DC4481
G8/2
Insc. *Ninestone Rig*
Annotated plan of stone circle with
    elevation of stones *A* and *C* and
    hillside sketches.
$\frac{1}{4}$ in : 1 ft
Blue ink and pencil
559 mm × 379 mm

SHEARERS (THE)
Old Dyke
NT71NE 49   NT79131927
DC4484
G8/9
Insc. *Eleven Shearers, Hownam*
Plan of dyke with chart showing the
    approximate height of the stones above
    the ground.
$\frac{3}{8}$ in : 10 ft
Black and blue ink and pencil
560 mm × 379 mm

SHEARERS (THE)
(see above)
DC4872/co
G8/9
Insc. *Eleven Shearers*
Annotated plan of stones with view along
    dyke to hillside.
$\frac{1}{2}$ in : 30 ft
Black ink
251 mm × 104 mm
Plastic film

'TRESTLE CAIRN'
Cairn
NT71NE 34   NT75181612
DC4483
G8/6
Insc. *Dere Street II*
Annotated plan of cairn.

$1\frac{1}{2}$ in : 10 ft
Blue ink and pencil
560 mm × 379 mm

Shetland

GIANTS' STONES, SHETLAND
Stone Circle
HU28SW 3   HU24308055
DC4658
Z3/2
Insc. *Giants' Stones*
Plan of stone circle.
No scale
Pencil
449 mm × 305 mm

GIANTS' STONES, SHETLAND
(see above)
DC4779
Z3/2
Insc. *The Giant's Stones*
Annotated plan of stone circle.
$\frac{1}{2}$ in : 10 ft
Black ink
415 mm × 323 mm
Plastic film

GIANTS' STONES, SHETLAND
(see above)
DC4780
Z3/2
Insc. *Giants' Stones*
Annotated plan of stone circle.
$\frac{1}{2}$ in : 10 ft
Black ink
287 mm × 282 mm
Plastic film

HALTADANS, FETLAR, SHETLAND
Cairn
HU69SW 4   HU62209240

DC4661
Z5/1
Insc. *Fetlar, Haltadans is a cairn.*
Plan showing location of cairn.
1 : 10000
Blue ink and pencil
331 mm × 294 mm
Tracing paper

DC4657
Z3/1
Insc. *Stanydale*
Annotated plan of the settlement and field
    system.
No scale
Pencil
500 mm × 380 mm
Tracing paper

HAMARS (THE), LOCH OF STROM,
SHETLAND
Cairn
HU45SW 2   HU40335020
DC4659
Z3/3
Insc. *Circle, on W side of Loch of Strome*
Annotated plan of circle.
1 : 50
Pencil
567 mm × 390 mm

STANYDALE, SHETLAND
(see above)
DC4783
Z3/1
Insc. *Stanydale, Shetland*
Annotated plan of 'temple'.
¹₄ in : 1 ft
Black ink
692 mm × 498 mm
Plastic film

HAMARS (THE), LOCH OF STROM,
SHETLAND
(see above)
DC4782
Z3/3
Annotated plan of circle.
¹₄ in : 1 ft
Black ink
240 mm × 205 mm
Plastic film

YAA FIELD, EAST BURRA,
SHETLAND
Standing Stone
HU33SE 16   HU37803285
DC4660
Z4/1
Insc. *East Burras*
Plan of standing stone and nearby stones
    forming a ring.
1 : 84
Black ink and pencil
438 mm × 337 mm

STANYDALE, SHETLAND
Settlement and Field System
HU25SE 1   HU285 502
DC4656
Z3/1
Insc. *Stanydale*
Plan of 'temple'.
1 : 50
Black ink and pencil
571 mm × 388 mm

YAA FIELD, EAST BURRA,
SHETLAND
(see above)
DC4781
Z4/1
Insc. *East Burra, Shetland*
Annotated plan of standing stone and
    other nearby stones.
¹₈ in : 1 ft
Black ink
377 mm × 366 mm
Plastic film

STANYDALE, SHETLAND
(see above)

Stirlingshire

DUNTREATH, DUMGOYACH
Standing Stones
NS58SW 3   NS53288072
DC4842
A11/1
Insc. *Near Blanefield*
Annotated elevation drawing of standing
  stones with hillside sketch.
${}^3\!s$ in : 10 ft
Black ink
99 mm × 80 mm

Sutherland

CLACH MHIC MHIOS
Standing Stone
NC91NW 11   NC94041508
DC4566
N2/4
Insc. *Clach Mhic Mhios, Glen Loth,
  Sutherland*
Elevation drawing of stone with horizon
  profiles.
No scale
Black ink and pencil
460 mm × 300 mm
Graph paper

CLACH MHIC MHIOS
(see above)
DC4861
N2/4
Insc. *Glen Loth*
Annotated elevation of standing stone
  with diagram of horizon. Annotated
  'Menhir Clach Mhic Mhios, showing
  moon setting with declination -($\varepsilon \pm i$),
  where $\varepsilon = 23° 53'.1$ '.

No scale
Black ink
201 mm × 137 mm
Plastic film

DRUIM BAILE FIUR
Stone Circle
NC50SE 13   NC56040295
DC4677
Tracing of stone circle from the RCAMS
  Inventory for Sutherland.
1 in : 25.3 ft
Black ink and pencil
556 mm × 376 mm
Tracing paper

LEARABLE HILL
Standing Stone, Stone Circle, Stone Rows
  and Cairns
NC82SE 1   NC89252349
NC82SE 4   NC89122351
NC82SE 6   NC89272350
NC82SE 9   NC89222351
DC5343
N2/1
Plan of standing stone, stone circle, stone
  rows and cairns with hillside sketch.
${}^1\!2$ in : 10 ft
Black ink
259 mm × 195 mm

LEARABLE HILL
Stone Circle
NC82SE 4   NC89162351
DC4677
Tracing of the stone circle from the
  RCAMS Inventory for Sutherland.
1 in : 25 ft
Black ink and pencil
556 mm × 376 mm
Tracing paper

LEARABLE HILL
Stone Rows
NC82SE 6   NC89272350

DC4562
N2/1
Insc. *Learable*
Plan of stone rows. Annotated '1st survey'.
$^1/_{10}$ in : 1 ft
Blue ink and pencil
558 mm × 382 mm

LEARABLE HILL
(see above)
DC4563
N2/1
Insc. *Learable Hill*
Plan of the stone rows showing in different colours of ink the surveys of 1957 and 1963.
No scale
Red and black ink
555 mm × 350 mm
Tracing paper

LEARABLE HILL
(see above)
DC4564
N2/1
Insc. *Learable Hill*
Plan of stone rows. Annotated '2nd (partial) survey'.
No scale
Black ink and pencil
559 mm × 379 mm

LEARABLE HILL
(see above)
DC4677
Tracing of the stone rows from the RCAMS Inventory for Sutherland.
No scale
Pencil
556 mm × 376 mm
Tracing paper

RIVER FLEET
Stone Circle

NH79NE 8   NH76909908
DC4565
N2/2
Insc. *Fleet, near the Mound*
Plan of stone circle.
$1^1/_2$ in : 10 ft
Blue ink and pencil
558 mm × 380 mm

RIVER SHIN
Stone Circle
NC50SE 22   NC58220493
DC4565
N2/3
Insc. *Shin River, Lairg*
Annotated plan of stone circle.
$1^1/_2$ in : 10 ft
Blue ink and pencil
558 mm × 380 mm

STRATH, ALLT BREAC
Stone Rows
NC91NE 6   NC95491854
DC4677
Tracing of the stone rows from RCAMS Inventory for Sutherland.
No scale
Pencil
556 mm × 376 mm
Tracing paper

West Lothian

CAIRNPAPPLE HILL
Henge Monument and Cairn
NS97SE 16   NS98727173
DC4601
P7/1
Insc. *Cairnpapple Hill*
Plan of the stone holes.
1 in : 10 ft
Red ink and pencil
558 mm × 380 mm

Wigtownshire

DRUMTRODDAN
Standing Stones
NX34SE 2   NX36444430
DC4457
G3/12
Insc. *Drumtroddan*
Annotated plan of standing stones.
⅛ in : 1 ft
Black ink and pencil
556 mm × 383 mm

DRUMTRODDAN
(see above)
DC4869
G3/12
Insc. *Drumtroddan*
Annotated plan and elevations of standing
    stones.
½ in : 10 ft
Black ink
100 mm × 91 mm

DRUMUILLIE LANE
Stone
NX27SW 7   NX20977234
DC4868
G3/3
Insc. *Site on Tarf Water*
Plan of stones in area with elevation of
    the Laggangarn standing stones.
¾ in : 1000 ft
Black ink
203 mm × 128 mm

EAST THORNEY
Stones
NX14SW   NX10 42
DC4801
G2/4
Insc. *Stones near Port Logan*
Annotated plan of stones.
1 in : 100 ft

Black ink and pencil
759 mm × 556 mm

EAST THORNEY
(see above)
DC4866
G2/4
Insc. *Stones near Port Logan*
Plan of stones.
⅝ in : 300 ft
Black ink
203 mm × 186 mm
1938 and 1939

GLENTERROW
Standing Stones
NX16SW 8   NX14536250
DC4454
G3/4
Annotated plan of the standing stones
    with astronomical events marked.
1 in : 200 ft
Pencil
557 mm × 355 mm

KNOCKING STONE
Stone
NX27SW M1   NX21887174
DC4868
G3/3
Insc. *Site on Tarf Water*
Plan of stones in area with elevation of
    the Laggangarn standing stones.
¾ in : 1000 ft
Black ink
203 mm × 126 mm

LAGGANGARN
Standing Stones
NX27SW 4   NX22237166
DC4868
G3/3
Insc. *Site on Tarf Water*

Elevation of the standing stones with plan
  of other stones in area.
³₄ in : 1000 ft
Black ink
203 mm × 126 mm

LONG STONE
Standing Stone
NX27SW 6   NX22727148
DC4868
G3/3
Insc. *Site on Tarf Water*
Plan of stones in area with elevation of
  the Laggangarn standing stones.
³₄ in : 1000 ft
Black ink
203 mm × 126 mm

LONG TOM
Standing Stone
NX07SE 3   NX08167183
DC4867
G3/2
Insc. *Haggstone Moor*
Plan of standing stone showing
  relationship with other monuments.
  Diagrams of the moon setting behind
  the Mull of Kintyre from the stone and
  Haggstone Moor standing stone.
No scale
Black ink
207 mm × 200 mm
Plastic film

TAXING STONE
Standing Stone
NX07SE 1   NX06237096
DC4867
G3/2
Insc. *Haggstone Moor*
Plan of standing stone showing
  relationship with other monuments.
  Diagrams of the moon setting behind
  the Mull of Kintyre from 'Long Tom'
  and Haggstone Moor standing stone.
No scale

Black ink
207 mm × 200 mm
Plastic film

TORHOUSEKIE
Stone Circle and Standing Stones
NX35NE 12   NX38375650
NX35NE 14   NX38255649
DC4455
G3/7
Insc. *Torhouse*
Annotated plan of stone circle and
  standing stones.
¹₈ in : 1 ft
Black ink and pencil
558 mm × 379 mm

TORHOUSEKIE
(see above)
DC4456
G3/7
Insc. *Torhouse*
Plan of the stone circle and standing
  stones.
No scale
Black ink
556 mm × 393 mm
Linen backed tracing paper

TORHOUSEKIE
(see above)
DC4784
G3/7
Insc. *Composite Survey of Four Circles.
  Each reduced to same diameter and
  orientated on major axis*
Plan.
No scale
Black ink
363 mm × 284 mm
Tracing paper

WREN'S EGG (THE)
Standing Stones
NX34SE 10   NX36104199

DC4458
G3/13
Insc. *Wren's Egg*
Annotated plan of standing stones.
$\frac{1}{8}$ in : 1 ft
Black ink and pencil
557 mm × 376 mm

WALES

Brecknockshire

CERRIG DUON
Stone Circle
SN82SE 1   SN85122063
DC4806
W11/3
Insc. *Maen Mawr*
Annotated plan of stone circle and nearby
    alignment of small stones.
$\frac{1}{8}$ in : 1 ft
Black ink and pencil
559 mm × 380 mm

DRYGARN
Stone Circle
SN93NW 1   SN92113826
DC4652
W11/5
Insc. *Stone Circle near Ynys Hir,
    Llanfihangel Nant Bran*
Annotated plans of stone circle.
1 in : 8.12 ft & 1 in : 16.15 ft
Pencil
556 mm × 375 mm
Tracing paper

NANT TARW E
Stone Circle
SN82NW 5   SN81872583
DC4648
W11/4
Insc. *River Usk E Circle*
Annotated plan of stone circle with
    elevation of outlier stone *C*.
$\frac{1}{8}$ in : 1 ft

Blue ink and pencil
556 mm × 379 mm

NANT TARW W
Stone Circle
SN82NW 5   SN81972579
DC4649
W11/4
Insc. *W. Circle River Usk*
Annotated plan of stone circle with inset
   showing position of E and W circles
   and outlying stone.
¹₈ in : 1 ft and ³₄ in : 100 ft
Green and red ink and pencil
556 mm × 381 mm

NANT TARW W
(see above)
DC4650
W11/4
Insc. *W Circle. River Usk*
Plan of stone circle.
¹₈ in : 1 ft
Black ink
555 mm × 379 mm
Linen backed tracing paper

SAITH-MAEN
Stone Alignment
SN96SW 3   SN94936030
DC4639
W11/1
Insc. *Saeth-Maen*
Plan of stone alignment.
¹₄ in : 1 ft
Blue ink and pencil
556 mm × 379 mm

SAITH-MAEN
(see above)
DC4852
W11/1
Insc. *Saeth-Maen*
Plan of stone alignment.

¹₄ in : 1 ft
Black ink
203 mm × 69 mm

TRECASTLE MOUNTAIN E
Stone Circle
SN83SW 6   SN83353109
DC4646
W11/2
Insc. *Y-Pigwn Trecastle, NW Circle*
Annotated plan of stone circle.
¹₈ in : 1 ft
Black ink and pencil
559 mm × 380 mm

TRECASTLE MOUNTAIN E & W
Stone Circles
SN83SW 6   SN83313106
                          & SN83353109
DC4647
W11/2
Insc. *Y-Pigwn, Trecastle*
Annotated plan of stone circles.
³₈ in : 10 ft
Pencil
555 mm × 380 mm

TRECASTLE MOUNTAIN W
Stone Circle
SN83SW 6   SN83313106
DC4807
W11/2
Insc. *Y-Pigwn, Trecastle SE Circle*
Plan of stone circle.
¹₈ in : 1 ft
Green ink and pencil
560 mm × 382 mm

YNYS-HIR
Stone Circle
SN93NW 1   SN92113826
DC4651
W11/5
Insc. *Ynys-Hir*

Annotated plan of stone circle.
$1\frac{1}{2}$ in : 10 ft
Black ink and pencil
559 mm × 380 mm

YNYS-HIR
(see above)
DC4768/co
W11/5
Insc. *Ynys-Hir*
Plan of stone circle.
$1\frac{1}{2}$ in : 10ft
Dyeline
327mm × 316mm

Caernarvonshire

CEFN COCH STONE
Cairn (possible)
SH77SW 10   SH72197463
DC4678
Tracing of cairn from the RCAMW
    Inventory for Caernarvonshire.
No scale
Pencil
558 mm × 376 mm
Tracing paper

CERRIG PRYFAID
Stone Circle
SH77SW 72   SH72457132
DC4678
Insc. *Cerrig Pryfaid*
Annotated tracing of stone circle from the
    RCAMW Inventory for Caernarvon-
    shire.
1 in : 48 ft
Pencil
558 mm × 376 mm
Tracing paper

DRUIDS CIRCLE
Stone Circle
SH77SW 1   SH72287464
DC4631
W2/1
Insc. *Penmaen-Mawr*
Plan of stone circle with elevation of one
    of the stones to NW.
1 in : 8 ft
Black ink
556 mm × 379 mm
Linen backed tracing paper

DRUIDS CIRCLE
(see above)
DC4632
W2/1
Insc. *Penmaen-Mawr*
Annotated plan of stone circle with
    elevation of stone to NW.
1 in : 8 ft
Black ink and pencil
558 mm × 380 mm

DRUIDS CIRCLE
(see above)
DC4678
W2/1
Insc. *The Druid's Circle*
Tracing of stone circle from the RCAMW
    Inventory for Caernarvonshire.
1 cm : 10 ft
Pencil
558 mm × 376 mm
Tracing paper

DRUIDS CIRCLE (West of)
Stone circle
SH77SW 10   SH72197463
DC4633
W2/1
Insc. *Penmaen-Mawr. Complex to W of
    Main Circle*
Annotated plan of stone circles.
1 in : 10 ft

Black ink and pencil
559 mm × 380 mm

DRUIDS CIRCLE (West of)
(see above)
DC5350
W2/1
Insc. *Penmaen-Mawr. Complex to W. of*
    Main Circle (W2/1A)
Annotated plan of stone circle.
1 in : 10 ft
Black ink
371 mm × 212 mm
Plastic film

GRAIG WEN
Stone Circle
SH77SW 77   SH72507476
DC4633
W2/1
Insc. *Circle to NE from main circle and*
    lower some 53 ft
Annotated plan of stone circle.
1 in : 4 ft
Blue ink and pencil
559 mm × 380 mm

Cardiganshire

HIRNANT
Cairn Circle
SN78SW 3   SN75328395
DC4654
W14/1
Insc. *Hirnant*
Plan of cairn. Annotated 'survey by J.R.
    Hoyle'.
⅝ in : 1ft
Black ink
465mm × 391mm
Plastic film

HIRNANT
(see above)
DC4655
W14/1
Insc. *Hirnant*
Plan of cairn.
No scale
Black ink
389 mm × 360 mm

Carmarthenshire

CASTELL-GARW
Cairn Circle
SN12NW 12   SN14572702
DC4642
W9/4
Insc. *Castell-Garw*
Annotated plan of stone circle.
1½ in : 10 ft
Black ink and pencil
556 mm × 381 mm

CASTELL-GARW
(see above)
DC4770/co
W9/4
Insc. *Castell Garw*
Annotated plan of stone circle.
1½ in : 10 ft
Dyeline
285 mm × 250 mm

Flintshire

PENBEDW PARK
Stone Circle
SJ16NE 6   SJ17126793
DC4636
W4/1
Insc. *Tardd-y-Dwr, Penbedw Hall*

Plan of stone circle.
1 in : 10 ft
Black ink and pencil
560 mm × 380 mm

PENBEDW PARK
(see above)
DC4678
W4/1
Tracing of stone circle from the RCAMW
Inventory for Flintshire.
No scale
Pencil
558 mm × 376 mm
Tracing paper

Merionethshire

MEINI HIRION
Standing Stones
SH52NE 3   SH58332700
DC4636
W5/3
Insc. *Meini Hirion*
Plan of standing stones.
¼ in : 1 ft
Blue ink and pencil
560 mm × 380 mm

MEINI HIRION
(see above)
DC5351
W5/3
Insc. *Meini Hirion*
Plan of standing stones.
¼ in : 1 ft
Black ink
220 mm × 209 mm
Tracing paper

MOEL TY UCHA
Cairn Circle

SJ03NE 1   SJ05593716
DC4634
W5/1
Insc. *Moel Ty Ucha*
Annotated plan of cairn with sketch of
hillsides.
1 in : 4 ft
Black ink and pencil
559 mm × 383 mm

TYFOS near LLANDRILLO
Cairn Circle
SJ03NW 5   SJ02853876
DC4635
W5/2
Insc. *Tyfos near Llandrillo*
Annotated plan of cairn.
1 in : 8 ft
Black ink and pencil
559 mm × 379 mm

Monmouthshire

GREY HILL
Stone Circle
ST49SW 6   ST43809353
DC4653
W13/1
Insc. *Gray Hill*
Plans of stone circle. One showing circle
in detail and the other showing the
position of the circle in relation to
outlying stones.
1 in : 4 ft and 1 in : 20 ft
Blue ink and pencil
555 mm × 376 mm

Montgomeryshire

KERRY HILL
Stone Circle

SO18NE 11   SO15768607
DC4637
W6/1
Insc. *Nantyrhynan*
Plan of stone circle.
$\frac{1}{8}$ in : 1 ft
Black ink and pencil
556 mm × 380 mm

RHOS-Y-BEDDAU
Stone Circle and Avenue
SJ03SE 3   SJ05773020
DC4638
W6/2
Insc. *Cerig Beddau (On 6 in OS), Rhos-Y-Beddau*
Annotated plan of stone circle and avenue.
1 in : 20 ft
Black ink and pencil
555 mm × 380 mm

RHOS-Y-BEDDAU
(see above)
DC4769/co
W6/2
Insc. *Rhos-Y-Beddau*
Plan of stone circle and avenue.
$\frac{1}{2}$ in : 10 ft
Dyeline
381 mm × 171 mm

RHOS-Y-BEDDAU
(see above)
DC4679
W6/2
Insc. *Rhs-Y-Beddau St. C. [sic] & Avenue*
Tracing of the stone circle, avenue and cairn from the RCAMW Inventory for Montgomeryshire.
1 in: 40 ft
Pencil
558 mm × 378 mm
Tracing paper

Pembrokeshire

CERRIG MEIBION ARTHUR
Standing Stones
SN13SW 6   SN11813102
DC4641
W9/3
Insc. *Cwm-Garw*
Plan of standing stones.
$\frac{1}{4}$ in : 1 ft
Pencil
556 mm × 380 mm

CERRIG MEIBION ARTHUR
(see above)
DC5352
W9/3
Insc. *Cwm Garw*
Plan of standing stones.
$\frac{1}{4}$ in : 1 ft
Black ink
172mm × 146mm
Plastic film

GORS-FAWR
Stone Circle
SN12NW 1   SN13462937
DC4640
W9/2
Insc. *Gors-Fawr*
Plan of stone circle and outliers.
$1\frac{1}{2}$ in : 10 ft
Blue ink and pencil
556 mm × 379 mm

PARC-Y-MEIRW
Standing Stones
SM93NE 12   SM99883591
DC4644
W9/7
Insc. *Parc-Y-Meirw (Fishguard)*
Annotated plan of standing stones.
$\frac{1}{8}$ in : 1 ft

Blue ink and pencil
555 mm × 385 mm

PARC-Y-MEIRW
(see above)
DC5353
W9/7
Plan of standing stones with profile of
    Mount Leinster showing the setting
    moon.
¾ in : 50 ft
Black ink
201 mm × 199 mm

RHOS-Y-CLEGYRN
Cairn Circle
SM93NW 24   SM91303543
DC4643
W9/5
Insc. *St. Nicholas*
Annotated plan of cairn circle.
1½ in : 10 ft
Blue ink and pencil
556 mm × 381 mm

WAUN LWYD
Standing Stones
SN13SE 2   SN15773126
DC4645
W9/8
Insc. *Dolau-Main*
Plan of standing stones.
¼ in : 1 ft
Pencil
559 mm × 380 mm

WAUN LWYD
(see above)
DC5354
W9/8
Insc. *Dolau Main*
Plan of standing stones.
¼ in : 1 ft
Black ink

248 mm × 157 mm
Plastic film

Radnorshire

FEDW
Stone Circle
SO15NW 2   SO14325797
DC4678
W8/2
Insc. *Rhos Maen*
Tracing of stone circle. Annotated
    'Visited 1960 found destroyed'.
No scale
Pencil
558 mm × 376 mm
Tracing paper

FOUR STONES
Standing stones
SO26SW 3   SO24586079
DC4642
W8/3
Insc. *Four Stones*
Annotated plan of stone circle.
¼ in : 1 ft
Black ink and pencil
556 mm × 381 mm

RHOS-Y-GELYNEN
Stone Alignment
SN96SW 1   SN90536308
DC4639
W8/1
Insc. *Rhosygelynnen*
Plan of stone alignment.
1½ in : 10 ft
Blue ink and pencil
556 mm × 379 mm

RHOS-Y-GELYNEN
(see above)

DC4851
W8/1
Insc. *Rhosygelynnen*
Plan of stone alignment.
$\frac{1}{4}$ in : 1 ft
Black ink
187 mm × 64 mm

MISCELLANEOUS

DC4664
Map of Islay, Jura, Lorn, Knapdale and
   Kintyre showing possible solstitial
   sites.
No scale
Black ink and pencil
439 mm × 358 mm
Tracing paper

DC4665
Plan illustrating positions of celestial
   bodies in 1700 B.C.
Red and black ink and pencil
279 mm × 252 mm

DC4670
Plan of ellipses.
Blue ink and pencil
476mm × 298mm

DC4671
Insc. *Suns declination 1800 B.C.*
Sketch diagram.
Pencil
476mm × 298mm
Graph paper

DC4672
Contour plan traced from 1″ OS survey
   of Ireland, 3rd edition.
Red ink and pencil
337 mm × 256 mm
Tracing paper

DC4673
Plan showing the geometry of a flattened
circle.
Black ink
308 mm × 293 mm
Linen backed tracing paper

DC4674
Plan showing flattened and egg-shaped
rings.
Black ink and pencil
455 mm × 301 mm

DC4675
Insc. *Types of Rings found in Standing
Stone Circles and in Cup and Ring
marks*
Annotated diagram.
No scale
410 mm × 402 mm
Plastic film

DC4676
Unidentified plan of cup and ring marks.
Black ink
584 mm × 388mm
Plastic film

DC4748
Unidentified plan of stones.
Annotated 'This I do not recognise.
A.S.T. 13/4/86. It seems to be done by
pantograph. Is it Gavrinis? But Thoms
did not survey Gavrinis?. Acha-
vanich?'.
Black ink
758 mm × 558 mm

DC4804
Unidentified plan.
Insc. *New Grange*
Plan showing the geometry of cairn.
Annotated 'A.T. never visited New

Grange in Ireland is it not? A.S.T.
20.1.85'.
No scale
Black ink
316 mm × 315 mm
Linen backed tracing paper

DC4805
Unidentified plan.
Insc. *Hanging Stones*
Geometrical plan of stone circle.
Pencil
343 mm × 287 mm
Tracing paper

DC4722
Unidentified plan.
Annotated 'This webplotted from 1922
DATA and used over 1973 plan to fix
line *RN* as well as possible'.
1 : 250
Pencil
588 mm × 478 mm
Tracing paper

DC4723
Unidentified annotated plan.
1 in : 40 ft
Pencil
478 mm × 413 mm
Tracing paper

DC4724
Unidentified annotated plan.
1 : 500
Pencil
560 mm × 371 mm
Tracing paper

DC4846
Histograms showing equinoxes and
solstices at the major and minor
standstills.
Black ink

369 mm × 317 mm
Plastic film

DC4857
Insc. *Temperature differences to be applied to mean annual site temperature of equinoxes and solstices. Based on a diagram of Kirkwall temperatures prepared by the late Alexander Strang Thom in 1976.*
Chart
Black ink
308 mm × 216 mm
Plastic film

5

The metrology and geometry of Megalithic Man

ALEXANDER THOM and ARCHIE THOM

Many people do not agree with the conclusions we have drawn from our analyses of the surveys of standing stones and cup-and-ring marks. This is probably because they have not taken the time to appreciate and to understand the underlying theories. It is of course impossible to prove any of our ideas in the way that for example a proposition in Euclid can be proved, but we consider we have given sufficient detail in our published books and papers to make it very difficult to think that our hypotheses are wrong. Our theories are based on our surveys and these we know are reliable and good; by a good survey we mean one from which any distance can be scaled accurately.

We propose to explain some of the points in greater detail and we are sure that anyone who takes the trouble to follow these through will be convinced about Megalithic Man's ability and knowledge.

Cup-and-ring marks

We shall start with the metrology of the cup-and-ring marks. Consider the histogram shown in Fig. 5.1. This is a plot of the various measurements that were made of circular rings by R.W.B. Morris and D.C. Bailey (Thom & Thom 1978a: 46). Instead of representing each measured diameter by a dot we represent it by a little gaussian curve. Two different types of these are used. Diameters which are reasonably well known are shown by the higher and narrower curve whereas those which are somewhat uncertain are shown by the wider and flatter curve. The advantage of using this method of plotting is that we can add the ordinates together and so get an enveloping line which presumably represents the lot. The measurements by Morris and Bailey were probably made by using a graduated scale but when they were made neither

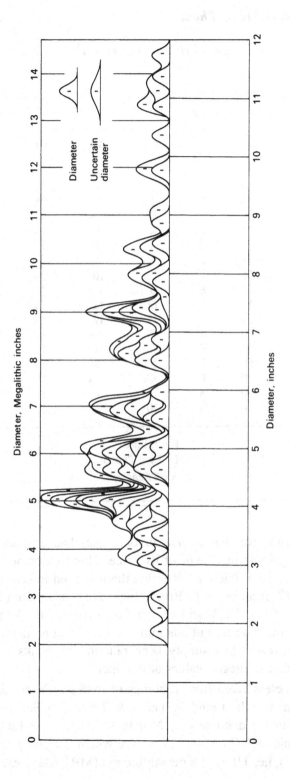

Fig. 5.1. Cup-and-ring marks. Histogram of measured diameters for the circular rings.

Table 5.1. *Triangles from cup-and-ring marks recorded in Thom & Thom (1978a)*

Site	Page	Type	a	b	c	Angle A		
			MI	MI	MI	°	′	″
Gallows Outon	51	E	5	4	3	90		
Moss Yard	52	I	$2\frac{1}{2}$	2	$1\frac{1}{2}$	90		
Cardrones House	53	I	$1\frac{1}{4}$	1	$\frac{3}{4}$	90		
Douchray	54	E	9	$7\frac{1}{2}$	5	89	48	32
Douchray	54	I	$4\frac{3}{4}$	$1\frac{1}{2}$	$4\frac{1}{2}$	90	15	55
Glasserton Mains	55	E	6	$4\frac{1}{2}$	4	89	36	08
Glasserton Mains	55	I	$3\frac{1}{4}$	3	$1\frac{1}{4}$	90		
Knappers Farm	56	I	$10\frac{1}{4}$	10	$2\frac{1}{4}$	90		
Panorama Stone	58	E	6	4	$4\frac{1}{2}$	89	36	08
Panorama Stone	58	E	12	11	$4\frac{3}{4}$	90	14	24
Panorama Stone	58	E	14	12	$7\frac{1}{4}$	89	48	53
Knock	60	E	$7\frac{1}{2}$	$6\frac{1}{2}$	$3\frac{3}{4}$	89	55	36
Knock	60	E	$6\frac{1}{2}$	6	$2\frac{1}{2}$	90		
Knock	60	E	$5\frac{1}{2}$	$4\frac{3}{4}$	$2\frac{3}{4}$	90	16	27
Knock	60	E	$4\frac{1}{2}$	$4\frac{1}{4}$	$1\frac{1}{2}$	89	43	09
Gourock Golf Course	50	Δ	5	4	3	90		
Gourock Golf Course	50	Δ	10	8	6	90		

Mean $A = 89° 57' 22''$

E = Ellipse
I = Egg, Type I
Δ = Triangle

Morris nor Bailey (nor for that matter anyone else) had any idea of the existence of the 'megalithic inch', nor of its value. This first came from Fig. 5.1 itself. It is, we are sure, fairly evident that the unit used in setting out the rings was about 0.817 of an ordinary British inch. We mark these megalithic inches along the top of Fig. 5.1. It will be seen how often the marks pick up a peak or a measurement. If we try out other units we find that none but this one fits at all. If the diameters had simply been random the peaks would not necessarily have landed at integral values of the 'inch'.

Particulars of diameters taken from rubbings of marks in Yorkshire, made by Evan Hadingham, will be found in Thom & Thom (1978a: Table 5.2). Combining these with the diameters by Morris and Bailey we find that the value of the megalithic inch (MI) is 0.817 inches, which is exactly 1/40 of the megalithic yard (MY), i.e. 1/100 of a megalithic rod (MR). Many examples of

Table 5.2. *Triangles from stone rings recorded in Thom, Thom & Burl (1980)*

	Site	Page	Type	a	b	c	Angle A		
				MY	MY	MY	°	′	″
L6/4	How Tallon	72	E	35	28	21	90		
S1/1	The Hurlers	74	II	12	9	8	89	36	08
S2/8	Postbridge	114	E	21	20	6	91	11	36
S3/1	Stanton Drew	116	E	39	36	15	90		
S3/1	Stanton Drew	116	E	52	48	20	90		
S4/1	Winterbourne Abbas	118	E	22	19	11	90	16	27
S5/1	Stonehenge	122	E	27	17	21	89	55	11
S5/3	Avebury	126	S	125	100	75	90		
S5/4	Woodhenge	130	I	37	35	12	90		
A2/8	Temple Wood	147	E	32	30	11	90	15	38
A7/22	Ardnave	150	I	5	4	3	90		
A8/6	Machrie Moor	152	E	2	√3	1	90		
A9/2	Ettrick Bay	154	E	37	29	23	89	59	25
B1/9	South Ythsie	168	E	22	16	15	90	21	29
B1/24	Blackhill	186	E	20	18	9	89	06	57
B1/27	Sands of Forvie	192	E	16½	15³⁄₈	6	89	57	23
B2/4	Esslie the Greater	200	I	5	4	3	90		
B2/7	Cullerlie	206	E	8½	7³⁄₈	4¼	89	48	52
B2/7	Cullerlie	206	E	3¾	3³⁄₈	1⁵⁄₈	90	09	48
B7/1	Clava	246	I	5	4	3	90		
B7/5	Daviot	252	E	18½	17½	6	90		
B7/13	Loch Nan Carraigean	260	E	22½	22	5	89	17	02
B7/18	Druid Temple	270	I	5	4	3	90		
G4/1	Carsphairn	276	E	30	22³⁄₈	20	89	57	32
G6/1	Twelve Apostles	288	I	5	4	3	90		
G7/6	Whitcastles	300	E	38	34	17	89	57	02
G9/10	Borrowston Rigg	304	II	15½	12¼	9½	89	59	05
G9/15	Allan Water	308	I	17	13	11	89	47	59
P1/19	Croft Moraig	348	E	11	8	7½	90	21	29
P2/2	Ballinluig	352	E	9½	8	5⅛	89	59	21
P2/9	Guildtown	360	E	12	8½	8½	89	48	06
Z3/1	Stanydale	364	I	25	24	7	90		
Z3/1	Stanydale	364	I	13	11	7	89	37	41
Z3/3	Loch of Strom	368	I	5	4	3	90		
W11/3	Maen Mawr	392	I	13	12	5	90		
W11/3	Maen Mawr	392	II	12	9	8	89	36	08
B7/4	Boat of Garten	250	E	17½	16	7	90	19	11
W2/14	Penmaen Mawr	373	E	31	29½	9½	90	03	04

Mean A = 89° 59′ 01″

E = Ellipse
I = Egg, Type I
II = Egg, Type II

ellipses and semi-ellipses are also shown (*ibid.*: Ch. 5). A particularly good example is the spiral at Knock which is built up from half-ellipses, all based megalithic yard (MY), i.e. 1/100 of a megalithic rod (MR). Many examples of ellipses and semi-ellipses are also shown (*ibid.*: Ch. 5). A particularly good example is the spiral at Knock which is built up from half-ellipses, all based on the megalithic inch (*ibid.*: Fig. 5.11).

A method of drawing ellipses on stone is also explained (*ibid.*). Working on a smooth stone surface without metal points must have been very difficult, and Dr J.C. Orkney has suggested that the geometrical constructions (see below, and Fig. 5.3 and Table 5.1) were perhaps made on a piece of half-cured leather stuck to the surface. Points on the lines could then have been pecked through and the construction on the rock scribed by a flint point. This would be a more satisfactory method than trying to set out the figure directly on to the polished stone with flint points fastened to pieces of hard wood. The geometry would have been developed beforehand perhaps working on smooth sand with sticks and ropes possibly to a scale of 1 MI : 1 MY.

Triangles (from cup-and-ring marks and from stone rings)

Table 5.1 gives the triangles for the shapes of cup-and-ring marks given in Thom & Thom (1978a), and Table 5.2 gives those from the stone rings detailed in Thom, Thom & Burl (1980). The mean value of angle A in Table 5.1 is 89° 57′ 22″. The mean A in Table 5.2 is 89° 59′ 01″. We do not know how Megalithic Man set out an accurate right angle and so we say that it is coincidental that the two means are so nearly equal.

It is evident that megalithic people knew about the 3-4-5 triangle. Look at the set of cup-marks on the rock on the golf course above Gourock (Thom & Thom 1978a: Fig. 5.2); the little rings shown here in Fig. 5.2 were carefully set out by us on tracing paper using two triangles, one $3 \times 4 \times 5$ MI and one $6 \times 8 \times 10$ MI and the tracing paper was placed over the rubbing of the cups. The agreement obtained between the cups and the plotted rings is obvious. It is perhaps difficult to find a probability level for the commonsense point of view that these cups were cut to suit 3-4-5 triangles in megalithic inches, but it seems obvious. Now $3^2 + 4^2 = 25 = 5^2$ and so 'the square on the hypotenuse of a right-angled triangle is equal to the sum of the squares on the other two sides'. We incline to the idea that Megalithic Man did not know this theorem but that he knew about the 3-4-5 triangle and about a number of other triangles which also gave a right angle. In Tables 5.1 and 5.2 we see a large number of these triangles which have been picked up from megalithic designs

- cup-and-ring markings and stone rings. We see that some satisfy the theorem exactly but that a number are only close approximations. We consider that if Megalithic Man had known the theorem in its mathematical form he would have tried examples which satisfied it exactly in integral units, but perhaps this is asking too much.

It is becoming more and more evident that Megalithic Man always preferred to use lengths which were integral numbers of whatever unit he was using. This explains the interest in right-angled triangles with integral sides. When these were incorporated in a geometrical design more of the lines could be integral. (Think of using rods of definite lengths to set out designs. Obviously it is much easier if the lengths are integral.) Megalithic Man sometimes used halves and on occasion quarters of his unit. It is easy to obtain a half of a length on the ground.

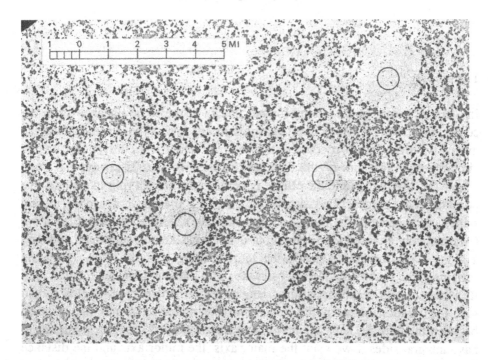

Fig. 5.2. Cup marks on a rock on Gourock Golf Course.

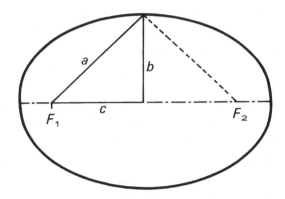

Ellipse   Major axis=2*a*
Minor axis=2*b*
a,b, and *c* all integral

Fig. 5.3. The ellipse.

The ellipse

In a circle, points on the circumference are all at the same distance from the centre. Probably for Megalithic Man this radius had to be a multiple of the unit. An ellipse has two centres, the foci, F_1 and F_2. At any point on the circumference the sum of the distances to the two foci is always the same (Fig. 5.3). After the circle the ellipse is perhaps the easiest ring to set out. The periphery can be described simply by attaching the ends of a cord to two stakes in the ground (the foci), and then running a third stake round the loop. (Alternatively a closed loop can be put over two stakes in the ground and then a loose stake run round inside the loop keeping it tight.) Now bring the scribing stake to the position opposite the point midway between the two foci. It is immediately obvious that the triangles so formed are right-angled and so give the relation between the major and the minor axes and the distance between the foci. So here again we have a right-angled triangle which has to have integral sides if we want the major axis, the minor axis and the distance between the foci to be integral multiples of the unit.

Today we know that some people have the greatest difficulty in mastering the most elementary arithmetic and geometry, whereas others have no difficulty whatever in picking up both. It must have been the same in megalithic times, only some 130 generations ago. Maybe someone in the latter category was experimenting one day with lengths of cord and stakes and discovered that he could describe an elliptical shape. Others again using a cord may have shown this man, or perhaps someone later, how to find out the elementary properties mentioned above.

It was natural for these people to measure the circumference. Could this be made integral if the major and minor axes were to be integral? We know today that is is impossible but these people found many close approximations.

Megalithic stone rings

We have dealt above with cup-and-ring marks and the same line of argument applies to circles and rings of stones. A histogram for the diameters of these is shown in Thom (1967: Fig. 5.1). The dimensions given for all the circles were obtained from careful field surveys.

The main classes of megalithic rings are (1) circular rings; (2) flattened circles of two types; (3) egg-shaped rings of two types; and (4) ellipses. There are also one or two other special types of ring, for instance the ring at Avebury and the ring at Moel ty Ucha.

Stonehenge

We made a particularly accurate survey of Stonehenge. Professor Atkinson has pointed out that the rain water soaking through the soil has evidently dissolved the top of the underlying chalk and so reduced ground level by about a foot and a half. Accordingly we measured to the bases of the stones from 6 in to 2 ft above the ground level.

Our original survey was plotted to a scale of 1:84. A very much reduced copy of this is shown in Thom, Thom & Thom (1974: 75). The sarsen circle is obviously contained between two circles with perimeters of 45 MR and 48 MR. Evidently each stone had a width on the inner perimeter of one rod and there is half a rod between each. Since there are 30 stones this is how a dimension of 45 rods was obtained.

The mean perimeter of the sarsen stone circle is 46^{1}2 MR and this circle through the hypothetical or idealised centres of the stones has a diameter of 37.004 MY. In all our work on rings we have built up our hypothesis on arcs passing through stone centres, except on this sarsen circle with the inner faces of the stones smoothed, and on the trilithon ellipse.

The trilithons are accurately contained between two ellipses 30 × 20 and 27 × 17 MY. Since the inner faces of these stones had been smoothed we shall concentrate on the smaller ellipse. This nearly satisfies the Pythagorean convention in integers: $2a = 27$, $2b = 17$ and $2c = 21$. The square of the hypotenuse is 729 and the sum of the other two squares is 730. We assume that $2a$ is exactly integral in megalithic rods and then take the two cases when either $2b$ or $2c$ is integral. In either method the residual from the whole number 28 MR for the perimeter is so small that we may well ask how it was possible to measure round the perimeter 200 ft long with an error of only an inch or two. Only one closer approximation is known and that is the ellipse at Pen Maen Mawr. For futher details see Thom & Thom (1978a: 145).

It is perhaps worth while pointing out here that our attempt to explain the geometry inside of the trilithon ellipse (*ibid.*) leads us to propose the 22-14-17 ellipse shown. This ellipse has a distance of 17 MY between foci, this being the minor axis of the trilithon ellipse and the diameter of the proposed inner circle. Things do not fit here, however, with the precision we found for the sarsens and trilithons.

Diameters and perimeters

In Thom, Thom & Burl (1980) we have collected reproductions of all our surveys of rings in Britain. We have several times described the way in which rings have been disturbed over the millenia by such things as frost, ground movement, growing trees and human intervention. In the present paper we have used the above-mentioned data but have omitted all those rings which seem to be so badly disturbed that we could not possibly obtain a reasonably accurate diameter or reliable dimensions. We measure diameters in megalithic yards and perimeters in megalithic rods. We take the value of the megalithic yard to be 2.722 ft and the corresponding value of the megalithic rod, $2\frac{1}{2}$ times this, namely 6.805 ft. We shall write the diameter as

$$dn + df$$

where dn is the integral part and df the fractional part. Similarly we shall write the perimeter as

$$pn + pf$$

where pn is the integral part and pf the fractional part. Thus neither df nor pf is ever greater than unity.

Elliptical rings

Starting with elliptical stone rings, we take from Thom, Thom & Burl (1980) the 16 ellipses given in Table 5.3. Here we use only sites where there are sufficient stones showing to enable us to determine dimensions with reasonable accuracy. It is possible that excavation will reveal more. There are five or six other sites in Thom, Thom & Burl (1980) which are there called 'ellipses' but in each of these cases an examination of the survey shows that we could not rely on the dimensions and so we omitted these.

With the major axis of the ellipse 2*a*, the minor axis 2*b* and the distance between the foci 2*c*, we have the relation that

$$a^2 = b^2 + c^2 \ .$$

The values of *a*, *b* and *c* finally used are given in Table 5.3 and the reader must judge for himself how successful we have been in finding an ellipse which fits the stones. In about four cases the values of *a*, *b* and *c* fitted the relation in whole numbers. In the other cases we do not know whether the erectors preferred to keep *c* an integral whole number or to keep *b* integral. Accordingly we have taken the two cases and for each we have calculated and tabulated the perimeter. It will be seen that there is very little difference between the perimeters in the two cases.

In this paper we divide the fractional part *df* or *pf* into four equal parts. Any measurement falling into the lower quartile or the upper quartile is considered favourable and any measurement falling between $^1\!_4$ and $^3\!_4$ unfavourable.

We have set up a hypothesis that major and minor axis diameters were intended to be integral in megalithic yards with the perimeters integral in megalithic rods, and see that in only one case (Daviot) does our hypothesis fail.

We can also consider the five cases in the middle of our table of ellipses (Thom 1967: Table 6.4) under the heading '(b) Definite ellipses from other sources'. It appears that Professor A.E. Roy's survey of the Tormore ellipse shows this ring to be 'unfavourable'. Yet it was this ring with its symmetry and proportions which first showed Roy that megalithic elliptic rings existed. Carefully recalculated particulars of the other four rings will be found in Table 5.4. It is worthwhile examining this table carefully, along with Table 5.3, and to note that the ellipse perimeters are always very close to an integral number of megalithic rods. We ask anyone who is sceptical about this hypothesis to try to produce such an ellipse for himself. Only by trying will he find out how difficult it really is to get the axes and the perimeter integral at the same time. And yet Megalithic Man produced the examples we have given without mathematical tables and without calculating machines.

Table 5.3. *Ellipses recorded in Thom, Thom & Burl (1980)*

Page	Site		2a	2b	2c	P	P	Deviation *pf*
			MY	MY	MY	MY	MR	
73	L6/4	How Tallon	$17\frac{1}{2}$	14	$10\frac{1}{2}$	49.63	19.85	−0.15
115	S2/8	Post Bridge	$10\frac{1}{2}$	10	3.20	32.14	12.86	−0.14
115	S2/8	Post Bridge	$10\frac{1}{2}$	10.06	3	32.30	12.92	−0.08
117	S3/1	Stanton Drew	39	36	15	117.86	47.14	+0.14
117	S3/1	Stanton Drew	52	48	20	157.14	62.86	−0.14
119	S4/1	Winterbourne Abbas	11	$9\frac{1}{2}$	5.54	32.24	12.90	−0.10
119	S4/1	Winterbourne Abbas	11	9.53	$5\frac{1}{2}$	32.28	12.91	−0.09
123	S5/1	Stonehenge	27	17	20.98	70.01	28.00	0.00
123	S5/1	Stonehenge	27	16.97	21	69.97	27.99	−0.01
155	A9/2	Ettrick Bay	$18\frac{1}{2}$	$14\frac{1}{2}$	11.49	52.03	20.81	−0.19
155	A9/2	Ettrick Bay	$18\frac{1}{2}$	14.49	$11\frac{1}{2}$	52.02	20.81	−0.19
169	B1/9	South Ythsie	11	8	7.55	30.03	12.01	+0.01
169	B1/9	South Ythsie	11	8.07	$7\frac{1}{2}$	30.10	12.04	+0.04
187	B1/24	Black Hill of Drachlaw	10	9	4.36	29.86	11.95	−0.05
187	B1/24	Black Hill of Drachlaw	10	8.93	$4\frac{1}{2}$	29.85	11.94	−0.06
193	B1/27	Sands of Forvie	$16\frac{1}{2}$	15.32	6	50.07	20.03	+0.03
193	B1/27	Sands of Forvie	$16\frac{1}{2}$	$15\frac{3}{8}$	5.99	50.08	20.03	+0.03
251	B7/4	Boat of Garten	$17\frac{1}{2}$	16	7.06	52.63	21.05	+0.05
251	B7/4	Boat of Garten	$17\frac{1}{2}$	16.04	7	52.71	21.08	+0.08
252	B7/5	Daviot	$18\frac{1}{2}$	$17\frac{1}{2}$	6	56.56	22.62	−0.38
277	G4/1	Carsphairn	30	24	18	85.08	34.03	+0.03
353	P2/2	Ballinluig	$9\frac{1}{2}$	8	5.13	27.54	11.02	+0.02
349	P1/19	Croft Morag	11	8	7.55	36.03	12.01	+0.01
349	P1/19	Croft Morag	11	8.05	$7\frac{1}{2}$	30.10	12.04	+0.04
373	W2/14	Penmaen-Mawr	31	$29\frac{1}{2}$	9.53	95.05	38.02	+0.02
373	W2/14	Penmaen-Mawr	31	29.51	$9\frac{1}{2}$	95.06	38.03	+0.03

Without Daviot, RMS(*pf*) = 0.09 MR.   With Daviot, RMS(*pf*) = 0.12 MR.

Table 5.4. *Ellipses from other Sources (Thom 1967: 82 & Table 6.4(b))*

Site	2a	2b	2c	P	P	Deviation pf
	MY	MY	MY	MY	MR	
Auchangallan	18	17	5.97	58.99	23.00	0.00
Auchangallan	18	16.97	6	54.94	22.97	0.03
Clauchried	13	11	6.93	37.76	15.11	0.11
Clauchried	13	10.95	7	37.70	15.08	0.08
Braemore	34	29.5	16.97	99.87	39.95	0.05
Braemore	34	29.45	17	99.77	39.91	0.09
Learable Hill	24	20.5	12.48	70.01	28.00	0.00
Learable Hill	24	20.49	12.5	69.99	28.00	0.00

RMS(*pf*) = 0.06 MR.

Flattened circles and eggs

On going through the flattened circles given in Thom, Thom & Burl (1980) we found 15 with the perimeters favourable and seven with the perimeters unfavourable. Similarly for the Type I and II eggs together we found eight favourable and four unfavourable.

In addition to the above flattened circles and eggs there are two Type I megalithic eggs at Beaghmore, Co. Tyrone, the larger based on a triangle which is $6 \times 8 \times 10$ MY and the smaller on a triangle $3 \times 4 \times 5$ MY (A.S. Thom 1980: 15).

Circles

On going through Thom, Thom & Burl (1980) again we found 78 circles of which 47 had the diameter favourable and 31 unfavourable. Amongst the 78 we found 49 with perimeters favourable and 29 with perimeters unfavourable.

On our information it appears that the ellipses are in a different category. The other rings certainly show a tendency to have the perimeters integral but in the 16 ellipses only one is unfavourable. It is a pity that we do not have archaeological dates for all these rings.

Thus we see that amongst all our surveys, in all the main types of ring, the erectors liked to have the perimeters integral. Obviously we did not measure the perimeters by tape, but calculated them from the measured dimensions, that is from measured diameters, or major and minor axes. We do not know why they wanted these quantities to be integral, but especially in the ellipses there can be no doubt about it. The probability level works out to have a value of about 1 in 2000. It is all the more surprising to find that that the rule does not apply exactly to the ring at Tormore, Arran.

The use of 10 MY

This unit appears strongly in the geometry of Avebury (see below) and a look at our survey of Woodhenge and its interpretation shows that the rings had perimeters of 40, 60, 80, 100, 140 and 160 MY (Thom, Thom & Burl 1980: 130; Thom 1967: 47). The use of 10 MY does not otherwise obtrude itself.

Which ring came first?

The first rings to be set out were probably circles. When the erectors began to be more particular the diameter was probably specified as a multiple of whatever unit was being used or perhaps the size was specified by the length of the perimeter. In *attempting* to get both the diameter and the circumference integral they probably invented the flattened circles and the egg-shaped rings.

We suggest that the last shape to be designed was the ellipse. We base this opinion on the fact that practically all ellipses had major and minor axes nearly integral and the perimeter also integral. For the other shapes, flattened circles and eggs, certainly a preponderance had favourable dimensions, but it was only a preponderance whereas in the ellipses it was practically all.

Avebury

The ring at Avebury is perhaps one of the most instructive sites in Britain. Unfortunately it has been very badly damaged but there is enough left to give us a very definite idea of the geometrical layout. This is shown to a small scale in Fig. 5.4. Details of the geometry from which this stems will be found in Thom & Thom (1978a: Fig. 4.1). Without going into any of the mathematics we shall here merely illustrate in Fig. 5.4 the kind of geometry which was used. It will be seen that the construction is based on the 3-4-5

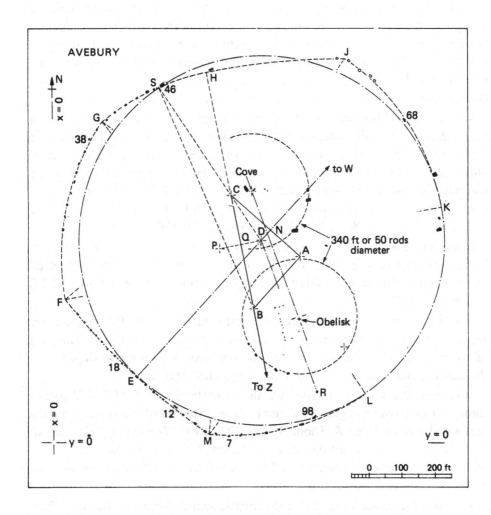

Fig. 5.4. The Ring of Avebury.

triangle *ABC* with *AB* = 75, *AC* = 100 and *CB* = 125 MY. Now from *C* a line *CD* is drawn 60 MY long with *DN* (at right angles to *AC*) 15 MY long. This makes *DN* parallel to *BA*. *D* is considered as the centre of the whole construction. With centre *D* and radius 200 MY, describe a circle. Find a point *W* on *ED* with *EW* = 750 MY, and with centre *W* draw the arc *FEM*. Arc *GSH* has centre *B* and radius 260 MY; similarly arc *FG* has radius 260 MY and centre at *A*, and arc *ML* has the same radius and the centre at *C*. Arc *HJ* has radius 750 MY and centre on *CB* produced, at *Z*. Make *QP* = *QD* both at right angles to *CB* and with *P* as centre complete the ring with arc *JK*.

Now let us look at what we have drawn. All the lengths used with the exception of the radius *PJ* are integral in megalithic rods and five are integral in multiples of 10 MR. Clever as the erectors were it is too much to expect that they could have every dimension integral and so we need not be surprised that the arc *JK* had a non-integral radius. Why did they use this peculiar construction?

We have seen how very interested the erectors were in getting as many as possible of the lengths integral and we believe that this controls the whole setting out of the design. We see that the perimeter consisted of a number of arcs meeting at what we might call corners. Using the dimensions shown we carefully calculated the exact length of each arc, round the curve between the corners. In megalithic rods these arcs are *ME* 38.89; *EF* 46.97; *FG* 79.95; *GH* 51.87; *HJ* 59.89; *JK* 78.30; *KM* 165.03; total 520.90 MR.

It will be seen that all the arc lengths except *JK* are reasonably near to an integral number of megalithic rods. Dr D. Heggie (priv. comm.) has estimated that the probability level of this happening by chance is between 0.1 and 1.0 per cent.

It will be seen that the stones on the west side of the ring fit the geometry exactly. In setting out the design, radius *PK* had to be non-integral. This side of the ring has never been excavated and our assumed geometry depends on the burning pits near *J* and one or two other stones near *K*.

It is most important to realise that the theoretical arcs *LMEFGSH* passed through, or almost through, all the stones. The distances of the stones from the arcs are shown in Thom & Thom (1978a: Table 4.1). The average distance of the middle of a stone from the arc is 1.7 ft, small in comparison with the size of a stone. Obviously Avebury could not have been set out by pacing, yet this has been suggested several times.

It is also important to note that the arcs shown in Thom & Thom (1978a: Fig. 4.3) are not independent arcs drawn to fit the stones in each individual arc, but that they all form part of the design shown in Fig. 5.4. Thus if any one of the arcs is moved, say to the north-east, then every one of the arcs in Thom & Thom (1978a: Fig. 4.3) must also be moved by the same amount to the north-east.

The theodolite traverse on which our survey depends had its azimuth checked at three points astronomically and it closed to within a few inches. Somehow or another the megalithic builders of this ring managed to obtain the same kind of accuracy, otherwise the discrepancies listed in Thom & Thom (1978a: Table 4.1) would have been very much greater. This makes one realise how accurately the builders really worked.

We might note that a 'least squares' solution of the geometry of the stone positions given in Thom & Thom (1978a) shows that the value of the megalithic yard is about 2.723 ft or 0.830 m. This can be compared with the value found in Brogar (2.725 ft) and the values found at Le Ménec, France (2.721 ft) and that at Kermario, France (2.724 ft).

It would be very interesting to have the ground at the apices *A, B, C* and *D* excavated to see if there is any trace remaining of these points. It may of course be that as at Stonehenge, the surface of the chalk has been lowered by the action of rain on it.

The effect of the difference between the arc and the chord

For a short arc the amount by which the length of the arc round the curve between its ends is greater than the chord is $c^3/24r^2$, where c is the length of the chord and r is the radius of the arc.

We shall assume that Megalithic Man measured the distance round the curves by using relatively short straight rods. If he were setting out a circle he presumably would arrange the diameter so that there was an integral number of chords in the perimeter. We come along but we do not attempt to measure the perimeter at all. We calculate it and as our calculations of course allow for going round the arc we obtain a greater perimeter than Megalithic Man had used or thought he was using. We should in fact expect the perimeter which we get to be very slightly greater than integral. In fact we find that there is very little difference between the two. If Megalithic Man had worked perfectly and we had worked equally perfectly, then the difference should indicate to us the length of the measuring rod which had been used by Megalithic Man, i.e. by using the above formula and solving it for c. This would be properly reasoned if our value for the megalithic yard had been obtained entirely from measurements between stones in straight lines such as at Carnac; but we must remember that the majority of our measurements used for determining the megalithic yard came from rings and circles.

The Avebury calendar

Lees (1984) shows how, by using movable markers, the complex of rings, stones, post holes, etc. at The Sanctuary could have been used as an accurate solar and lunar calendar. At the same time another marker was used on the inner ring to enable the months to be recorded. Owing to the incommensurability of the solar and lunar periods no calendar can be perfect but the Sanctuary calendar was as perfect as it could possibly be. An analysis of the stones showed that the Sanctuary calendar was in some ways superior to the present-day calendar which was ruined by the jealousy of the Emperor Augustus. He wanted to make his month (August) have as many days as Julius Ceasar's month (July). To do this he had to take a day from another month, and hence February is in the peculiar position in which we find it today.

According to Lees, the Sanctuary Calendar can be operated like a circular abacus with 42 stones in the outer circumference. Other rings in Britain may have been used in the same way. We have indicated here and elsewhere that the erectors evidently considered the circumference to be more important than the diameter. If the circumference was to be used as an abacus then we have a reasonable explanation for this preference.

Dr A. Burl has pointed out that there are some circles on the west coast of England which have 42 stones in their circumference.

Stone spheres

The most perfect proof of the ability of Megalithic Man to understand and use solid geometry (the geometry of the sphere) is provided by the fact that over 300 balls made from hard stone, each with a surface divided into triangles, squares, pentagons or hexagons have been found. The distribution of the 'find-spots', that are chiefly in Scotland, is shown by Marshall (1977: Figs. 10-14). These can be compared with the positions of the geometrically constructed rings of standing stones (Fig. 5.5).

It can be shown that there are only five so-called platonic solids, viz.:

(1)   the tetrahedron consisting of 4 equilateral triangles;
(2)   the cube consisting of 6 squares;
(3)   the octahedron consisting of 8 hexagons;
(4)   the dodecahedron consisting of 12 pentagons; and
(5)   the icosahedron consisting of 20 equilateral triangles.

These were described by Plato perhaps a thousand years *after* the above-mentioned stone balls were made. Among the stone balls found, all five of the platonic solids are represented on one form or another.

In Fig. 5.6 we show a photograph of one of the platonic solids, a dodecahedron, in the form of a dovetailed box made by the senior author in the late 1950s when he was very interested in the mathematics involved in constructing geodesic domes. The corresponding stone ball is shown in

Critchlow (1979: 149, Fig. 146). Many illustrations of these solids are shown in the same book including a photograph of the tetrahedral sphere (*ibid.*: 133, Fig. 115).

Possibly the makers of these balls made first a perfect sphere and then scribed out on it the figure they required. It is evident that they had a perfect mastery of their subject.

The date of these balls is not known, but only in the period when Megalithic Man was setting out the sophisticated stone rings has a sufficiently

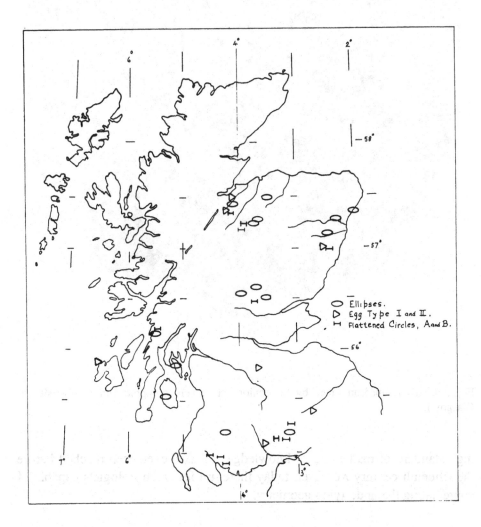

Fig. 5.5. The distribution of geometrically constructed rings of standing stones in Scotland.

Fig. 5.6. Dodecahedron made by the senior author (Photographic Unit, University of Glasgow).

high standard of mathematical knowledge and skill ever been reached before the fifteenth century AD. Even today there are few archaeologists capable of appreciating the underlying geometry.

Conclusion

When the geometrical ideas and measurements in this paper are combined with the radiocarbon dating obtained by archaeologists it is hoped that a complete picture of the prehistoric development of the subject will emerge.

Acknowledgment

Thanks are due to Mrs H.M. Gustin for her invaluable help throughout, in the preparation of the tables, diagrams and manuscript.

Postscript

My father worked on the above paper with the help of Hilda Gustin and myself from the autumn of 1982 until the spring of 1983. Apart from some recent small alterations judged necessary to make it flow more easily, the paragraphs, tables and figures are as he wanted them to be. (He was becoming too frail to live by himself any more in Thalassa and had to go to live with his daughter, Beryl Austin, in Banavie in December 1983.) In the autumn of 1983 and the spring of 1984 he was working on what was to be his last and final paper, 'Moving and Erecting the Menhirs.'

A.S.T. 30 March, 1986

BRIDGING PIECE

6

Megalithic landscapes

CHRIS JENNINGS

Many artists have been impressed by the standing stones and stone rings of the British Isles not only because of their beauty in the landscape but because of the mystery of their role in prehistoric culture.

In making a photographic survey of megalithic sites my intention was to record and document the stones in their surrounding landscape. As an artist I looked at the stones with particular reference to their siting in the landscape; often the stones appeared to act as a focal point unifying sky, horizon and land. An art which is integral to society and incorporates geometric and astronomical science is a major concern for some contemporary artists. My own enthusiasm for the stones was reinforced by the theories of Professor Thom as proposed in his book *Megalithic Sites in Britain* (Thom 1967).

Encouraged by Professor Thom, and with the assistance of a Leverhulme Fellowship, I photographed over 300 sites; these photographs are my 'portraits' of the stones and I hope that they might go some way towards inspiring the preservation of megalithic sites in Britain.

Professor Thom's foreword to my photographic book (unpublished) is printed below and demonstrates his sense of the importance of the link between science and art, stressing not simply the preservation of our cultural heritage, but the continuation of the social and political relevance of art and science.

> Many times, perhaps after a long walk, as I approached a site which I had not previously seen I felt the sense of anticipation growing as I got nearer. Would it be a good site? Would it be worthwhile surveying and would it contribute anything to our knowledge of what these alignments

(Continued on page 172)

Fig. 6.1. Duloe, Cornwall (Chris Jennings).

Fig. 6.2. Staldon, Devon (Chris Jennings).

Fig. 6.3. Penmaen-mawr, Clwyd (Chris Jennings).

Fig. 6.4. Moel-ty-Uchaf, Clwyd (Chris Jennings).

Fig. 6.5. Swinside, Cumbria (Chris Jennings).

Fig. 6.6. Blakeley Moss, Cumbria (Chris Jennings).

Fig. 6.7. Ballochroy, Kintyre (Chris Jennings).

Fig. 6.8. Ballymeanach, Argyll (Chris Jennings).

Fig. 6.9. Ballinaby, Islay (Chris Jennings).

Fig. 6.10. Spittal of Glenshee, Perthshire (Chris Jennings).

Fig. 6.11. Castle Frazer, Aberdeenshire (Chris Jennings).

Fig. 6.12. Monymusk, Aberdeenshire (Chris Jennings).

Fig. 6.13. Callanish III, Lewis (Chris Jennings).

Fig. 6.14. Clach an Trushal, Lewis (Chris Jennings).

Fig. 6.15. Ring of Brogar, Orkney (Chris Jennings).

Fig. 6.16. Lund, Unst, Shetland (Chris Jennings).

(Continued from page 155)

were for? I needed every site I could find to add to the material in the statistical analysis, but there is much more than that. Where there were still several stones upright, I quickened my pace and got beside them, I suppose in an unconscious attempt to get in touch with these mysterious people who had been prepared to go to the enormous labour of erecting them.

I am sure Mr. Jennings experiences similar feelings as he approaches a site, but armed with his camera he is able to carry away the sensation and present it to others.

I tried to carry a camera but I found I had no time at all to use it. All my attention was on the theodolite to enable me to carry away my impression of the site. Both methods are necessary if we are to get a complete record, if we are to be able to pass on to others not only the details but the atmosphere of the place. We have probably done more damage to our precious archaeological sites since the coming of the internal combustion engine than had happened in the many thousands of years since they were built and used and, in most cases with official approval and help, the destruction is still going on. Jennings has done a great deal to preserve something for the future. I and many others believe that we are dealing with the remnants of an earlier civilisation; a civilisation which vanished and has left behind only these few records in stone. What will our civilisation leave behind? A polluted world covered with scars to show the rapacity and greed of mankind. If Jennings' work can do anything to warn us of the danger, it is indeed worthwhile.

Professor A. Thom 1978

Part 2

RESEARCH PAPERS

Archaeological research inspired by Alexander Thom

7

'Without Sharp North...'
Alexander Thom and the great
stone circles of Cumbria

AUBREY BURL

At Swineshead ... is a druidical temple, which the country people call
Sunkenkirk ... [The entrance] is nearly south-east ... This monument of
antiquity, when viewed within the circle, strikes you with astonishment,
how the massy stones could be placed in such regular order, either by
human strength or mechanical power.

W. Hutchinson. The History & Antiquities of Cumberland I . 1794

L1/3. Sunkenkirk. C. 93.7 ft. (B) Az = 128°.8. $h = +0°.5$. $\delta = -21°.5$.
Sun. 'Entrance'.

Alexander Thom. Megalithic Sites in Britain. 1967

Sunkenkirk, or Swinside, was only one of 19 Cumbrian rings surveyed and
planned by Thom who realised how informative they were about the methods
and needs of their builders. 'Two of the most important circles in the north of
England are Castle Rigg and Long Meg and Her Daughters' (Thom 1966: 22).
He gave evidence for this by providing accurate plans for both rings, by
commenting on their possible astronomical alignments and, at the end of his
paper, including a sketched panorama of Castlerigg that showed the heights of
the stones and their spacing against the mountainous background of an
horizon divided into 10° sections. On to this diagram Thom superimposed the
rising and setting positions of the sun, moon and 16 stars. Ten years earlier he
had defined Castlerigg as a flattened circle, 107 ft ± 0.3 (32.6 m ± 0.1) in
diameter, whose asymmetrical shape had been intentionally created by people
using geometrical principles in their design (Thom 1955: 281). It was an

objective analysis of a stone circle, the work of an engineer that revealed potentials for stone circle studies never before so clearly presented either by amateur investigators or by professional archaeologists.

It is strange how rarely advances in stone circle research have been made by an archaeologist. To the contrary, with little worthwhile artefactual evidence coming from rings that lacked stratigraphy and were deficient in pottery, human bone and charcoal, stone circles became the *personae non gratae* of British prehistory. In scholarly literature Stonehenge was described, Avebury was mentioned and a reluctant, vague line or two was written about the existence of other little-known sites. Pessimism about discovering their purpose was commonplace. It was characteristic of a book by a professional archaeologist that a chapter entitled 'Sacred Sites', concerning stone circles, could begin with 'what they signify no man can say' and conclude, 13 Stonehenge-heavy pages later, with 'in all essentials the great stone circles retain their mystery' (Clark 1948: 103-16).

Most of the discoveries about the rings have been made by non-archaeologists, surveyors, engineers, clerics, astronomers, solicitors, from the seventeenth century onwards, an illustrious pageant of true amateurs amongst whom Alexander Thom occupies an honourable place. He stands with John Aubrey and William Stukeley as a person who investigated outside the boundaries of archaeological convention, less concerned with pots and flints than with the neglected aspects of shape, design and orientation. These had been disregarded, quite deliberately, by a majority of archaeologists as illusions fostered by romantics who had no understanding of the realities of prehistoric existence.

As late as 1976, stone circles were still believed to have been erected around 1700 BC as temples by the continental Beaker Folk (Branigan 1976: 74-6) whose megalithic rings, Stonehenge being the most impressive example, contained stones imprecisely set up in line with the sun. 'What rites were there performed cannot be guessed, but in a general way midsummer and midwinter festivals belong to high latitudes where the sun's annual journey from north to south and back is conspicuous; in Mediterranean lands orientation by the meridian is more normal and more natural' (Childe 1940: 109).

Much has changed. Now it is accepted that the earliest circles were built by natives around 3200 BC, long before any beaker pots were known in these islands, and the rings themselves were laid out according to strict rules concerning cardinal points. Many of these megalithic enclosures had calendrical alignments in their designs. Perhaps nowhere in the British Isles are these features more apparent than in the great stone circles of Cumbria, and it is been largely through the plans and studies of Alexander Thom that they have been recognised. His work has compelled other students to consider the implications of his data. No longer is it possible to dismiss a 'circle' as devoid of any clue as to its function simply because excavation has produced nothing tangible. Even without artefacts or dating evidence a ring possesses size, shape and design and it was Thom who made archaeologists think about such matters.

There are over 50 Cumbrian stone circles in what was once Cumberland, Westmorland and Lancashire North-of-the-Sands (Burl 1976a: 342-8) but it is the nine very large rings in and around the Lake District that are the subject of this paper: Brats Hill; Castlerigg; Elva Plain; Grey Croft; Grey Yauds; Gunnerkeld; Long Meg and Her Daughters; Studfold and Swinside. Outside Cumbria there are three related sites, the Twelve Apostles and the Girdle Stanes in south-west Scotland, and Ballynoe in Ireland. With the exception of Grey Yauds, now removed, and Ballynoe the others have been planned by Thom and his comments about them, from 1954 onwards, are summarised in the Appendix. Included also in this paper, in the context of ritual enclosures as depôts for the distribution of stone implements from the Lake District 'axe-factories', are the henges of Mayburgh and King Arthur's Round Table near Penrith, Broadlee in Dumfriesshire, and the destroyed stone circle called the Lochmaben Stone on the coast of the Solway Firth.

These are huge rings, 100 ft (30 m) or more across, the henges with spacious central plateaux, the megalithic circles with rough stones closely-set around an interior much larger than most others in the British Isles. Examination of their designs reveals subtleties not obvious on the ground and a sceptic might decide that these had been subconsciously imposed by the wish of a modern enthusiast to 'rediscover' geometries and alignments unconsidered by prehistoric communities. Fortunately, Thom's plans can be checked against those of another meticulous surveyor, C.W. Dymond, who plotted five rings and two henges in the late 19th century (Dymond 1880; 1881; 1891). There is little variation in size, shape or orientation between Thom's work and his.

The first of the Cumbrian stone circles to be described was Long Meg and Her Daughters mentioned by Camden. 'This the common people call *Long-Megg*, and the rest *her daughters*; and within the circle are two heaps of stones, under which they say there are dead bodies bury'd' (Camden 1695: 831). Perceptively for his time he added a footnote that the cairns were 'no part of it; but have been gather'd off the plough'd-land adjoyning and ... have been thrown up here together in a waste corner of the field'. John Aubrey knew of these 'Tumuli, or Barrows of cobble-stones, nine or ten foot high', and of a 'Giants bone, and Body found there' (Aubrey 1665-1693, I: 115-16). He meant to ask for more information from the local minister but if he did the reply has been lost.

The fable that the outlying pillar of Long Meg was the body of a petrified witch has fascinated generations of visitors but it is another circle, Castlerigg, at the mountainous heart of the region, that has attracted most attention, perhaps because it is so close to the tourist centre of Keswick. Wordsworth and Keats saw it. Stukeley described it. Ley-hunters found shadow-lines in it. And Alexander Thom was intrigued by the complex alignments, the axis and the perfect setting for the ring.

It is sometimes wrongly called the Carles as though the stones were 'ceorles' or husbandmen transformed into stone for some forgotten sin, but the name comes from a misreading of Stukeley. Having been there in 1725 he wrote, 'They call it the Carsles, and corruptly I suppose, Castle-rig. There seemed to be another larger circle in the next pasture toward the town' (Stukeley 1776: 48). He believed it to be a temple of the druids. So did the poet, Thomas Grey, who found it in a cornfield, and the belief in druids lingered into the present century, gradually being replaced by more scientific claims that celestial alignments had been built into many stone circles.

Thomas Pennant was the first to suggest that the rectangle of low stones in the south-east quadrant of Castlerigg had an astronomical significance, put there 'for the respect paid by the antient natives of this isle to that beneficent luminary, the *Sun*' (Pennant 1774: 58). Lewis, who was convinced that some hills had been sacred to people of prehistoric times, thought that the situation of the ring had been chosen so that it would lie with the mountain of Skiddaw to its north-west and the three peaks of Blencathra to the north-east in 'the grandest position in which I have ever seen a stone circle placed' (Lewis 1886: 473). Morrow, following the researches of Sir Norman Lockyer at other British stone circles, considered that there were three important alignments at Castlerigg. The major line was from an outlying stone to the circle centre. This marked the sunrises in late April and early August and the builders of the ring had been warned of these forthcoming events by the risings, respectively, of the star Arcturus in 1400 BC and the Pleiades in 1650 BC (Morrow 1909). A few years later Anderson, in contradiction, claimed that there were several solar orientations, with hill-summits acting as foresights, to the midsummer and midwinter solstices and to the 'Celtic' festivals of Beltane at the beginning of May and Samain in early November (Anderson 1915). Archaeologists continued to say little about the circle which 'was thought to have been built primarily for ceremonial purposes, though the beliefs held at the time remain unknown' (Fell 1972: 36).

It is the lonest tract in all the realm
Where lived a people once among their crags,
Our race and blood, a remnant that were left
Pagan among their circles, and the stones
They pitch straight up to heaven.

C.A. Parker (Fell 1972: 36)

It was against this background that Alexander Thom made his own survey of Castlerigg which was, according to him, a flattened circle of his Type A, its diameter of 107.1 ft being only a very little shorter than 40 of his megalithic yards. In the ring the largest stone at the south-east had been erected to define opposing azimuths to the midsummer sunset and to sunrise at Candlemas in early February. There were also axial alignments to three lunar and two solar phenomena. The outlier to the WSW seemed to be important and 'in fact this was one of the lines which convinced the author of the necessity to examine the calendar hypothesis in detail' (Thom 1967: 151). As it happened, the research which followed was based on a misapprehension but, ironically, Thom was right for the wrong reason. The outlier never had been an astronomical marker but had been moved to the edge of the field in recent times (Burl 1976a: 77). Even so, its apparent significance led to conclusions about a prehistoric calendar that had never previously been considered. The existence of such a calendar is relevant to the question of the origins of four supposed 'Iron Age' or 'Celtic' festivals which may, in reality, have started three thousand years earlier in the Neolithic period (Thom 1967: 107-17; Burl 1983: 34).

To Thom, Castlerigg was a symbolic observatory, a marvel of design in which the alignments were never intended for refined observations. 'We might guess that all these [alignments] were built early in the age, that is when an interest was first being taken in the solstice and equinox without any attempt being made to make accurate measurements' (Thom & Thom 1978a: 178). He added that he did not know what the stone circle was intended for but he believed that the siting of its stones was 'controlled by the desire to indicate the rising or setting positions of the sun at important times' (*ibid.*: 22). In this his opinion was not far from some archaeologists less scornful than many of their colleagues about astronomical lines. 'Whilst it is not impossible that an astronomical alignment was built into the circles this does not appear to have been of major importance and may perhaps be best compared with the orientation of later religious buildings' (Clare 1975: 13).

The problem then remains of the purpose of rings such as Castlerigg. The druids may have gone but the hills and the sightlines remain, some of them, like Morrow's, of an improbably late date in the Late Bronze Age. Thom himself offered no date for Castlerigg although he did propose 1900 and 1800 BC for the setting of Arcturus at Brats Hill, 1600 BC for the setting, perhaps of Pollux, between two pairs of neighbouring stone circles, and 1700 BC for the rising of Antares in the same megalithic complex (Thom 1967: 99). Whether such dates are feasible and whether rings like Brats Hill and Castlerigg were raised predominantly as calendrical observatories can be tested by examining the evidence for the time of their erection and then considering what associations these Cumbrian circles possess that might tell us something of their purpose.

Sadly, the few restricted excavations of the Cumbrian stone circles and henges have done little to determine their date and purpose. In 1826 a Mr. Wright of Keswick dug out some burnt bones, antlers and animal remains from two of the five cairns inside the Brats Hill circle (Williams 1856). These finds, of limited value in themselves, were even less helpful because the cairns were almost certainly later additions to the ring. In 1875 an unpolished stone axe was found at Castlerigg and seven years later an exploration inside the unusual rectangle of stones at the south-east of the same circle located a 3 ft (1 m)-deep crude pit at its western end filled with black soil and stones with traces of burnt wood or charcoal near its bottom (Dover 1882). In 1901 Dymond dug two narrow trenches across Swinside, at right-angles to each other like a hot-cross bun, and unearthed only a small lump of charcoal and a bit of decayed bone near the centre (Dymond 1902). Little more was discovered in the two seasons of 1937 and 1938 at Ballynoe in Co. Down when Van Giffen obtained some burnt bone, flints and some Loughcrew ware sherds from the secondary cairn inside the ring (Waateringe & Butler 1976: 83-84).

The first excavation of a henge in Cumbria was hardly more productive. Begun by Collingwood in 1937 (Collingwood 1938), it was completed by Bersu in 1939. A long trench lay on the axis. In it the remains of a corpse consumed by a fire of hazel wood rested under a collapsed cairn of stone slabs (Bersu 1940). Even the complete reconstruction of the Grey Croft stone circle in 1949 recovered very few artefacts. A broken Cumbrian stone axe was picked up near a disturbed stonehole at the east, and the team of carefully supervised schoolboys collected some charcoal, bracken, six hawthorn berries, flints and a jet or lignite ring from the internal cairn inside the circle. Helpful though this was in showing that the mound had been heaped up in a prehistoric autumn, the finds from it gave only the vaguest indication of the age or meaning of the ring into which it had been inserted.

Fortunately, although the results of excavations have been disappointing, the distribution of the great stone circles is much more informative because, paradoxically, several of them concentrate in what seems to be an inhospitable area. Cumbria is an isolated region, bordered by the Solway Firth and the Irish Sea to north and west, by Morecambe Bay and the Lancashire marshes to the south and by the bleak, windblown Pennines to the east. Constricted by these barriers, this part of England is further constrained by the dome of fells and mountains that swells from its heart and from the air it is like a hacked and dented tin helmet with only its narrow, sloping rim capable of supporting settlement. Even this strip is limited in its appeal. The west coast is a mere few miles wide, confined between the sea and the mountains. At the south-west is a low-lying tract of limestones, good for farming but cut off from the Lake District by the Scafell massif and by Black Combe to the west 'and the sense of cultural isolation of the region from the north is embodied in the old Furness saying "Nowt good ever came round Black Combe"' (Hogg 1972: 31).

To the north of the mountains the hills flatten into the 300 square miles of the Carlisle Plain whose heavy boulder-clays were once thought to have been covered in a dense prehistoric oak forest inimical to any settlement. Although this view is now much modfed, the absence of good stone precluded any megalithic ring being erected there.

The most fertile area in Cumbria is the Eden Valley to the east with the river rising near the Stainmore Gap and flowing northwards past Carlisle into the Solway Firth. For much of its course it is bounded by gentle hillsides with rich sandstones and shales, abundant in timber and wildlife. Although flanked to its west by the harsh ridge of the Lazonby Fells, it was this well-watered and sheltered valley with its access to north-eastern Britain through the Tyne-Tees Gap to the north, and to Yorkshire through the Stainmore Gap at its south, that was most densely populated. Later prehistoric settlements, burial-places and finds of Bronze Age pottery and implements testify to its continuing popularity. Yet this pleasant, easily-farmed landscape was not the one chosen by the builders of several large stone circles. Instead, they inhabited the superficially daunting region of the central mountains. It is here that Castlerigg and Elva Plain, Brats Hill and Swinside were put up in a part of Cumbria where virtually no other Neolithic site exists. Their presence gives the impression of an 'island culture', sufficient in itself and largely unaffected by outside traditions.

To an outsider this central zone of the Lake District with its hard, sedimentary rocks was a forbidding place of steep-sided mountains, scree slopes, boulders, tumbling streams and thin, acidic soils. It would seem that no-one could live there but the landscape was misleading. In it, in the valley bottoms, there were tracts of sweet grassland where post-glacial silts had produced areas of rich loam. 'The presence of Castlerigg stone circle near Keswick shows that organised settlement was established deep within the region ... , a fact which can be explained by the inducements offered by the fertile, glaciated valleys' (Hogg 1972: 29-30). This, and the axe-factories nearby in the Langdales, account for several early stone circles there.

Other Neolithic monuments were built only on the outskirts of the mountains. There is no room for a detailed discussion of them but it seems that both the long burial cairns and the henges had eastern rather than Cumbrian origins. The Yorkshire affinities of long mounds such as Raiset Pike and Skelmore Heads have been discussed by Powell (1972). The situation of others to the east of the mountains (Druids Grove; two at Mossthorn, Newton Reigny; Trainford Brow, Cow Green and Ewe Close) suggests that the building of such tombs was the work of people from the far side of the Pennines (Manby 1970). Stukeley remarked on several of these long mounds, the burial-places of his 'arch-druids', near Penrith in the Eden Valley. Another, the Currick, one of the long 'cairns or carracks as the Scotch call them' (Stukeley 1776: 45), lies to the north near the Tyne-Tees gap. Only Sampson's Bratfull and Miterdale are to the west of the Lake District and even these are well away from the mountains. Interestingly, however, Sampson's Bratfull is no more than five miles north-east of Grey Croft stone circle on the coast and the same distance from the axe-finishing site of Ehenside Tarn with its Grimston pottery from Yorkshire and with an early date of 3010 ± 300 bc (c. 3800 BC). It is possible, therefore, that the stone circle was a native depôt from which unfinished axes were traded or exchanged, passed on to an intrusive group from whose base the polished axes would eventually be sent across the Pennines to the Yorkshire Wolds.

Why long cairns such as Raiset Pike or 'Yorkshire' henges such as King Arthur's Round Table with its elliptical shape, two entrances and NW-SE axis should never be found close to the factory sources is, at first, surprising. The majority of the Group VI Langdale products appear to have been taken to the area around Bridlington on the east coast for redistribution (Cummins 1974: 204; 1980: 53), and it might be expected that a 'trading centre' such as King Arthur's Round Table would have been constructed near Keswick and the Langdales rather than near Penrith almost 20 miles to the east.

One possibility is that the region was forbidden to outsiders. Certainly the great henge of Mayburgh, unlike any known in Yorkshire, stands directly in front of King Arthur's Round Table as though blocking the way westwards past the hills and along the valley of the River Greta. It is also feasible that the mountains were dreaded by the easterners. Around Bridlington the land rises only 50 ft in ten miles, and a mere 12 ft rise five miles east of Hull is called Hedon, 'the hill of heather'. The towering slopes of the Lake District may have astonished and daunted the inhabitants of the lowlands. Even in the eighteenth century Daniel Defoe could write of 'a chain of almost impassable mountains' and 'unpassable hills, whose tops, covered with snow, seemed to tell us all the pleasant part of England was at an end' (Defoe c.1700: 550-1). Nor would the trails have been easy. A little before Defoe, the diarist, Celia Fiennes, grumbled at being able to ride only eight miles in four hours, and in prehistoric times the wildness of the unaccustomed heights and the hostility of the natives may well have persuaded strangers that the boundaries of the region were safer than its unknown heart.

The mountains were avoided in later times also. No avenues of standing stones were put up there. Finds of beaker pottery concentrate in the Eden Valley (Clough 1968) and so do the food-vessels and urns of the Early Bronze Age. Given such lacunae it is reasonable to suppose that the presence together at the centre of the Lake District of axe-factories and great stone circles is not accidental but is an indication that the two were both contemporary and functionally connected. It is a hypothesis supported by the distribution of the rings, by finds of axes and by radiocarbon dates. Two of these came from Great Langdale itself, one of 2730 ± 135 bc from charcoal associated with stone tools and chippings and another of 2524 ± 52 bc from a similar context. These average about 3400 BC in recalibrated years, showing that the axe-factory sites are nearly five and half thousand years old. It is possible that rings such as Castlerigg, so close to the stone sources, are of the same general period, much earlier than previously suspected.

The idea of the factories and the circles being interconnected is not new. Over 50 years ago Collingwood noticed how the location of the larger rings and the stray finds of axes tended to coincide. 'The circles and the axes thus hang together, and seem to demand explanation as the relics of a single people' (Collingwood 1933: 178). Axes, roughly shaped, have been found at the factories themselves, at Pike of Stickle, Harrison Stickle, Scafell Pike and others (Waterhouse 1985: 6). They have been found along the routes taken by the miners down from the high scree slopes, and they have been found in the vicinity of the great stone circles. In 1901, preserved in the peat at Portinscale just west of Keswick and Castlerigg, four rough-outs and a polished axe were found near a pile of chippings and a thick log stump with a battered top, a 'Celt maker's manufactory' (Cowper 1934). Axes have been found near Long Meg and Her Daughters and near Swinside. Others have been recovered from the circles themselves, part of a polished stone axe lying just under the turf in the entrance to Mayburgh henge. Stukeley, moreover, recalled that 'in ploughing at Mayborough they dug up a brass Celt' (Stukeley 1776: 44) which suggests that the henge continued to function as a distribution centre well into the second millennium when bronze axes had come into fashion. Another stone axe lay in a pile of stones taken from the Hird Wood Circle (Waterhouse 1985: 145). Three have come from Castlerigg. Williams (1856) mentioned a rough-out and a polished axe discovered there and a third unpolished one, $8\frac{1}{4}$ in (21 cm) long was found in 1875 (Cowper 1934: 95). Yet another, broken, lay near the disturbed stonehole at the exact east of Grey Croft stone circle (Fletcher 1958: 6).

Axes such as these were in production for over a thousand years and their discovery inside a stone circle offers no more than a probability that the ring was put up nearer to 3500 than to 2200 BC. Happily, the radiocarbon date from the Lochmaben Stone ring is more definite. All but two of the stones have gone from this ring only a few hundred yards from the modern banks of the Solway Firth near Gretna Green. More properly called the Clochmabane-stane, one great granite boulder 9 ft 6 in (2.9 m) high stands at the edge of a plateau with a smaller stone 80 ft (24 m) to its NNE. In 1841 the site was described as 'a number of white stones placed upright circling half an acre of ground in an oval form' which, if correct, would give the ring a mean diameter of about 160 ft (50 m), well in keeping in shape and size with other great rings in south-west Scotland.

That a 'circle' did exist there is indirectly confirmed by its identification as the *Locus Maponi* of the Roman *Ravenna Cosmography*, a site by a loch which clearly was well-known. In the fourteenth century it was used as a meeting-place for the settlement of disputes over land boundaries.

The Lochmaben Stone fell in 1982. During its re-erection oak charcoal was obtained from the bottom of its stonehole and gave a date of 2525 ± 85 bc (Crone 1983: 18). When corrected, this is the approximate equivalent of 3275 BC and quite close to the beginning of axe production in the Lake District. The shape of this vast oval, like that of the Twelve Apostles 24 miles to the west, and its distance from Great Langdale suggest that it may have been put up somewhat later than the circles of the central regions and at a time when Cumbrian axes were reaching south-west Scotland (Williams 1970).

It is unlikely that any stone circle, great or small, was only an axe-trading centre. It is more probable that each was the focus of a community, sometimes

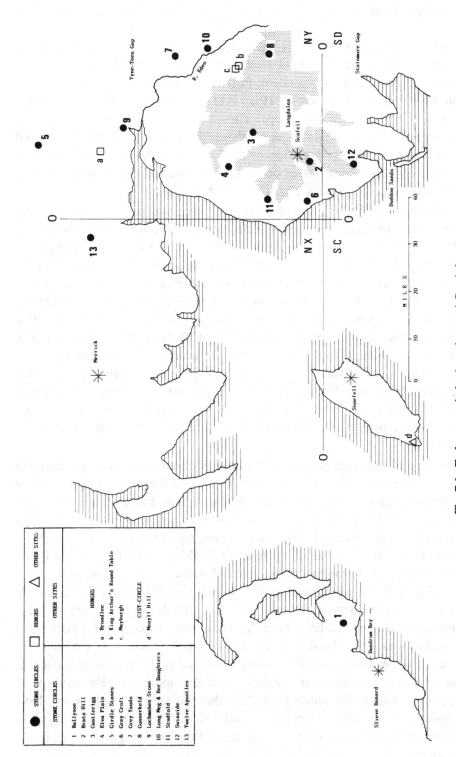

STONE CIRCLES ● □ **HEDGES** △ **OTHER SITES**

STONE CIRCLES
1 Ballynoe
2 Brats Hill
3 Castlerigg
4 Elva Plain
5 Girdle Stanes
6 Grey Croft
7 Grey Yauds
8 Gunnerkeld
9 Lochmaben Stone
10 Long Meg & Her Daughters
11 Studfold
12 Swinside
13 Twelve Apostles

HEDGES
a Broadlee
b King Arthur's Round Table
c Mayburgh

CIST-CIRCLE
d Meayll Hill

Fig. 7.1. Early stone circles in and around Cumbria.

of a single family, sometimes, as with the large rings, of a sizeable population for which it acted as a communal meeting-place in which ceremony as well as trading was a necessary component. From this viewpoint the spacing of the great rings, as the map shows (Fig. 7.1), provides an indication of the 'territories' of the different groups. The circles are quite neatly separated, eight to ten miles apart, in tracts of about 50 to 80 square miles each and conveniently strung together, side by side almost without a break, showing how axes may have been passed outwards from one community to the next as gifts or exchanged in ritual barter. There are two gaps.

To the south of Swinside, on the Furness peninsula, there is no early ring. 'This... raises the question whether Furness may not have had timber circles; its heavier soil and scantier stone would make this not unlikely' (Collingwood 1933: 178). There is, in fact, a stone circle near Ulverston at the Druids' Circle on Birkrigg Common eleven miles south-east of Swinside. It consists of two concentric rings of low stones whose outer diameter of 85 ft (25.9 m) and two taller stones at the north link it in a general way with the circles that are the subject of this paper (Thom, Thom & Burl 1980: 70-71, L5/1). Although its distance from Swinside fits comfortably with the spacing of the other rings, its excavated contents of a tiny collared urn, artefacts and cremations point to it being a later monument than the others and it is best visualised as a circle erected at a time when more distant areas of Cumbria were being exploited and when people had traversed the obstacle of Black Combe.

The one other noticeable gap in the lines of circles wriggling out from the central mountains is that between Grey Yauds and the devastated ring of the Lochmaben Stone. As the site of the missing 'circle' lies precisely where the modern city of Carlisle now sprawls (NY 4055) it will never be known if any such structure existed there. There is, however, a place-name hint. 'Carlisle' is derived from Caer Luguvalos, 'the town of Lugh' (Gelling 1978: 48), a Celtic god in whose honour the harvest festival of Lughnasa was held annually at the beginning of August. Lugh means 'the shining one' (MacCana 1970: 28). He was a deity 'with distinctively solar attributes' (Rolleston 1911: 88), and it is just possible that in 'Carlisle' there lingers a tenuous record of a time when rituals and feasts associated with the sun were performed in an enclosure by the banks of the River Eden. If there were such a ring it was probably a henge. Stone is scarce in north-west Cumbria. Near Brampton even Hadrian's Wall was constructed of turf, and it is unlikely that prehistoric people would have dragged stones for miles when the alluvial soil was ideal for the digging of an earthen enclosure just as other people did at Broadlee henge 15 miles to the north-west.

At first, all these Cumbrian rings were spacious and open, uncluttered by any internal feature but, over the centuries, several had cairns built inside them, a not uncommon practice in the Bronze Age when burials were introduced into ancient sacred sites. These cairns were secondary structures of no relevance to the intended use of the ring. A multi-phase cisted cairn was added to the Ballynoe stone circle (Waateringe & Butler 1976: 84), just as an oval mound containing Bronze Age material was heaped up inside Grey Croft (Fletcher 1958). John Aubrey was told of two massive tumuli near the middle of Long Meg. Camden believed them to be the result of field clearance. They had gone by the time Stukeley saw the ring in 1725. 'In the middle of the circle are two roundish plots of ground, of a different colour from the rest apparently, and more stony and barren' (Stukeley 1776: 47).

Increasing the likelihood that such mounds were additions is the fact that they are seldom in the exact centre of the circle. Inside Studfold the low cairn is almost 10 ft (3 m) out of place. An exploration before 1924 uncovered a damaged cist underneath it (Mason & Valentine 1925). Two cairns were reported inside Castlerigg. Thom plotted one on his plan and a recent resistivity survey detected the other. Both have been levelled but their bases can just be made out in the northern half of the circle, each about 14 ft (4 m) across and a full 20 ft (6 m) to the north-west and north-east of the centre respectively. About the same distance to the south-east is the corner of the unusual rectangle of stones.

An indication that such internal cairns were intrusive latecomers is most obvious at Brats Hill on Burn Moor in the Eskdale valley. Four other smaller rings there also contain cairns but Brats Hill has no fewer than five, four of them cramped together in the western half of the circle. Two were explored in 1829 by Mr Wright of Keswick. Living where he did, he would have known Castlerigg well and it is significant that he claimed that the fifth and easternmost cairn at Brats Hill, isolated from the others, was surrounded by a 'parallelogram of stones similar to that in the Keswick circle, very few of which remain' (Williams 1856: 226). The observation is doubly interesting. First it might explain the Castlerigg setting as an enclosure once bordering a small, now vanished cairn of which Dover found only the underlying pit. Secondly, Thom remarked that the Brats Hill cairn, which he called the 'fifth cell', lay on the major ESE axis. He noted 'that the circle at Burnmoor (L1/6E) is almost the same size and shape' as Castlerigg and that it was actually possible to superimpose their plans. Such a fact strengthens his conviction that the rings were carefully planned by people following rules of design and were not casually laid out as some recent researchers have supposed for Castlerigg (Barnatt & Moir 1984: 205).

There is virtually no sign of the Brats Hill rectangle today but the cairn is very evident 30 ft (9 m) ESE of the centre. In passing, there is an intriguing but possibly accidental arithmetical pattern at Brats Hill where the 42 stones of the circle stand around the cairns each of which has 14 kerbstones, a division of 3 which, as will be seen, is a mathematical feature in later Cumbrian rings (Burl 1976b: 28).

Table 7.1. *The sizes and shapes of rings in and around Cumbria*

	Site	Grid ref.	Thom ref.	Shape	Diameters (ft)	Area (sq ft)	(ratio)*
Stone circles							
1	Ballynoe, Co. Down	J 481404	-	C	100	7854	0.9
2	Brats Hill, Cumbria	NY 173023	L1/6E	F	105 × 97	8125	0.9
3	Castlerigg, Cumbria	NY 292236	L1/1	F	108 × 98	8596	1.0
4	Elva Plain, Cumbria	NY 176317	L1/2	C	113	10029	1.2
5	Girdle Stanes, Dumfriess.	NY 254961	G7/5	C?	128	12868	1.5
6	Grey Croft, Cumbria	NY 034024	L1/10	F	89 × 83	5961	0.7
7	Grey Yauds, Cumbria	NY 544486	-	C?	156?	19113?	2.2
8	Gunnerkeld, Cumbria	NY 568178	L2/10	C	102	8171	1.0
9	Lochmaben Stone, Dumfriess.	NY 312659	G6/3	O?	167?	21780?	2.5
10	Long Meg and Her Daughters, Cumbria	NY 571373	L1/7	F	359 × 311	87454	10.3
11	Studfold, Cumbria	NY 040224	L1/14	C	110	9503	1.1
12	Swinside, Cumbria	SD 172883	L1/3	C	94	6940	0.8
13	Twelve Apostles, Dumfriess.	NY 947794	G6/1	E	288 × 256	57293	6.7
Henges							
a	Broadlee,Dumfriess.	NY 216747	-	O	149 × 130	15213	1.8
b	King Arthur's Round Table, Cumbria	NY 523284	-	O	167 × 148	19412	2.3
c	Mayburgh, Cumbria	NY 518284	-	C	280	61575	7.2

C - Circle   F - Flattened
E - Egg   O - Oval

---

*   Under AREA the 'ratio' shows the area of the site relative to that of Castlerigg. The interior of Broadlee henge, for example, is almost twice as large as that at Castlerigg.

Once it is accepted that the great stone circles were originally planned as open rings several deductions can be made about their purpose. It can be presumed that they were intended to accommodate many people because otherwise the very spaciousness of their interiors would have diminished the spectacle of any ceremony performed there. How many people will never be known. The heaviest stone of all, the monstrous east boulder at Long Meg, weighs about 30 tons and could have been hauled and levered into place by a gang of about 150 men. Yet a ring such as Long Meg, enclosing some 87454 sq ft (8125 m²) may have been meant for many more than that. In their proportions the areas of these rings mirror the densities of modern populations and it may be that the area of a ring is a delicate indicator of the number of people who once used it (Table 7.1).

Castlerigg has an area of about 8596 sq ft (799 m²). Taking this as an arbitrary norm for the early Cumbrian rings, the Twelve Apostles is seven times more spacious and Long Meg is almost ten and half times as big. It is unlikely to be fortuitous that these are the rings closest to the biggest towns, Dumfriess and Penrith, with populations in 1971 of 29384 and 11299, whereas Keswick, near Castlerigg, had only 5169 inhabitants. The smallest ring, Grey Croft, three-quarters the size of Castlerigg, is not far from Gosforth with no more than 922 people (Bartholomew 1972: xv, xvii, xxii, 298). As might be expected, the circles along the Eden Valley, in a fertile, easily worked countryside, are about six times as capacious as those on the west coast and in the mountains even though those rings are nearly three times the national average of 3420 sq ft (318 m²).

A circle's area, therefore, may be taken as a rough guide to the site's population. Seventy men could have erected the heaviest stone at Castlerigg. One might assume an adult male community of about 100 and a total population of male and female, old and young, of 400 or more. Extrapolating from this Long Meg, ten times bigger, could have served a populace of over 4000 people, a number which a territory of 80 square miles could have supported quite comfortably. The existence of many villages, hamlets and farms in the neighbourhood of Long Meg today is a proof of the fertility of the land and is in gross contrast to the few, scattered habitations near Keswick and Castlerigg.

The social structure of these early farming and mining communities is still unclear. Nor is very much yet known about their more abstract skills. Despite the arithmetical compatibility of the stones at Brats Hill there is little evidence for numeracy amongst the builders of these rings. The number of stones and their spacing vary from site to site. The major requirement seems to have been to put stones close together regardless of how many were used. At Castlerigg the 48 or so stones are set 7 ft (2.1 m) apart, at Long Meg some 70 stones 15 ft (4.6 m) apart, and at Swinside the 60 stones were separated by 5 ft (1.5 m) gaps. This is a quite different picture from that in later rings such as the recumbent stone circles of north-east Scotland. There, irrespective of the length of the perimeter, convention preferred ten or eleven stones in many of the rings. In consequence, the small circle of the Hill of Fiddes and the larger Sunhoney, with perimeters of 157 ft and 261 ft (47.9 m and 79.6 m) respectively, each had eleven stones.

Some elementary form of counting may have been known but it is not obvious in the early prehistory of Cumbria. It is only later, in the Early Bronze Age, that semi-numeracy is suggested by the preference for twelve-stone rings. There is a strong likelihood that counting-systems based on either 6 or 3 were used in Cumbria (Burl 1976b: 24). With a radix of 3 a number such as seven would have been calculated 3+3+1, and with a radix of 6, simply 6+1. These were systems which would have worked well for small numbers but which, when amounts of above 20 were to be computed, could only have worked efficiently when recorded on tally-sticks or some other form of record. If such tallies existed in the British Isles they have either not survived or have not been interpreted correctly.

In Cumbria the rings presumed to be earliest can be recognised by their numerous stones. They are circles or flattened circles, of local stones, big and unshaped boulders whose relative heights were a matter of indifference except for those that were chosen for an entrance or an astronomical marker.

Over the centuries the designs of rings became more formalised tending towards a particular number of stones that were more regularly spaced and arranged in an ovoid rather than a circular pattern. As with everything else in prehistory there was no chronological chest of drawers with one megalithic phase closing before another opened. The process of change was more akin to the drifting and merging of slow clouds. Yet sites such as Castlerigg, Long Meg, Ballynoe and the Girdle Stanes are likely, for architectural reasons, to have been erected before the more orderly and regimented Grey Croft and the Twelve Apostles.

The most distinctive feature in the earliest rings and one which does not occur in the later ones is the entrance. Resembling an embryonic avenue, it is formed of two extra stones just outside the circle, side by side, perhaps 10 ft (3 m) apart, and in line with two others in the ring itself. Simple though the construction was, this short and narrow rectangle of four stones created a neat and effective approach to the interior and this may have been the sole wish of the builders. Yet from the centre of the circle these 'portals' stand, more often than chance would predict, in line with some calendrical or cardinal position:

Fig. 7.2. The outlier and south circle stones at Long Meg from the west (Aubrey Burl).

at the south-west and south-east respectively at Long Meg (Fig. 7.2) and Swinside (Fig. 7.3), at the north in Castlerigg and the west in Ballynoe. The possible significance of these orientations will be discussed.

As well as the megalithic rings, the two henges at Penrith also had stone-lined entrances. At Mayburgh William Stukeley recorded stones lying in the wide gap on the east and John Aubrey copied the plan that Sir William Dugdale had made showing two stones standing there (Aubrey 1665-93, I: 114). The Class II henge of King Arthur's Round Table, just to the east of Mayburgh, had opposing causeways at the NNW and SSE of its ditch, 'that to the North having two huge [stones] (viz: on each side one) of about five foot in thicknesse' (*ibid.*). It is of interest to note that this was the entrance nearer to Mayburgh and the mountains. The southern entrance, looking towards the Stainmore Gap and Yorkshire, had no stones in it (Bersu 1940: 201).

Even Castlerigg, where two imposing pillars line a wide gap at the north, may have had an additional pair outside them (Fig. 7.4). Stukeley wrote that 'at the north end is the kistvaen of great stones' which might have referred to a former rectangle, and Anderson (1915: 102) observed that 'it seems possible that they originally formed a gateway to a short avenue, such as exists in the

Circles of Swinside, Long Meg, Stanton Drew and others...'. It is feasible that it was from this tradition of putting up four stones to make an impressive entrance that the later custom of erecting long avenues of standing stones developed. Such a derivation of avenues from simpler Cumbrian prototypes would explain why avenues are more numerous around the Lake District than elsewhere in Britain: at Broomrigg, Lacra, Moor Divock and Shap; and there is even a vestigial line of little stones once connecting the idiosyncratic oval of the Loupin' Stanes to the great circle of the Girdle Stanes a third of a mile away by the bank of the River White Esk. It might also account for the presence of avenues at Avebury and Stanton Drew in Wessex, introduced into southern Britain as an idea accompanying the physical distribution of stone axes from Cumbria.

Yet other than this the architectural influence of the early stone circles was slight. Northwards, beyond the destroyed site of the Grey Yauds with its 88 stones of local granite (Waterhouse 1985: 151), only the Girdle Stanes has any close likeness to the Lake District rings. The appearance of the Lochmaben Stone oval is not known, and the Twelve Apostles oval, with no entrance and with a mere dozen widely-spaced stones seems to belong to a

Fig. 7.3. The south-east entrance at Swinside (Aubrey Burl).

Fig. 7.4. The north entrance at Castlerigg (Aubrey Burl).

rather later period. Eastwards, on the far side of Long Meg and across the Pennines, there is nothing comparable to Castlerigg or Brats Hill even though so many Group VI axes have been found in Yorkshire. To the south, only 16 miles from Penrith, the ring of Gamelands (Thom's Orton, L2/14, Flattened Circle, Type *A*) is more similar to north Welsh rings such as the Druids' Circle on Penmaenmawr near the Group VII axe-factory of Graig Lwyd.

There is an impression of isolationism as if the communities in and around the Lake District mountains were content to stay in their homeland, never venturing far from it, permitting outsiders only to settle at its edges. If this were so then it was not the miners but the foreigners who collected the axes and took them out of Cumbria.

One unexpected exception to this reclusive existence was the link between south-west Cumbria and north-eastern Ireland where the circle of Ballynoe is almost identical to Swinside in all its features. Although 80 miles of Irish Sea separate the rings, the crossing was made easier by the haven of the Isle of Man halfway between the Duddon Sands near Swinside and Dundrum Bay a few miles from Ballynoe. On the island the landmark of Snaefell, 2034 ft

Fig. 7.5. View looking north at Ballynoe, Co. Down (Crown Copyright, reproduced with the permission of the Controller of HMSO).

(620 m) high, rose like a mountain in the sea and, in fair weather, experienced navigators could have reached it in a day, paddling their long, skin-lined boats across the almost tideless waters (Davies 1946: 42). From the Isle of Man the silhouette of Slieve Donard, 2796 ft (852 m), in the Mourne mountains guided the seafarers westwards and, returning, they would have steered towards the even more dominant outline of Scafell, 3210 ft (978 m) in the Lake District mountains (Bowen 1972: 40-41).

It is perhaps not coincidence that the only site on the Isle of Man anything like a stone circle is the unique ring at Meayll Hill (SC 189677), a Neolithic oval of cists with entrances at north and south as though its builders had adopted some of the traits of the Cumbrian rings (Kermode & Herdman 1914: 40-53). There is proof of contacts. Nearly a third of the stone axes on the island came from the Langdales and it is likely that travellers from the Lake District used the Isle of Man as a staging-post on their way to Ireland.

On a clear day the Irish coast can be seen from Meayll Hill and there, not far from the modern town of Downpatrick, a 'Cumbrian' ring was erected. 'Ballynoe in coastal Co. Down has a peripheral great stone circle which may have been built by people influenced from Cumbria' (Davies 1945: 142) (Fig. 7.5). So alike is this ring and Swinside that it is more probable that it was Cumbrian people themselves who put up Ballynoe. The presence in northern Ireland of Group VI axes from the Langdales is well attested (Jope & Preston 1953) and more recently others have been identified in Co. Down itself (P.C. Woodman, pers. comm.) strengthening the belief that Ballynoe, dissimilar to any known Irish stone circle, was constructed by stone-axe traders from the area around Swinside who reproduced their own 'temple' in Irish stone.

Both are almost perfect circles, Ballynoe about 108 ft (32.9 m) in diameter, Swinside nearly 94 ft (29 m) (see Fig. 6.5 of Jennings' contribution in this volume). Both are constructed of local stones, Ordovician grit at Ballynoe, porphyritic slate at Swinside, and each ring may originally have had some 64 closely-set stones on its circumference, higher, lower, thick or slender, unshaped and coarse, but both at Ballynoe and Swinside there is a taller stone at the north. The rings share two other features. Both have characteristic entrances of four stones. At Swinside these stand at the south-east lining a gap 7 ft (2.1 m) wide and, probably by accident, the same width occurs at Ballynoe where the portals are at the WSW. Another facet of the circles, generally uncommented on, is the possibility that both sites were levelled before the erection of the stones, at Ballynoe by people heaping up an earthen platform (Chart 1940: 120), at Swinside by cutting into the slope to make a shelf on which the circle would stand.

This was not uncommon. Where possible the people who put up the early Cumbrian circles avoided steep gradients and uneven ground. The Girdle Stanes were raised at the foot of a hillside where the ground levelled out. Brats Hill lies between knolls and hillocks on an even stretch of moorland. Thom himself commented on how Castlerigg was 'beautifully situated on a flat level part of the field' (1967: 150). The Grey Yauds, dismantled for the walls of the nineteenth century common, stood on 'a dark and dreary waste' on level ground just west of the one surviving stone (Waterhouse 1985: 151). Studfold is splendidly situated at the edge of a moor from which the land falls to east, west and south.

Only Long Meg and Her Daughters does not conform to this custom, standing as it does on a slope of some 3° that falls nearly 13 ft (4 m) from south to north across the ring's western side. Yet just across the lane to the east there is a wide, level field, and immediately to the west of the ring the slope flattens out at its crest. It is unlikely that there were settlement sites to both east and west on flat ground where the circle might have stood. A plausible explanation for such a perverse choice of site is that there was an older stone that the builders wished to incorporate into their new circle, the elegant outlier of Long Meg. This thin pillar of sandstone was different in size and shape from the stones of the ring which are squat and rounded porphyritic boulders and it may have been erected in the early Neolithic as a territorial marker or funerary memorial. Many standing stones are known to predate more elaborate megalithic structures in Brittany where early Neolithic menhirs were broken to be re-used as capstones in passage-graves and allées-couvertes (Mohen 1984: 1537). In the British Isles there is less evidence, although at the Rollright Stones in Oxfordshire Neolithic people deliberately put up a megalithic ring against an older standing stone (Lambrick 1983: 34-6). This custom was known also in Brittany where chambered tombs and huge burial-mounds were placed as close as possible to towering menhirs.

The builders of the circle against Long Meg's emaciated pillar may have attempted to arrange the stones of the ring and its entrance so that the outlier would be in line with the midwinter sunset, and this introduces the problem of what significance particular stones had to the Neolithic mind. There have been anthropological interpretations of them as fertility symbols like the 'male' and 'female' avenue stones at Avebury or as guardian goddesses in Breton tombs (Burl 1985b: 15). At Long Meg 'the vulgar notion that the largest of these stones has breasts, and resembles the remainder of a female statue, is caused by the whimsical irregularity of the figure, in which a fervid imagination may discover a resemblance of almost anything' (Smith 1752, repr. 1886: 72), a criticism which could be applied equally well to many more recent statements about stone circles and their sexual symbolism.

There are carvings on Long Meg. The lower half of this outlier is rectangular in section. Three of its sides are badly weathered, raddled by rain, but the east face is less exposed and on it there are incised concentric circles, a series of arcs, a spiral and some eroded motifs almost impossible to discern even in the most favourable light (Simpson 1867: 19-21). Such decoration is usually considered indecipherable by archaeologists but art-historians and anthropologists have been less pessimistic. From a study of megalithic art in the passage-graves of eastern Ireland Brennan concluded that the spiral was a representation of the sun at its 'standstills', the anti-clockwise spiral portraying the winter solstice when the shadows were longest and the coils of the spiral were reversed and widely spaced (Brennan 1983: 190).

The spiral at Long Meg is an anti-clockwise one, and although the carving does not face the south-west it is interesting that from the centre of the circle this tall, tapering outlier does stand in line with the midwinter sunset (Thom 1967: 99. 'CO'), a fact which suggests that the circle-builders were perpetuating a long-established Neolithic orientation.

At first sight it seems odd that the Long Meg pillar is not framed between the monstrous stones of the entrance. Twelve feet (3.7 m) high, its top rises off-centre like a shrivelled finger above the line of the two western portals as though they were backsights. This arrangement may have been planned for the benefit of future generations. The centre of Long Meg's vast and open ring, 359 ft (109.4 m) across, would have been difficult to locate exactly and an observer only 5 ft (1.5 m) to east or west of it would have been a full degree out, a matter of two or more days, when attempting to celebrate the midwinter solstice.

It is arguable, therefore, that the ring of Long Meg had a calendrical line built into it, employing an ancient carved stone as a sighting-device. Caution, however, is needed when using a term such as 'calendrical'. It could mean any time of the year that a couple of stones happened to define, early December, the middle of October or April the first. With 365 days in a year and with several opposed pairs of stones in a circle, an enthusiast could construct a megalithic calendar out of almost any ring. Consequently, there are two constraints in this paper. The lines are limited physically by accepting only those from a circle's centre - and even this is a problematical position - to the tallest stone, to an entrance or to any outlying stone standing close to the circle. Temporally, the alignments are restricted to the most important solar events. (There is little evidence, as yet, for lunar observations in Cumbria.) Such positions are defined by the solstices, the equinoxes and by the occasions of the four great 'Celtic' festivals of Imbolc, Beltane, Lughnasa and Samain at the beginnings of February, May, August and November (Burl 1983: 34). It may not be by chance that these solar phenomena correspond closely to the declinations of the even-numbered 'Epochs' of Thom's 'Megalithic Calendar' (Thom 1967: 110). In Table 7.2 the azimuths, with a 0° horizon, are cited as a rough guide. They are for the latitude of Castlerigg.

Despite the ruin of several circles in which stones have fallen or have been taken away there is a correlation between these calendrical positions and the three architectural features of the rings. As well as the destroyed Grey Yauds and Lochmaben Stone, other sites such as Elva Plain, Gunnerkeld and Studfold are in too poor a condition for anything calendrically certain to be claimed for them. Elsewhere the evidence is clearer and, except where cited, has been obtained from the surveys of Thom (1967). At Ballynoe the midpoint of the entrance at 264° (Jope 1966: Fig. 58), is nearly in line with the equinoctial sunset. It is actually directed towards it if the northern pair of

Table 7.2. *Declinations and azimuths for the Cumbrian Rings*

Epoch	Date	Sunrise declination °	azimuth °	Sunset declination °	azimuth °
0	Vernal equinox, March 21	+0.4	89.4	+0.6	271.0
2	Beltane, early May	+16.6	60.6	+16.7	299.8
4	Summer solstice, June 21	+23.9	45.6	+23.9	314.4
6	Lughnasa, early August	+16.8	60.1	+16.6	299.6
8	Autumn equinox, September 21	+0.5	89.1	+0.3	270.6
10	Samain, early November	-16.2	118.9	-16.4	240.5
12	Winter solstice, December 21	-23.9	134.4	-23.9	215.6
14	Imbolc, early February	-16.3	118.9	-16.2	241.3

entrance stones had been a backsight as seems to be the case at Long Meg. Thom noticed how the tallest stone at Castlerigg, a huge pillar over 8 ft (2.4 m) high and set radially to the circumference, had a declination of 16°.0 in line with sunrise in early February and November. At Brats Hill (Dymond 1881: 56) the outlier has a similar declination, -16°.2, again looking towards the sunrise in early November. The midwinter solstice alignment at Long Meg has already been discussed. At the Twelve Apostles the tallest stone is at the south-west (Thom, Thom & Burl 1980: 288). With an azimuth of 241° and a declination of -16°.1 it also appears to have been a Samain alignment. At Swinside Thom assumed a line through the middle of the south-east entrance (*ibid.*: 34-5). The azimuth of 128°.8 and a horizon altitude of 0°.5 produced a declination of -21°.5 of little astronomical importance. If, instead, the line had passed through the southern pair of portal stones the azimuth would have been 134°.5 and the declination would have been -24°.6, close to midwinter sunrise.

These are not precise sightlines. But they indicate a concern with the sun and especially with the sun at the darker quarters of the year, at midwinter and at the time of Samain, the most feared of all the festivals, when the ghosts of the dead rose from their graves to threaten the living (MacCana 1970: 127) and 'when the Otherworld became visible to mankind, and all the forces of the supernatural were let loose upon the human world' (Ross 1970: 200).

Important though these observations are, the Cumbrian rings contain other properties. They not only possess alignments which correlate with calendrical declinations but they also have features which mark the cardinal points of north, east, south or west, a fact surprisingly recognised two hundred years ago. Smith, who had been so scornful about Long Meg as a megalithic deity, added that 'the four [stones] facing the cardinal points are by far the largest and most bulky of the whole ring. They contain at least 648 solid feet, or about 13 London cartloads' (Smith 1752, repr. 1886: 71). He was correct. The east boulder in particular is enormous, the western is almost as big, and those at north and south are almost unbelievably heavy (Dymond 1881: 40).

> Where can we find two better hemispheres,
> Without sharp North, without declining West?

John Donne. The Good Morrow

During the compilation of his data Thom noticed how many megalithic circles and rows contained meridional lines. 'There are a great many sites with very definite indications of a north/south line' (1967: 95). This observation, with east and west included, can be applied to all the stone circles under consideration.

The low outlier at Grey Croft is at the north, 354°, and in that ring a broken stone axe was unearthed by the disturbed hole of a stone at the exact east. Votive deposits of axes are known in other ritual centres in Britain and Brittany. Two chalk axes lay in postholes at Woodhenge in line with the south and with midsummer sunrise. It will be recalled that a stone axe was picked up in the east entrance at Mayburgh henge, apparently deliberately buried there.

Cardinal points are marked by some entrances to circles, north at Castlerigg, 0°, and Gunnerkeld, 358°, and east, 92°, at Mayburgh (Dymond 1891: plan). In many rings the tallest stone is the indicator: north at Ballynoe, 357°; Studfold (Waterhouse 1985: 71) and Swinside, 6°; south at Brats Hill, 178°; west, 262°, at Elva Plain where it is the longest of the fallen stones in the ring; and east at Long Meg, 86° and perhaps also at the Girdle Stanes where the western third of the circle has tumbled into the undercutting waters of the River White Esk making any conclusions about orientation debatable. Even in some later Cumbrian rings such as Oddendale (Thom's site, L2/13) and the Druids' Temple on Birkrigg Common (L5/1) there are tall stones at east and north and this pervasive concentration on cardinal points must have been purposeful.

At the Twelve Apostles a north-south line, 8°-188°, was defined by the long axis of the oval (Thom, Thom & Burl 1980: 288). This technique of establishing an orientation by a long or a short axis may explain why so many non-circular rings were erected. It was a method used at Woodhenge where the axis of the ovals pointed to midsummer sunrise and it appears also in several rings in north Wales (Burl 1985a: 79).

There is hardly a precise alignment in the Cumbrian rings, and if any one site were taken on its own one would be justified in thinking that a supposed calendrical or cardinal orientation was accidental, even imaginary. Single sites can be misleading. Instead, it is the repetition, in ring after ring, of comparable alignments that reinforces the belief that the lines were intended and needed by prehistoric people. General patterns, rather than individual sightlines, buttress the argument in favour of archaeoastronomy.

For nearly 20 years the writer has been urging archaeologists and archaeoastronomers to turn from the study of single sites, searching for high-precision alignments, and, instead, to look for 'a group of similar monuments with the same orientation in a restricted locality' (Burl, MacKie & Selkirk 1970: 27). Where this has been done, as with the Cumbrian circles, then generalised and approximate but recognisable alignments have been demonstrated: in chambered tombs (Burl 1981b: 67-8); in the recumbent stone circles of north-eastern Scotland (Ruggles & Burl 1985); and in the free-standing megaliths of western Scotland (Ruggles 1984a: 303-5). The lines discovered are never so fine that they could have been used for celestial prediction but they are consistent enough to show that in the Neolithic and Bronze Ages of the British Isles people had an obvious interest in the sun and the moon.

More research, coupled with scepticism but unimpeded by prejudice, will bring us closer to an understanding of what such alignments, like those in the Cumbrian stone circles, meant to the communities that laid out them out. It is, after all, the pursuit of the whole past, not just the comfortably preferred elements of it, that should be the preoccupation of all who profess an interest in antiquity. No less than architecture and artefacts, astronomy and its alignments are a legitimate part of that pursuit.

How such positions were fixed is conjectural. They are not exact but they are so nearly correct that Neolithic people must have tried to calculate where they should be. There was no Pole Star (α Ursae Minoris) to assist them. Even today it is some 50' out of line with the North Pole. Two thousand years ago it was 12° away and in 3500 BC this second magnitude star would have been nearly 30° from True North (Hawkins 1966: 20). Only Thuban (α Draconis) in the irregular constellation of Draco glimmered near the north at that time and it is unlikely that prehistoric people could have picked it out from the other stars, especially as one of them, Eltanin (γ Draconis), was brighter.

With their semi-numeracy it is improbable that people were competent to compute the days of a year, dividing the number into two, 182 or 183, and then counting daily from a solstice to the equinox. Thom suggested that observers might have bisected the angle between the east and west elongations of a circumpolar star (1967: 96). It is just as possible that they simply halved the distance between two solar events such as midwinter and midsummer sunsets or midwinter sunrise and sunset, setting up posts at these extremes, stretching a rope from one to the other and then finding the midpoint by folding the rope in half. Being unaware that the hills and valleys of the skyline would distort this 'centre', their alignments inevitably had imprecisions built into them.

Vital to an understanding of what stone circles were to their builders is the question of why such emphasis was placed on these cardinal positions. It is easy to believe that calendrical lines were laid out to commemorate and record the occasions of a festival or the turning of the year. The equinoctial alignments at Mayburgh and Long Meg, both in the Eden Valley and close to the Stainmore Gap and Yorkshire, are in keeping with the idea of autumnal gatherings with outsiders when the summer work in the axe-factories was at an end. In places where articles such as stone axes were exchanged this would have been of extreme importance, involving as it did the presence of strangers. There is a vivid parallel with the stone axe trade of Australian aborigines. 'The tools were traded along known routes by stages and ... these journeys were arranged to coincide with seasonal festivals of magical and social significance' (Bunch & Fell 1949: 15). Gift exchange and periods of non-hostility required both sanctuaries and rituals at accepted times of the year to sustain them. It is also arguable that the winter alignments in the remoter Cumbrian circles were for local rituals to which no foreigner would have been admitted.

In contrast to these calendrical lines, the north-south alignments seem meaningless. That it was not so is shown by the ways in which other people in the world have stressed such directions. In 210 BC the Chinese emperor, Qin shi Huang died and was interred beneath a gigantic earthen pyramid. Each side of the pyramid faced a cardinal position so that it would symbolise the cosmos, the world as an ordered system. The pyramid stood for the 'Apex of Heaven', the north pole of the sky. 'The heavens revolved about the Apex of Heaven and the world revolved around the emperor. He was the steady spot at the center that provided stability and order to the world' (Krupp 1983: 111).

The poet T.S. Eliot, who, in *East Coker*, wrote 'of old stones that cannot be deciphered' also wrote in *Burnt Norton* of 'the still point of the turning world'

> Where past and future are gathered. Neither movement from nor towards,
> Neither ascent, nor decline. Except for the point, the still point...

T.S. Eliot. *Four Quartets*

This was the essence of the north. The sun and moon and stars revolved around it but the North was always still, unmoving, eternal. Krupp gave other examples of north-south alignments based on the belief in the immutability of the north. The holy city of Beijing was planned on a north-south axis 'where earth and sky meet' (Krupp 1983: 261). The Mississippian Indian village of Cahokia was oriented north-south and the burials in it were similarly oriented. And with the kivas or subterranean temples of prehistoric south-western American Indians, the Anasazis, it is possible to penetrate the thinking behind these alignments.

Kivas were usually circular, roofed buildings with doorways to north and south. One of the largest at Casa Rinconada had a niche high in the wall to admit the rays of the midsummer sun. Another niche was aligned towards the rising sun at the equinoxes. Casa Rinconada, however, was not an observatory. It was a place where the earth and the sky met and where the four directions of the world were reproduced. Myths were told of the first kiva which was circular because it duplicated the circle of the sky. A kiva was an 'attempt to image the cosmos in an earthly building' (Williamson 1982: 216), 'a metaphor of the cosmos in stone... an image of time and space' (Williamson 1984: 112).

> At the round earth's imagin'd corners, blow
> Your trumpets, Angells, and arise, arise
> From death, you numberlesse infinities
> Of souls...

John Donne. *Holy Sonnets. VII*

This may have been what a stone circle was to its people, a place where axes and gifts were exchanged, a place where annual gatherings were held, a place to which the bodies of the dead were brought before burial, but, above all, a place that was the symbol of the cosmos, the living world made everlasting in stone, its circle the circle of the skyline, its North point the token of the unchangingness of life, a microcosm of the world in stone, the most sacred of places to its men and women.

Should this interpretation be correct, then it will not have come out of the work of excavators but from the plans and analyses of Alexander Thom and others before him without whose information and stimulus such research would not have begun. Years ago John Aubrey wrote that he had brought the stone circles 'from an utter darkness to a thin Mist' (1665-1693, I: 25). The mist remains, a little thinner today, but through it, with the work of Alexander Thom, the sun is rising.

APPENDIX

Bibliographical references to work by Alexander Thom on Cumbrian stone circles

1 Brats Hill, Cumbria. L1/6E (Burnmoor)   *TTB: 30-31, 40-41*

Construction. 1966: 35; 1967: 40. Flattened *A. c.*105 ft diameter.

Astronomy. 1966: 7, 13; 1967: 99.

   (A). Circle to circle (Low Longrigg NE, L1/6A). Az = 348°. *h* = 7°.5. decl. = 42°.1. Arcturus 1900 BC.

2 Castlerigg, Cumbria. L1/1.   *TTB: 28-29, 30-31*

Construction. 1955: 281; 1966: 22, 35, 55, Fig. 39; 1967: 39, 64; 1971: 12. Flattened *A.* 108 ft diameter.

Astronomy. 1955: 284; 1966: 7, 13, Fig. 39; 1967: 99, 114; 1971: 11; TT78: 22, 178.

   (A). Stone to stone. Az = 127°.0. *h* = 5°.2. decl. = -16°.0. Epoch 1. November sunrise.

   (A). Stone to stone. Az = 307°.0. *h* = 4 .6. decl. = +24°.3. Midsummer sunrise.

   (A). Stone to stone. Az = 157°.1. *h* = 2 .8. decl. = -29°.8. Major moonrise.

3 Elva Plain, Cumbria. L1/2   *TTB: 32-33*

Construction. 1955: 281; 1967: 40. Circle. *c.*113 ft diameter.

Astronomy. -

4 Girdle Stanes, Dumfriess. G7/5   *TTB: 298-299*

Construction. 1955: 281; 1967: 40, 137; TT78: 4. Circle. Part only. 128 ft diameter.

Astronomy. -

5 Grey Croft, Cumbria. L1/10 (Seascale)   *TTB: 46-47*

Construction. 1967: 39, 139. Flattened *D.* 89 ft diameter.

Astronomy. 1966: 7, 13; 1967: 99.

   (A). Circle to outlier. Az = 354°.0. h = 1°.0. decl. = +36°.3. Deneb.

6 Gunnerkeld, Cumbria. L2/10   *TTB: 56-57*

Construction. 1967: 40, 139. Concentric circles. *c.*102 ft and 50 ft diameters.

Astronomy. -

7 Long Meg and Her Daughters, Cumbria L1/7   *TTB: 42-43*

Construction. 1955: 281; 1966: 22, 38; 1967: 40, 139; TT78: 2-3. Flattened *B. c.*358.8 ft
    diameter.

Astronomy. 1954: 390, 400; 1967: 99.

   (A). Circle to outlier. Az = 223°.4. h = 1°.1. decl. = -24°.2. Midwinter sunset.

   (A). Circle to Little Meg (L1/8). Az = 65°.1. h = 3°.4. decl. = +16°.7. Epoch 2. May
       sunrise.

8 Studfold, Cumbria. L1/14 (Dean Moor)   *TTB: 50-51*

Construction. 1967: 40. Circle. 110 ft diameter.

Astronomy. -

9 Swinside, Cumbria. L1/3 (Sunkenkirk)   *TTB: 34-35*

Construction. 1955: 281; 1967: 39; TT78: 3. Circle. 93.7 ft diameter.

Astronomy. -

10 Twelve Apostles, Dumfriess. G6/1   *TTB: 288-289*

Construction. 1955: 281; 1967: 40. Flattened *B. c.*288.4 ft diameter. TTB: 288-289. Could be Egg, Type I, based on 3, 4, 5 sides of 18, 24, 30 MY. Perimeter = 125.3 MY.

Astronomy. -

Notes

1. In this list 'TT78' refers to Thom & Thom (1978a) and 'TTB' refers to Thom, Thom & Burl (1980). Other citations are to works in the relevant year by A. Thom alone.

2. The main reference, where a plan and notes are printed together, (1980), is cited to the right of the title.

3. For astronomical alignments (1967: Table 8.1, 97-101) only the (A) lines are listed.

8

Investigating the prehistoric solar calendar

EUAN MACKIE

Introduction and Background

Origin of the writer's interest

After the work of Sir Norman Lockyer (1894; 1906) and Admiral T.B. Somerville (1912; 1923; 1927) early this century, British archaeology seems to have more or less ignored the question of astronomical practices in ancient times, having inevitably been more concerned until recently with basic problems like dating, excavation techniques and the reconstruction of sequences of material cultures. In the second half of the twentieth century the possibility of discovering the calendrical and astronomical potential of Neolithic sites - and of obtaining in this way a new insight into prehistoric ceremonial and intellectual activities - re-emerged in 1965. At this date was published Gerald Hawkins' *Stonehenge Decoded* (Hawkins & White 1965), an ambitious work which tried to show that Neolithic Stonehenge was a form of calculating device which could keep track of the solar calendar, of the moon's movements and could even predict eclipses. On the whole it was not well received by the profession in Britain which found its arguments short on archaeological expertise and rather too full of special pleading (Atkinson 1966; Hawkes 1967).

The features which Hawkins claimed to be astronomical in purpose were many pairs of stones and post-holes on the site, each of which, he assumed, had been deliberately arranged to form a straight line pointing to a place on the horizon where the celestial bodies concerned rose or set at significant points in their cycles of movements. The essential argument was that this happened so often that it could not be explained by chance. Of course, since the horizon seen from Stonhenge is mainly low and featureless, most of these 'significant' rising and setting points were only indicated by the features on the site themselves, and not by any natural horizon marks. Thus it was not possible to use the argument that the artificial lines gained greater credibility by consistently pointing at such natural marks. Since Hawkins was concerned with a single site it was difficult for anyone other than mathematicians and astronomers - and those few archaeologists who knew Stonehenge thoroughly - to make much of a contribution to the debate. However, the climate of archaeological thought was altered to some extent by *Stonehenge Decoded*, if only in the sense that it revived a mainly sceptical interest in an almost forgotten aspect of the discipline.

Two years later there appeared a very different work - *Megalithic Sites in Britain* by Alexander Thom (Thom 1967). This also dealt with possibly deliberately arranged, astronomically significant orientations in prehistoric monuments but, deliberately avoiding Stonehenge, it considered scores of related but visually much less impressive standing stone sites throughout England, Wales and Scotland. The statistical arguments were correspondingly more formidable than Hawkins', and Thom was the first to propose the existence in rugged terrain of very long, potentially accurate alignments. These consisted of a standing stone *backsight* pointing in some way to a distant natural *foresight* - a cleft or notch or mountain slope on the horizon. Some of the long ones would have been capable, for example, of defining the solstices to the day. At first this book was reviewed more favourably than Hawkins' (Atkinson 1968).

The writer had in fact been made aware a few years earlier of some of the possibilities held by 'archaeo-astronomical' techniques for shedding light on the intellectual development of prehistoric man; this due to Dr A.E. Roy, who showed him a draft of a paper on his work on the stone circles on Machrie Moor on the Isle of Arran (Roy *et al.* 1963). This dealt with the geometry of several of the circles, pointing out that one was clearly a deliberately arranged ellipse with an eccentricity of one half while others seemed to have more complex shapes, each with a distinct axis of symmetry. The astronomical potential of the site was not described at that time.

In spite of this early introduction, and of the appearance of Thom's work four years later, the writer remained only mildly interested and intrigued by the new evidence - deterred from more positive action partly because of the uproar that followed the appearance of Hawkins' book and partly because he had no training in the sophisticated surveying and statistical techniques which were evidently needed to carry out this kind of work. Probably his long-standing interest in astronomy, dating back to childhood, kept the slight momentum going. Yet the geometrical qualities, and the calendrical and astronomical potential, of standing stone sites still seemed an academic problem to a traditionally trained field archaeologist; Thom's book was hard going for one not rigorously versed in statistics and astronomy, being densely packed with data and making few concessions to the layman. The chances of matching Thom's expertise in theodolite surveying, mathematics and astronomy seemed so remote as to be positively discouraging. Doubtless this was, and perhaps still is, a widespread feeling in the British archaeological profession. There was then, for this writer at least, quite a long period of latent and inactive interest in 'archaeoastronomy' before he decided to try to make a positive contribution. An analogy drawn from evolutionary theory seems appropriate here. New mutations in a species sometimes survive for a long while in a rather precarious manner, without increasing or improving substantially, until some drastic change in the environment removes rivals and allows them to expand and diversify. The long and tenuous existence of the first small, warm-blooded mammals before the extinction of the dinosaurs is a good example, made more poignant by the current possibility that the giant reptilian rivals were removed by an extra-terrestrially caused cataclysm.

Another is provided by the development of the steam engine - the eventual driving force of the industrial revolution - which after its invention by Thomas Newcomen in 1712 existed for several decades as a crude and extremely inefficient atmospheric beam engine, used only for pumping out mines. Finally James Watt realised its potential for driving machinery and devised the rotary version; thereafter its development into the high pressure form and its subsequent diversification proceeded rapidly, and the mechanisation of the industrial revolution took fire. In a similar way the new ideas propounded in the 1960s by Hawkins, Thom and Roy on the basis of surface fieldwork lay dormant in the author's mind for several years until an opportunity for a constructive contribution finally appeared; his 'intellectual environment' then changed sufficiently for the idea to multiply and diversify. The evolutionary analogy can be pursued further; it is perhaps still too early to decide whether these 'intellectual mutations' will have a permanent and important place in the archaeological landscape of the start of the twenty-first century.

The start of active participation

Two events galvanised this mild interest and turned it into an active desire to pursue constructive fieldwork aimed at testing Thom's controversial new hypotheses, and both occurred in 1969. One was the first academic conference in Britain organised by an archaeologist and devoted to examining the impact of Thom's ideas on our discipline. This was arranged by Lionel Masters on behalf of the Department of Adult Education in the University of Glasgow and was addressed by four speakers. H.A.W. Burl described 'The archaeology of stone circles and henge monuments', A.E. Roy spoke on 'The astronomical background to stone circles and henge monuments' and A. Thom described 'The astronomy and geometry of stone circles and henge monuments'. For this event the writer had to prepare a lecture called 'Archaeology, geometry and astronomy - a prehistorian's view' and thus to work out his attitude as an archaeologist to Thom's work, to assess the new evidence and how it fitted in with our existing picture of late Neolithic Britain. Amongst other things he considered the plausibility of the idea of a Neolithic solar calendar of 16 'months' when considered against the fact that the apparatus for defining the 'months' appeared to be randomly scattered over the country, and rarely to be clustered in useful groups. A successor conference took place, also in Glasgow, in 1976 (MacKie 1976).

A little earlier in 1969 the writer contributed a short piece to *Current Archaeology* about the new ideas on stone circles and standing stones being put forward by Alexander Thom (MacKie 1969). Though perhaps slightly uncritical this was mainly a summary of these new ideas, and a request that they be considered and investigated by archaeologists. A response was provoked from H.A.W. Burl which forcibly articulated what might be termed the traditonal, or resolutely sceptical archaeological reaction (Burl, MacKie & Selkirk 1970); it is pleasant to report that, in the intervening years, the protagonists of that debate have both learned a great deal and that their positions are now much closer.

The second provocative event was the offer to the writer by Alexander Thom of a sight of the manuscript of his second book *Megalithic Lunar Observatories* (Thom 1971). Despite an early favourable review of *Megalithic Sites in Britain* (Atkinson 1968), it quickly became clear that Thom's first book presented a formidable problem for British archaeologists. The evidence was mainly statistical, and was based on the claimed existence of large numbers of long alignments defined by various combinations of standing stones which seemed to have been deliberately arranged to mark specific directions. These turned out to point consistently at places on the horizon where the sun, moon and some bright stars rose and set at significant times in their cycles of movements; often there was a peak or notch at these places. Yet still there seemed to be little a field archaeologist could do to test these hypotheses, apart from re-visiting all the sites and checking for himself whether they were plausible on the ground - a mammoth task which Clive Ruggles eventually carried out in the later 1970s (Ruggles 1984a).

However, in the new manuscript was some fresh information about the Kintraw standing stone in Argyllshire, diagnosed in *Megalithic Sites in Britain* as the backsight for a potentially highly accurate midwinter sunset alignment (Thom 1967: 54-6). The foresight was a conspicuous V-shaped notch between two of the Paps of Jura, 28 miles away from the stone to the south-west and in theory quite capable - atmospheric conditions being suitable - of indicating the shortest day exactly. However, a tree-covered ridge about a mile in front of the backsight stone in fact just obscured the notch from a position beside it, and Thom deduced from his general hypothesis that an observing position must therefore have existed slightly higher up on the steep hill slope to the north-east; from this spot, where there was a narrow, more level terrace, an unobstructed view to Jura was obtained.

This idea could obviously be interpreted in two ways. Either it was outrageous special pleading to get over an apparently fatal difficulty for the long alignment hypothesis, or it was a bold prediction made from that hypothesis and capable of being tested. The writer's realisation that the latter could be true - that at Kintraw there existed what might be an unique opportunity to test the astronomical long alignment hypothesis by standard archaeological excavation techniques - was a turning point in his interest in Thom's work. For if an artificial platform was found under the turf on this otherwise unpromising hillside, and could be dated suitably early, this would be a remarkable verification of Thom's interpretation of the site, and therefore of his general ideas. The discovery of such a platform, albeit undatable, in 1970 and 1971 and the subsequent controversies have been described several times (MacKie 1974; 1981).

The main results for the writer were, first, to increase his confidence in the alignment hypothesis and, second, to confirm that practical tests for Thom's ideas were not only possible but extremely desirable if the archaeological profession was to absorb anything of the new views, or even to consider them properly. Further work at a stone circle on the island of Islay, Argyllshire, in 1973 and 1974 (undertaken for quite different reasons) seemed to provide unexpected and strong support for Thom's geometrical hypothesis (MacKie

1977a: 92-4; 1981: 116-28) - that many if not most stone circles had been designed and laid out round carefully planned geometrical shapes and using a standardised unit of length. The long axis of the elliptical ring was apparently also aimed at a distant mountain peak in Ireland, at the winter solstice sunset position (MacKie 1981: Fig. 3.6).

A third potentially testable solstitial alignment was identified for investigation at Brainport Bay, Loch Fyne in Argyllshire, in 1977 and is the subject of the rest of this paper.

The solar calendar

The background to this research into the prehistoric solar calendar has already been fully explained (MacKie, Gladwin & Roy 1986), but a few further points are noted here. The crucial question seems to be whether in prehistoric times structures which were oriented in a simple way towards solar events, presumably for ceremonial reasons, evolved into long alignments capable of making useful calendrical observations.

If the existence of these could be proved this might well indicate the presence in late Neolithic Britain of a full-time class of 'wise men' including trained celestial observers - rather like, and perhaps ancestral to, the orders of Druids in Iron Age times (Piggott 1968; MacKie 1977a: 226). This is an important point which must be explained.

The identification of such long prehistoric alignments - usable as accurate observing instruments - should imply not only a definite desire for an exact calendar, well beyond the needs of a simple agricultural population, but also considerable mental skill. This is because more than one intellectual leap must be made from the simple and basic idea - which is obvious to every countryman and presumably was to Neolithic farmers also - that the approximate time of the year can be determined by watching the regular and endlessly repeated changes of the sun's position against a hilly horizon, and observing how these coincide with the seasons.

The next stage might arrive when a more precise calendar was needed for some reason. Then might a systematic attempt be made to define the exact length of the year in the way described earlier - by using one of the horizon features (or a distant artificial mark) to count the days in the year. Such investigations would inevitably lead on to the construction of a reasonably accurate solar calendar, and probably to interest in, and the defining of, the directions of the four cardinal points. South would be suggested by the position of the noonday sun and north perhaps by the circling of the stars round the celestial pole; the observation of the moving shadow cast by a post or a standing stone could easily lead to the exact definition of the north-south line. As soon as the equinoxes were defined by subdividing the year from the solstices, ideas on the importance of east and west would surely follow.

When this level of calendrical competence had been achieved it is easy to see how ancient buildings could have been erected so that their axes were orientated towards one of the calendrically important astronomical phenomena, or towards a cardinal point. Such a building need not itself have performed any precise calendrical function but would be oriented for ceremonial or religious reasons. The great Newgrange passage grave in Ireland is a classic example (Patrick 1974). Its builders must already have had the level of competence being described, and the orientation of the passage towards the sunrise on the shortest day would have provided a dramatic midwinter spectacle for those inside the temple/tomb; it would have been quite unable to define midwinter's day by itself but must have relied on an already-existing calendar.

Moreover the whole mound seems to have an axis of symmetry pointing in the same direction and along which the passage was built. All this activity, and especially the development in middle and late Neolithic times of really large and architecturally elaborate structures, surely suggests the existence by then of a class of priests and wise men of some kind; a society consisting mainly of farmers and herdsmen seems unlikely to have gone to all that trouble.

The third and highest level of Neolithic observing activity must certainly indicate such a class; indeed it could be argued that the existence of phenomena indicating the second level of activity must in fact imply the third. This stage would have involved the realisation that the observer's position (the backsight) can itself be adjusted in relation to a specific peak or notch on the horizon to create a long alignment giving fine measurements of the sun's position, or that of the moon and stars. Such a mental leap - aiming at great precision - is far beyond what is needed for a simple agricultural calendar, and must surely imply full-time, professional concern with such problems.

Having made this leap it would then have been necessary to spend many years first searching for suitable sites and then patiently observing to establish the exact position of the required backsights (usually marked by a standing stone). While not proving the existence of the very advanced lunar observatories claimed by Thom, the demonstrable existence of accurate solar alignments would make it more likely that the moon was watched systematically and that a more developed type of Neolithic society existed than the primitive rural ones usually envisaged by archaeologists (Hawkes 1967; Thorpe 1981).

The most puzzling aspect of all this as far as Neolithic Britain is concerned would seem to be why all this had to be done, and in such detail, after the colonisation of these islands by the first farmers. These had a long history of gradual movement over several millennia from the Neolithic heartland in the Near East during which time a calendar was surely developed. If the builders of the megalithic collective tombs of the early Neolithic period of Atlantic Europe were an élite group of priests and wise men, as the writer is still inclined to believe (MacKie 1977a; 1977b), this problem becomes even more acute.

The Brainport Bay Site

The existence of a third site where Thom's long alignment hypothesis could be tested was made known to the writer in 1976 by Col. P.F. Gladwin of Minard on Loch Fyne in Argyllshire (Gladwin 1978; 1985). As explained earlier, there were gaps in the evidence from the first two investigated - Kintraw and Cultoon - which it was hoped to fill. In particular what was needed was clear evidence for the artificiality of a supposed alignment, and of its primary purpose as a calendrical marker, as well as unequivocal dating evidence.

The layout of the site has been described in detail elsewhere (MacKie 1981: 128-37; MacKie, Gladwin & Roy 1986). In essence it is now seen to consist of two areas, namely the main linear site on the low ground next to the bay and the fallen Oak Bank stone, with associated rock carvings, on higher ground about 240 m to the north-west. There is also a cairn and an earthen bank at the south-western end of the main linear site. The various parts of the latter form a dramatic 'rifle barrel' sighting device, comprising a small standing stone seen through a notch arranged in the main outcrop, which points directly at the midsummer sunrise (MacKie 1981: Plate 3.5; MacKie, Gladwin & Roy 1986: Plate 1). Previous work suggested that this could have been an important ceremonial site for sun worship, and it raised the possibility that it might also have been used for something more sophisticated calendrically (MacKie 1981).

Problems for solution

The elements forming the main linear site at Brainport Bay can be seen to be largely artificial, and they seem well explained either as a deliberately-arranged set of oriented structures or as an accurate calendrical alignment aimed at the midsummer sunrise; it seems difficult in particular to doubt that the long stones were planned to fit into the sockets to form an unique sighting device. However, although flint implements of probable Neolithic date had been found on the old ground surface in most parts of the alignment, by 1982 the date of the paved areas and other artificial features was still not clear. Two C14 dates, in the second and fifth centuries AD were obtained for two separate deposits of charcoal on the old ground surface near the front (north-eastern) stone socket, but another much later one in the tenth century came from charcoal in a pit apparently dug through the paving on top of the main outcrop. Another date in the fourth century AD came from charcoal excavated from a trench full of ash and carbonised material next to one of the vertical rock faces on top of the Main Outcrop (the 'fire trench') (Table 8.1).

If the original construction could be accurately dated to late Neolithic times - presumably somewhere in the late third or early second millennia BC (Megaw & Simpson 1979: 130 ff.) - and if their purpose as an accurate calendrical alignment could be proved beyond reasonable doubt with evidence from the site itself, then the Thom alignment hypothesis would be strongly supported. If they were dated suitably early but could not be shown to be an accurate alignment, they could only be interpreted as a site designed for ceremonies at the summer solstice.

The indicated mountain foresights posed another problem. Neither has the kind of slope seen from sites claimed as classic solar obervatories like Ballochroy in Kintyre (Thom 1967: 151-5 & Fig. 12.2), parallel to the sun's diurnal movement and against which the upper edge of its disc can in theory be precisely marked (Thom 1971: 14, Fig. 1.1). Indeed in the third millennium BC the sun would have risen a little to the left of the notch where there is nothing to mark the spot (MacKie, Gladwin & Roy 1986: Fig. 5; MacKie 1981: Fig. 3.8). This notch has a declination of +23° 22', or 23° 06' for the sun rising with its upper limb exactly in it.

Despite this the date of the solstice could have been determined exactly by using an indirect method (Hoyle 1966; MacKie, Gladwin & Roy 1986). In about 1800 BC the sun's upper edge would have risen in the notch on two occasions 32 days apart and the day of the solstice could easily have been discovered by halving this interval. However since it is possible to use any notch to the right of the midsummer rising position in this way, such an indirect solstitial alignment cannot be regarded as convincing unless supported by independent evidence. Ideally this would take the form of another clear calendrical line nearby, or of clear archaeological indications from the site itself that a midsummer alignment, as opposed to an orientation, was intended. However, it is hard to imagine what form the latter could take here.

On the other hand the midsummer dawn over the notched peaks still provides a dramatic spectacle, and the design of the site - particularly the arrangement of the back platform - surely means that ceremonial activity at the summer solstice for a large group was also important here - perhaps indeed of primary importance. Numerous broken quartz pebbles were found on all three parts of the site and may imply the same thing (Gladwin 1985: 28), as also may the evidence cited of the continued use of the site right through to the tenth century (Table 8.1).

Finally, it was hoped that knowledge of the date and purpose of the adjacent features - standing stones, earth bank and cairn - would help with the interpretation of the main site. In particular, if convincing subsidiary astronomical alignments could be identified, the calendrical interpretation of the main alignment (as an indirect but potentially accurate alignment for the determination of the date of the summer solstice), only moderately plausible from the evidence available thus far, would be much strengthened. For all these reasons the writer has carried out, with the encouragement and full co-operation of Colonel Gladwin, planning and surveying at the site every year since 1977 as well as two spells of excavation in 1982 and 1983, with the results now described. Every new discovery by surface fieldwork at the site has, however, been made by Colonel Gladwin.

New results at Brainport Bay

The main linear site

The 1982 excavations quickly established that the cairn west of the main alignment was modern; it lies against an old estate wall and on top of surface rubbish which itself lies on a soil layer into which this wall had been set. The earth bank proved to be older but still relatively recent; it was composed of the same brown soil which forms the topsoil over all the higher ground south-west of the main site and which overlies a quite different, sandy soil. This buried ground surface was also found just behind the back platform, and both there and under the bank there were flint flakes lying on and just within it. This old soil, which was evidently formed in climatic conditions quite different from modern ones, was thus the ground surface at a time when flint was commonly used - that is down to late Neolithic times or even into the

Bronze Age. Since freshly-struck flints were lying on its surface, it may be that the later brown soil began to accumulate not long after they were deposited.

Though previous excavations had removed most of the topsoil from the main outcrop, it was possible in June 1983 to cut a fresh section across the revetted north-western edge of the large semicircular terrace (in squares Q39 and R39 - MacKie, Gladwin & Roy 1985: Fig. 2; 1986: Fig. 4); the stones of the edge could then be seen clearly to lie on the buried sandy soil and under the brown topsoil. The two terraces immediately south-west of the rock notch have thus been shown to be stratigraphically earlier than the formation of the modern brown topsoil and very probably to be contemporary with the flint-working horizon. By inference all the other primary artificial features on the Main Outcrop ought to be of the same age, and they should therefore be early enough to be contemporary with other standing stone sites in Scotland. In February 1984 a radiocarbon date in the eleventh century bc was obtained for charcoal lying with flints on the old ground surface under the Bank (Table 8.1); it is probably equivalent to a calendar year time-span centred on about 1520-1590 bc (Ralph, Michael & Han 1973).

On the other hand excavations around the bases of the 'observation boulders' showed clearly that these were sunk solidly into water-laid deposits - presumably those left by a prehistoric raised sea; the rear boulder seems to have split off the front one at that early time and fallen backwards. The rock is evidently jammed in a small cleft or slot of some kind in the bedrock, which all around the boulders is fairly level and only about 20 cm below the modern turf. Thus, contrary to earlier impressions (MacKie 1981: 136-7 & Plate 3.6)

Table 8.1. *C14 dates from Brainport Bay*

Lab. no.	Date	Origin
GU 1705	3010 ± 80 (1060 bc)	Charcoal on old ground surface under Bank
GU 1434	1815 ± 55 (ad 135)	Charcoal on paved area on Main Outcrop, just in front of front stone socket
GU 1703	1615 ± 55 (ad 335)	Charcoal from 'fire trench' through paving on Main Outcrop
GU 1595	1510 ± 60 (ad 440)	Charcoal in ash on old ground surface on Main Outcrop, close to north-eastern stone socket
GU 1000	976 ± 74 (ad 974)	Charcoal from an old pit dug through paving on top of the Main Outcrop

the boulders do not seem to have been moved by man at all, though the 'paving' between them may be partly artificial. The buried sandy soil was not found on the surrounding bedrock here, and evidently only existed locally on the lower terraces of the Main Outcrop nearby. Presumably these were raised just above high-water mark when the raised post-glacial sea washed around the Main Outcrop and the two boulders.

The alignments at the Oak Bank stone

The large prone stone on Oak Bank seemed to be the most promising outlier for investigation, since when upright it would have been plainly visible against the sky from the Main Outcrop. Direct observation of the setting sun on 21 June 1981 showed that this set well to the left of the stone as seen from the large terrace on the Main Outcrop. One of the observers then moved rapidly north-eastwards until the vanishing sun appeared to be more or less over the stone's position, and at this spot was found a small, low semi-circular platform of drystone masonry - unnoted hitherto - built against a low rock outcrop.

The presence of many trees on Oak Bank made observation of the exact position of the sunset difficult. At first Oak Bank seemed to be a good example of the discovery of an artificial backsight platform in the place predicted by a specific astronomical alignment hypothesis. This hypothesis was that the Oak Bank stone was set up to act as a foresight for midsummer sunset as seen from the large terrace on the main linear site. However, excavation of the North Platform in 1982 showed it to be relatively modern and probably to be some kind of kiln; ash, charcoal and iron fragments were recovered from the small square chamber inside. This particular alignment hypothesis was thus decisively disproved - perhaps not surprisingly in retrospect as it does seem rather pointless to define both sunrise and sunset on the same day at the same site.

It then seemed appropriate to test the hypothesis that the Oak Bank stone itself had been not an artificial foresight but a backsight for an entirely separate alignment of its own. Colonel Gladwin had previously discovered two unusual carvings on natural rock outcrops a few metres south-east of the fallen stone. Each, though much worn, clearly consists of a pecked cup-mark of standard early Bronze Age type but with a straight, shallow, pecked groove running through it. The groove of no. 1 is about 60 cm long and has an azimuth of 130°/310°; in one direction it points back directly at the Main Outcrop below. Carving no. 2 has a groove about 40 cm long lying along an azimuth of about 81°/261°, some 2° more than the line formed by the two cup-marks themselves (MacKie, Gladwin & Roy 1985: Fig. 3; 1986: Fig. 7); in the east this groove points to a level, featureless horizon across Loch Fyne.

Because of the fir trees little of the western and north-western horizons could be seen from the cup-marks, so in June 1983 the two azimuths were transferred 20 m on to the base of the fallen Oak Bank stone from where parts of the skyline could just be observed; since then the fir trees have grown up and nothing can now be seen. The groove of no. 1 was then found to point at a smooth hilltop, well to the right of the midsummer sunset position (MacKie, Gladwin & Roy 1986: Fig. 6). However the groove of no. 2 proved to point almost exactly at a conspicuous V-shaped notch - the only such in a long stretch of fairly level skyline about 1 km away (*ibid.*). In September enough vegetation was cleared to make it possible to see the western horizon from cup-and-line mark no. 2. Laying a straight metal tube in the groove and chalking along its sides showed that the pecked line does indeed point straight to the notch (Fig. 8.1).

Sets of observations of the sun were made in August 1983 to determine the azimuth and declination of the base of this notch. When reduced they give an azimuth of 260° 32′.5 and a declination of +0° 8′.1, almost exactly at the equinox; another computation from the same data gives a declination of +0° 9′.2 (MacKie & Roy 1985). The declination of the true base of the notch is perhaps 4′ less if one allows for the trees on the left slope, some of which are at least 40 ft high. This declination of *c*.+0°.1 is thus within the range of the equinoctial alignments defined by Thom, i.e. from -0°.5 to +1°.0 (Thom 1967: 113, Fig. 9.2).

Subsequently Colonel Gladwin, experimenting with other possible calendrical alignments from the Oak Bank stone, discovered that there is another clearly-defined cleft in the horizon about a mile away in the south-west, approximately in the midwinter sunset position; again it is the only such conspicuous notch in several degrees of uneven horizon. Because of the trees it is quite impossible to see this south-western horizon from the fallen stone or from the rock carvings, but a line from notch to stone was laid out through the wood with the help of a magnetic compass. A photograph was taken at sunset on 23 Dec. 1983 from the point where this line emerged from the wood and showed the sun setting exactly in this notch with its lower limb about level with the base.

Impressive though this phenomenon was, and hard to dismiss as pure coincidence, there was at that time no artificial direction indicator on Oak Bank to show that a south-westerly alignment was planned by the erectors of the standing stone, or that this second notch was being utilised as a calendar

Fig. 8.1. Brainport Bay, Oak Bank stone: the equinoctial alignment, formed by a notch on the western horizon and cup-and-line mark no. 2 (E.W. MacKie).

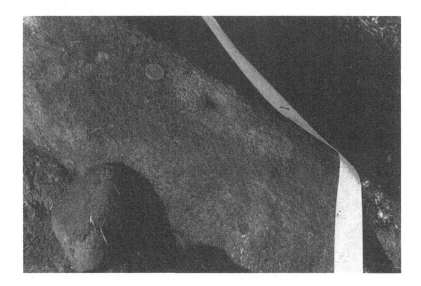

Fig. 8.2. Brainport Bay: New cup-mark on midwinter sunset alignment (E.W. MacKie).

marker for the shortest day. It was therefore not mentioned in the 1985 and 1986 accounts of the site (MacKie, Gladwin & Roy 1985; 1986). However in February 1986 Colonel Gladwin, while searching along the south-western line from the stone through the wood (previously marked with a tape), discovered a single cup-mark pecked on to a low outcrop on the line (Fig. 8.2).

Subsequent measurements were made with a theodolite positioned at the edge of the wood and as closely as possible on the projection of the line from the base of the fallen stone (just visible through the tree trunks) to the cup-mark; assuming this line to have an azimuth of 180°, the azimuth of the south-western notch was then found to be 179° 25′. The line through the wood from stone to theodolite is 63.3 m long and descends 5.55 m towards the south-west, a fact which could explain the discrepancy in azimuth of 35′. However, the south-western notch is certainly clearly indicated from the stone by an outlying cup-mark, and further measurements showed that the base of the notch as it would be seen from the stone has a declination of -23° 23′. Allowing for refraction the declination of the sun setting with its upper limb at the base of the notch would be -23° 35′, giving a figure for the centre of the disc of -23° 51′. This is very close to the position indicated by the megalithic alignments detected by Thom, of about -23° 54′, and may well be exactly at that figure if one allows for the trees on the left side of the notch (Fig. 8.3).

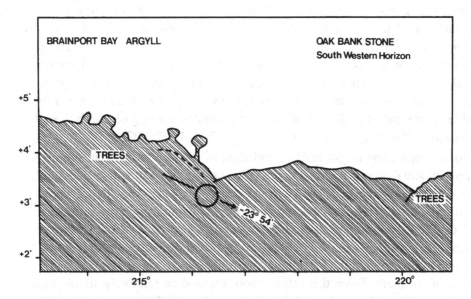

Fig. 8.3. The south-west notch seen from the Oak Bank stone. The horizon was drawn from the edge of the wood but the position of the sun is shown as it would have appeared from beside the stone.

Discussion of the Oak Bank alignments

The discovery that there are two exceptionally clearly and unambiguously marked calendrical alignments at the Oak Bank stone appears to have transformed the interpretation of the whole Brainport Bay site. It is important that both notches were unnoticed at first because of the dense fir plantation, and were only sought as tests, first for the modified astronomical alignment hypothesis as applied to the Oak Bank stone and the nearby cup-and-line carvings, and then specifically to check whether a midwinter line might also be present there. It is also of interest that the ridge containing the equinox notch is called 'Siaradh Druim' in Gaelic (Ordnance Survey 6 in map (2nd edn., 1899), sheet CL 16); this means 'the western ridge' and it can only be so regarded from the region of Brainport Bay (it is difficult to observe sunset positions from the main site because of the low ground). It is suggested that the reality of these two alignments can thus be regarded as proved beyond reasonable doubt, and this has important implications.

How could the equinoctial alignment have been established? Clearly not directly since in the spring and autumn the sun's real position changes rapidly in the same direction - at a rate of 24′ of declination (or, at this latitude, 43′ of azimuth) in 24 hours (HMSO 1983) - so there is nothing in the sky then to suggest any important calendar date. Clearly the concept of a calendrical equinox can only arise after the length of the year has been established and the dates of the solstices fixed, perhaps with an accurate direct alignment. However, as noted earlier, it is much easier to set up any solar alignment first, and to fix the solstitial dates from it by the simple counting of the days in the two parts of the year so defined (Thom 1954: 396; 1967: 107).

For example, one might mark a backsight in relation to a notch to indicate sunset on any day, say 10 April in our calendar, and therefore also (inevitably) on about 1 September. It would immediately be clear that there were on average 144 days from one marked sunset to the next through the summer and 221 or 222 between them through the winter. A very few years of observing would also reveal that the year contained just over 365 days and that an extra day had to be added to the calendar every four years to keep it correct with the sunset at the mark. From this information it would be relatively simple to fix the dates of the solstices by halving these totals, and then to define the dates midway between them - the equinoxes - by counting the days of the two intervals between the solstices and halving them. Indeed at this site the equinoctial line - being shorter, much easier to use and not susceptible to obsolescence over the centuries - might well have been established first and found to be more useful in the long term.

The equinox notch itself shows another interesting feature (MacKie, Gladwin & Roy 1985: Fig. 4; 1986: Fig. 8). One of the more convincing aspects of the Thom solar calendar hypothesis is the way the consistent 'off-setting' of the equinoctial markers by about +0°.5 in declination was explained (Thom 1967: Ch. 9, Fig. 9.2). Because of the eccentricity of the Earth's orbit the number of days between the spring and autumn astronomical equinoxes - defined as the times at which the sun reaches declination 0° - is actually greater than that between those of autumn and spring, on average 187 as opposed to 181. Thus prehistoric observers attempting to set up equinoctial alignments by counting the days between sunrises or sets at any alignment and then dividing would not establish horizon marks pointing at the true, astronomical equinoxes. Inevitably they would mark times slightly after it in the spring and before it in the autumn - compensating unwittingly for the greater number of days in the summer half of the year. These offset 'megalithic equinoxes' are exactly what the histogram of declinations seems to be showing (Thom 1967: Fig. 9.2).

The other point to remember is that, because of the rapid change of the sun's real position in the spring and the autumn, in successive half years it will always rise and set at slightly different positions on the days defined as the equinoxes; its positions at these times can vary up to 12′ of declination either way and thus over a 'zone of fluctuation' 24′ wide (MacKie, Gladwin & Roy 1985: Fig. 4; 1986: Fig. 8). Thus there can be no such thing as an 'accurate' equinoctial alignment because there is no single place on the horizon which marks it.

The Oak Bank western alignment, however, points not at $\delta = +0°.5$ but almost exactly at $\delta = 0°$ and, making the usual assumption that the upper limb of the sun was observed against the notch (to cut down the glare), the 'zone of fluctuation' might be thought to mean that this alignment was indeed aimed at the true astronomical equinox. In this case the first evening in spring, and the last in the autumn, that the sun's rim appeared in the notch would mark the date concerned. However knowledge of the true equinox depends on understanding that the Earth is suspended in space and that its axis is tilted. Such knowledge was apparently present in the east Mediterranean world at about 2000 BC; the description of the celestial sphere by the third century BC poet Aratus makes this fairly clear and he seems to have been using information dating from a much earlier time (Ovenden 1966: 10-12).

> Now some stars in numbers, going divers ways,
> Are ever drawn by heaven and ever joined;
> Yet never does the axis change the least,
> Aye fixed even as it is; and has in midst
> The earth in equipoise, and carries round the sky.
> This axis forms on either side a pole;
> The one we see but not the opposite.

(Mair 1921: 383)

Yet it seems safer to assume that this was not known in contemporary late Neolithic Argyllshire and that the alignment was aimed - after much trial and error by counting the days - at the 'megalithic' equinox rather than at the astronomical one. It could have worked in this way if the whole sun was observed in the notch instead of only the upper edge, and the same 'zone of fluctuation' would apply.

The interpretation of the south-west alignment at Oak Bank is more straightforward. Since the horizon is only 1.8 km away, the notch could not have been used to define the exact day of the winter solstice. To make the upper edge of the disc disappear exactly at the base of the notch on 20 December, and again on 21 December, the observer's eye would only be able to move sideways about 18 cm (6 in) between one evening and the next (MacKie 1977a: 75, Table 6b); this is much too short a distance to pick out midwinter with an accuracy of better than a few days. This line must therefore have been one which might be called a 'reminder' instrument - a device to give warning of the approach of the shortest day and to show when it had arrived, more or less. It seems that an accurate calendar must already have existed when it was set up.

There are several interesting general points about the Oak Bank alignments. Firstly, they are clearly working instruments; there are no signs so far of any ceremonial elaboration to these simplest of indicated lines and the situation of the Oak Bank stone - on a ridge well away from the main linear site - seems to confirm this. Secondly, the fact that one standing stone serves as the backsight for two distinct and indicated solar alignments not only provides a strong argument against this having come about by chance, but also must have been quite a difficult thing to arrange on the ground.

To achieve this the organisers would first have had to have found a promising position from which two conspicuous notches - both preferably by themselves - could be seen, one more or less in the west and the other towards the south-west. Presumably the equinoctial line was established first, for the reasons and in the manner already suggested, but in this case the backsight position would have to have been laid out as a line on the ground over a considerable distance, so that many alternative final positions were available. A long line of people could have done this if each person adjusted himself individually to the same solar phenomenon in the notch in Siaradh Druim and left a stake in the ground at his final position (Thom 1971: Fig. 1.1). Three months later at midwinter (the date of which, as already explained, must already have been known) it should have been comparatively easy for one or two people to move along this line of markers for the equinoctial sunset while facing the south-western notch and thus determine the one backsight position which would do for both. Considering the unevenness of the ground on top of Oak Bank, and the slopes of varying steepness nearby, this was no mean achievement.

Conclusions

Survivals of the old solar calendar

Our knowledge of the nature of the calendar in use in late Neolithic times derives from the work of Thom. It is firmly supported by the results from Brainport Bay, which are based entirely on autonomous archaeological evidence. It is therefore of considerable interest that the eight most important subdivisions of the year so detected still survive. Alexander Thom seems to have been the first to notice this but he did not go into any details (1967: 107, 109). Though a general one, it is a topic well worth pursuing here.

Our modern calendar is clearly a combination of two separate ones - a solar year but with subdivisions derived from the lunar calendar. The length of the year, and the need for a Leap Year, are obviously solar, and necessarily so if a calendar linked with the seasons is needed. The modern attempts to perfect such a link start with Julius Caesar. It also seems clear that important dates in the ancient solar calendar - like midwinter, and the spring and autumn equinoxes - have been transformed into Christian festivals like Christmas, Lady Day and Michaelmas.

However, the subdivisions are lunar; the very word 'month' indicates this although these no longer, with one exception, correspond to the lunar cycle of 29 days. The weeks are normally thought to derive ultimately from the ancient Jewish calendar (in which none of the days except the Sabbath was named) but there was also apparently an independent western European tradition of an astrological week of seven days. These were named after the seven planets (sun and moon included), originally with the Roman names but acquiring the Germanic equivalents in northern Europe (see *Oxford English Dictionary*, 1971 edition). However, it is surely possible that the week originates very far back in the seven day periods clearly signalled by the four quarters of the waxing and waning moon, as a logical subdivision of the original lunar month.

The study of prehistoric calendrical sites by Thom has shown convincingly that a solar calendar was already in use by about 2000 BC in which the year was subdivided quite differently (Thom 1967: 109 ff.). It was certainly divided into four quarters, and very probably further subdivided into 16 solar 'months' of from 22 to 24 days; it may even perhaps have been split into 32 solar 'weeks' of eleven or twelve days. It is interesting that the eighths of the solar year - which must, like the equinoxes, have been calculated by subdividing the intervals between the solstices - still survive in the modern British 'quarter days'. These are four days fixed by custom to mark quarters of the year and on which tenancies of houses begin and end, payment of rents are due and so on.

The English quarter days correspond in a straightforward manner with the solstices and equinoxes, with minor adjustments to fit with Christian festivals. They are Lady Day (25 March), Midsummer Day (24 June, also the day of St. John the Baptist), Michaelmas (29 September) and Christmas Day (25 December). Lady Day commemorates the Annunciation of Our Lady, the Virgin Mary and, until 1752 and the introduction of the Gregorian calendar, was the legal beginning of the year. Michaelmas is the Festival of St. Michael and all Angels. It seems plain that these two festivals as well as Christmas are Christian celebrations which were originally grafted on to important dates in the old pagan solar calendar, on the well-understood principle that it was easier to do this than to get the newly converted to give up their old feast and holy days.

It is of considerable interest that the quarter days in Scotland are different from the English ones and correspond well with the eighths of the solar year which fall half way between the solstices and equinoxes. These are Candlemas (2 February), Whitsunday (15 May), Lammas (1 August) and Martinmas (11 November). Using the sequence of solar 'months' deduced by Thom from an analysis of the standing stone alignments (Thom 1967: 113, Fig. 9.2) - and using the 'megalithic' rather than the true equinoxes - one obtains the following 16 'month' dates from spring 1985 (the eighths in italics): *22 March*, 14 April, *7 May*, 31 May, *23 June*, 16 July, *8 August*, 30 August, *21 September*, 13 October, *4 November*, 27 November, *20 December*, 11 January, *4 February*, 27 February and *22 March*. The correspondence with the English and Scottish quarter days is close.

The fact that the quarter days are still important legally, and that until recently they marked important seasonal and other festivities, suggests that they were inherited from a very old solar calendar with well-established feast days and religious ceremonies on the eight major subdivisions of the year. In ancient times the intermediate eighths were doubtless useful as seasonal indicators, and could be regarded as marking the beginning of spring (and the time to sow), of summer, autumn (and the time for harvest) and winter respectively. As with the English ones, three of the Scottish Quarter Days are now Christianised, Candlemas formerly being the Feast of Purification of the Virgin Mary (now the Presentation of Our Lord in the Temple), Lammas Day the Feast of St. Peter ad Vincula and Martinmas the Feast of St. Martin. Whitsunday, as described below, was different.

Candlemas might be said to mark the formal end of the first half of winter, and an old Scottish rhyme seems to contain a memory of its seasonal importance:

> If Candlemas Day be dry and fair,
> The half o' winter's come and mair;
> If Candlemas Day be wet and foul,
> The half o' winter was gone at Youl.

(Evans 1977)

There used to be a custom in Scotland called the 'Candlemas bleeze' (blaze) in which, in the latter part of the day, bonfires were made with any available furze.

Whitsunday (with the stress on 'whit') was in Scotland a quarter day without Christian religious significance. Whit Sunday (with the stress on 'sun'), or 'white Sunday', is the Christian festival, a fixed interval after Easter; the name is reputed to have originated in the white robes worn by those baptised at Pentecost (the seventh Sunday after the Crucifixion when the Holy Spirit descended to the disciples). The two different versions of the name seem to confirm that there are two quite different festivals here - the Christian one linked with the solemnity of the Crucifixion and the old pagan fertility rites of spring held at the beginning of May. A clear link with the May Day festivals is apparent in early English descriptions of 'Whitsuntide' with its maypoles and Morris dancing, and the 'Whitsun ale', a parish festival involving feasting, sports and general merrymaking (see *Oxford English Dictionary*, 1971 edition). The fertility rites survived until quite recently in various places in Scotland in egg-rolling ceremonies (Hole 1978: 94-95).

Lammas is most clearly a seasonal festival; the word comes from Old English 'hlaf maesse', or 'loaf mass', the festival of First Fruits when loaves baked from the first sheaves of the harvested wheat were consecrated, following a practice as old as that of the Hebrews' offering to Jehovah in Biblical times, and doubtless that of the Neolithic farmers of Europe.

Martinmas too had a strong seasonal element. In England fairs used to be held, servants were hired cattle were slaughtered and the meat salted to keep over winter; this 'Martinmas beef' recalls a time when conditions were more primitive and hard. All these 'quarter day' dates, with their old customs and added Christian celebrations, seem arbitrary in terms of the modern Gregorian calendar but fall close to the primary divisions of the solar calendar deduced from the megalithic alignments of four thousand years ago.

Likewise the important feast days of the ancient Celtic world also preserve the 'eighth' divisions of the prehistoric solar calendar. The old Celtic year was customarily divided into two - a winter half beginning on 1 November, the Feast of Samhain, and a summer half beginning on 1 May, Beltane. These were further divided by the quarter days of Imbolg on 1 February and Lughnasadh on 1 August (MacCana 1970). It is hard not to conclude that the modern Scottish Quarter Days are directly derived from these important ancient Celtic dates.

Imbolg on 1 February was the Irish pagan spring festival, and later became St. Brighid's day; it is evidently the equivalent of Candlemas and doubtless has the same origin in the ancient solar eighth-of-the-year date. It has been argued that Brighid herself is the principal goddess of the Celts only very thinly disguised, probably the equivalent of Minerva.

Beltane (Bealtuinne) fell on 1 May and has undertones of sun worship; the second part of the word means 'fire' while 'bel' could mean 'shining' or 'brilliant'. At that time the 'bel' fires were lit on hilltops and cattle were driven between the flames, either to protect them from disease or to prepare them for sacrifice. A connection with Apollo is possible.

The Irish harvest festival of Lughnasadh fell on 1 August and was dedicated to the god Lugh (Lud, Lugus), the 'shining one'; this is obviously the same as the Lammas festival. Lyon in southern France is named after Lugus, the Gaulish Mercury, and when Augustus Caesar chose it as the capital of Gaul, the annual festival to himself took place on August 1st, presumably fitting in with the most important local religious event.

Lastly, Samhain fell on 1 November and was the early winter festival approximately equivalent to Martinmas but closer to the 'megalithic' eighth date. However, in ancient Ireland this festival had a tremendous mythical significance, being the time of the year when the world of the spirits was closest, when the barriers between the natural and the supernatural were temporarily removed. Many important events in early Irish mythology took place on that date; famous kings and heroes died at that time, and unbridled carousing took place. Hallowe'en on 31 October, with its echoes of witches and the supernatural, is doubtless a modern survival of Samhain.

It is customary to doubt that sun worship was ever of much importance among the ancient Irish, the only significant hint of it being a remark of St. Patrick's (MacCana 1970: 32). However in view of the likelihood of the four important Celtic festivals originating in the late Neolithic calendrical subdivisions, and of the elaborate solar temples and observatories now known to have existed at that early time, the question should be reconsidered. It is accepted that the early Christian missionaries in Ireland converted the important pagan Celtic festivals to Christian ones. It now seems worth considering that the prehistoric Celts did the same to the even older solar religion they found there, and that four of the most prominent Neolithic festivals have survived in this way down to modern times.

Though no modern trace of the sixteenths and thirty-seconds of the solar year seems to survive now, it may be worth recalling the cache of 33 rounded quartz pebbles of similar size found on the main site at Brainport Bay (Gladwin 1985: 14). It may not be too fanciful to suppose that such a cache on an important ancient solar calendar site was used for telling off the solar 'weeks' of eleven or twelve days.

Discussion of the main linear site

The striking discovery of a double alignment on Oak Bank gives powerful independent support for the hypothesis that the main linear site was designed for the ceremonial observation of the summer solstice. This is particularly valuable as other data show quite clearly that the main site must originally have been a chance discovery by prehistoric man. The main outcrop is obviously natural, as is one side of the rock notch. The fact that there is a clear view from it up Loch Fyne to two distant peaks close to the midsummer sunrise position must also be due purely to chance and - even more extraordinary - the 'observation boulders' also seem to have arrived in their ideal sun-watching position by chance. Without the support provided by the Oak Bank alignments - sought and found because of deductions made after the main site had been interpreted - it might be argued that these chance features outweigh the clear signs of Neolithic or Bronze Age activity, and of the adaptation at that remote period of these natural features into a dramatic solstitial alignment.

If the position of the main site in the Bay was discovered rather than chosen, this would explain the less-than-ideal shape of the foresight mountains. Although precision could have been obtained with the indirect 'splitting the difference' method mentioned earlier, it is clear that the summer solstice alignment need not have been primarily for calendrical work. The equinox line could have served for this, and may indeed have been established first. The chance discovery of what must have seemed a miraculously suitable site might also explain the clear signs of ceremonial activity suggested by the design of the back platform and also, perhaps, by the scatters of broken quartz. The fairly clear evidence of late use of some kind - down to the tenth century at least - surely suggests the same.

Yet there is no getting away from the remarkable fact that the most prominent feature of this ceremonial site - adapted from fortuitously present natural rocks - is the 'rifle barrel' sighting device aimed at the only two really distant peaks visible. Why should the prehistoric builders have picked on this particular piece of ground from which Beinn Oss and Beinn Dubhchraig can just be seen, in the midsummer sunrise position and nearly 30 miles away? As explained, the peaks are quite unsuitable for the foresight of a precise solstitial alignment, and a variety of other and closer marks would do for an offset instrument.

There seems to be only one plausible conclusion. The arrangement closely resembles a long alignment like Kintraw but cannot be such a one because of subtle differences which are not obvious except to specialists (ancient and modern). It seems to the writer that the whole site only makes sense first if accurate long alignments already existed to be imitated in this way, and if they were thought to be important, and second if they were so important, and their operators had such prestige, that the discovery of the 'natural observatory' in Brainport Bay must have seemed sufficiently miraculous as to have endowed the site with a religious aura which persisted over as much as 3000 years, perhaps until the coming of Christianity.

In other words the writer would argue that Brainport Bay sheds a flood of light on the ceremonial practices of late Neolithic and Bronze Age Britain, shows with as clear evidence as is ever likely to be found in archaeology that short, functional calendrical alignments were in use then, and demonstrates to a high degree of probability that very long, potentially accurate alignments had also been devised. Brainport thus seems to be giving strong support to the essentials of what Alexander Thom suggested for many years. Finally the site also surely illustrates how the religion of the time was that blend of dramatic ceremonial and spectacle for the many, and esoteric knowledge for the few which seems eminently plausible in a British prehistoric context.

To assess what this means in terms of the social organisation of the time one must of course be more speculative. The writer has previously argued that the presence of accurate observing instruments in late Neolithic times would be best explained if there existed then a full-time class of specialist priests and wise men like the later Druids of Iron Age times (MacKie 1977a: 226 ff.), and the Brainport Bay evidence might be said to reinforce this hypothesis strongly. This in turn would make the extension of the hypothesis to the whole of Atlantic Europe in early Neolithic times more plausible (MacKie 1977b), although it could equally be argued that the Neolithic priestly class arose in Britain and was not imported. However these questions, though of fundamental importance, are beyond the scope of this paper.

Acknowledgements

Colonel Peter F. Gladwin first drew my attention to Brainport Bay and, since 1976, has given me constant help and encouragement in pursuing further work there. I have had the benefit of his previous work at the site, of numerous discussions about it and of the many discoveries he has made since 1976. The work there has been a joint effort, and someone living at a distance could not have worked a fraction as effectively without this help on the spot. Professor Archie E. Roy also discussed the site with me, gave advice on the astronomical aspects of the work, and undertook the basic calculations of the declinations of the Oak Bank alignments.

I thank the Forestry Commission (Scotland) and, since 1983 when he aquired the land, Lt. Col. R. Gayre of Gayre and Nigg, of Minard Castle, for ready permission to carry out surveys and excavations on the site. Dr H.A.W. Burl, Dr D.C. Heggie, Mrs Diana Reynolds and Dr J.N.G. Ritchie read early drafts of the chapter and made many useful suggestions. The following assisted me at the site in 1982 and are gratefully thanked: Dr and Mrs J.R. Baker, Mr J. Birdsey, Ms Helen Dyson, Mr M. Davis, Mr H.E. Kelly and Mr. S. White.

9

The stone alignments of Argyll and Mull: a perspective on the statistical approach in archaeoastronomy

CLIVE RUGGLES

Introduction

Alexander Thom was not the first to suggest that certain prehistoric stone structures might have been aligned upon the rising and setting positions of the sun, moon or stars. He *was*, however, the first to back up his conclusions with statistical evidence from many sites taken together - the product of extensive and high-quality fieldwork.

The value of what we can term a 'statistical' approach is easy to perceive. Any individual alignment of apparent astronomical significance could have arisen through factors quite unrelated to astronomy. There needs to be some attempt to demonstrate the intentionality of putative astronomical alignments. One way is through the analysis of a large quantity of data, whereby it can be shown that certain astronomical alignments are significantly more common than would have been expected by chance.

There are nonetheless dangers and drawbacks in applying a statistical approach to the study of alignments at archaeological sites. In this paper we attempt, through the example of a group of sites which has become of increasing interest to the author over some years, to illustrate some of the questions raised by applying statistical methods to a particular set of archaeoastronomical data.

A case study of the application of statistical methods to alignment data

We begin by describing the development, over several years, of a project to assess the possible astronomical significance of a sample of megalithic sites in western Scotland. The region was chosen because it was one in which Thom had a particular interest: a significant proportion of the data contributing to his statistical analyses came from this region, and several of his 'classic' astronomical sites, such as Ballochroy, Kintraw, and Kilmartin (Temple Wood) are located here.

Stage 1: A study of 300 western Scottish megalithic sites

The project in question arose from attempts to reassess Thom's statistical conclusions both with regard to lower-precision calendrical, lunar and stellar alignments (Thom 1967) and higher-precision lunar alignments (Thom 1971; Thom & Thom 1978b; 1980). Factors taken into account were (i) the implicit methodology of data selection; (ii) the archaeological status of the sites involved; (iii) the viability of using the sites for the observations proposed; and (in the case of the high-precision lunar alignments) (iv) the credibility (on purely astronomical and practical grounds) of making the observations proposed in the first place. These critiques, which themselves involved extensive fieldwork, resulted in the clear rejection of Thom's conclusions about high-precision lunar alignments (Ruggles 1981; 1982; 1983). The methodological improvements suggested by reassessing Thom's approach were implemented in a new independent survey of some 300 western Scottish sites (Ruggles 1984a).*

The independent survey was conducted under severe methodological constraints. For example, in selecting sites and structures for consideration, rigorous pre-defined criteria were adhered to. No survey data were reduced until the entire sample had been collected, in order to avoid the possibility of

---

*   This survey concentrated on the possibility of astronomical orientations of lower precision than those envisaged by Thom in his later work, and its results must be considered in conjunction with the earlier papers of this author where the high-precision alignments were comprehensively reassessed. It is, for example, quite misleading of MacKie (1986), in a review of this work, simply to ignore the earlier work and then criticise the survey of 300 sites on the grounds that it did not take into account possible high-precision alignments.

our selection strategy being influenced by the results obtained along the way. The conclusions of the study were as follows.

(1)   Indicated declinations manifested overall trends at three levels of precision. At the lowest level, declinations between about -15° and +15° were strongly avoided. At the second level, there was a marked preference for southern declinations between -31° and -19°, and for northern declinations above +27°. At the most precise level, there was marginal evidence of a preference for six particular declination values to within a precision of one or two degrees: -30°, -25°, -22°.5, +18°, +27° and +33°.

(2)   Certain coherent groups of sites were found to feature predominantly amongst the indications falling in particular 'preferred' declination intervals. These were sites in Mull and mainland Argyll in general, and the three-, four- and five-stone rows in these areas in particular.

(3)   When consideration was limited to the stone rows, pairs and single flat slabs in Mull and mainland Argyll, the overall declination trends noted above became more marked. Of those sites where measurements were obtained, the great majority, including every row of three or more stones and every pair of aligned slabs, were oriented in the south upon a declination between -31° and -19°. Only ten measured sites, non-aligned pairs of menhirs or single slabs, failed to fit the pattern.

The declination range -31° to -19° is of particular interest because it represents, to within a degree or so, the range of possible values of the southerly limit of the moon's monthly motions at different points in the 18.6-year cycle. The construction of deliberate orientations within this range need not have involved nightly observations of the moon in a given month, but could have been achieved simply by observing the rising or setting of the full moon nearest to the summer solstice. If such alignments were set up at arbitrary points in the 18.6-year cycle, then one would expect a scatter of declination values within the interval between about -30° and -19°.5, with more values occurring towards the edges of the range owing to the sinusoidal motion of the actual monthly limit within this range.

A preference for declinations near to -30°, over and above that accountable for by the sinusoidal effect, might indicate a specific interest in the southern major standstill moon, thereby implying that organised observations were undertaken over periods of at least 20 years. Similarly, a preference for declinations near to -19°.5 might indicate the same for the minor standstill moon. However, a preference for declinations near to -24° might indicate that there was also some interest in the winter solstitial sun.

The data available from the original project were insufficient to resolve these issues on purely statistical grounds. However, because the larger project singled out the settings of linear form in Mull and mainland Argyll, we felt we had some objective justification for examining these sites further in the light of the hypothesis of observations of the southern moon. In doing so we could began to relax our strict adherence to rigid, preconceived selection criteria, and to adopt a somewhat more interpretative approach. We still, however, felt it essential to document selection decisions and justify them in detail.

Stage 2: The linear settings of Argyll and Mull

Thus in a fresh project we concentrated upon the southern indications of the linear stone settings in Argyll and Mull - stone rows, aligned pairs, non-aligned pairs and single isolated flat slabs - making 92 sites in all. Our selection criteria could no longer be claimed to be pre-defined, since they were suggested directly by the previous exercise, and much of the data overlapped with that obtained previously. There was, in fact, one major source of fresh data, since the original data set, concentrating on alignments of the sort of precision noted by Thom (1967), had not included 'local' horizons, i.e. those closer than 1 km, where uncertainties due to possible changes in ground level since prehistoric times would be significant. In considering alignments accurate only to a degree or two, however, rather closer skylines could be taken into account. It would, however, be misleading to claim that these extra data constituted a valid 'test' of the lunar hypothesis, since had they not fitted the hypothesis we would have been prepared to suggest that only more distant horizons were used for astronomical purposes.

In May 1985 fieldwork was undertaken in an attempt to obtain as much as possible of the data required under the new selection criteria. A detailed report is given in Ruggles (1985).

The new data from stone rows and aligned pairs with local horizons to the south were found to fit the general pattern that indicated declinations falling between about -31° and -19°. However, the distribution of indicated declinations within this range was far from that to be expected if sites were merely oriented upon the limiting monthly moon at arbitrary points in the 18.6-year cycle. Instead, we found a grouping of indications within a degree or two of -30° and a second, rather wider, grouping centred upon -23°.

At this stage a pattern of some interest, described below, began to emerge amongst the two main geographical concentrations of this type of site: northern Mull and the Kilmartin area of Argyll. (Outside these concentrations the rows and aligned pairs of Argyll and Mull seem to fit a more general pattern of orientation between -30° and -19°.) It should be noted, however, that there was no formal justification for singling out this pattern, since the number of sites concerned was only 15 in total, selected from the larger set on the basis of the pattern in the first place.

Non-aligned pairs were found to fit no pattern at all, and the southern declinations obtained from single slabs, although giving some overall hint of similar preferences to those observed with the rows and aligned pairs, particularly in the Kilmartin area, were much more scattered. This is perhaps hardly surprising in view of the uncertainties inherent in determining an indication from the current disposition of such stones, together with the range of possible purposes for which such stones might have been erected.

Stage 3: The stone rows and aligned pairs in the two main concentrations

Following the analysis of the 92 sites as a whole we went on to explore the Kilmartin and northern Mull groupings in more interpretative manner. The northern Mull sites were revisited in May 1986. The aim now was to look at the sites in their geographical and archaeological context, in an attempt to start to accumulate independent information which might bear upon the astronomical interpretation of the sites. No attempt was made to define research strategies prior to fieldwork. Two resurveys were carried out, in one case because existing site data were found to be in error, and in the other to try to improve upon a difficult survey.

The project will continue. Only at this stage do we begin to feel free to reduce survey data as we go along, and to alter our survey strategy accordingly. It is hoped that during the next season of fieldwork computer reduction of survey data will take place on site, enabling more extensive and interpretative surveys to take place.

In the following section we shall briefly present the data relating to the two main groups of sites, proceeding to our current interpretations of them. Site reference numbers follow Ruggles (1984a), where the reader will find a full list together with cross-references.

The data

*Site descriptions**

The stone rows of Northern Mull

Glengorm (ML1). NM 4347 5715. Three standing stones, all around 2 m in height, in a triangular setting. Two of them (*A & C*), however, have been re-erected in recent times (RCAHMS 1980: no. 105). The remaining stone (*B*) is roughly rectangular, 0.8 m × 0.3 m, aligned NNW-SSE. Stone *A*, to the north, is almost in this alignment. Site plans: RCAHMS (1980: Fig. 44); TTB. Stone nomenclature follows RCAHMS.

Quinish (Mingary) (ML2). NM 4134 5524. A standing stone 2.8 m tall × 0.5 m × 0.4 m (*a*), together with three recumbent stones (*b, c, d*) which represent the remains of a group of five stones noted in the last century (RCAHMS 1980: no. 111). The remaining four stones appear probably to have formed an alignment some 10 m long, and the fifth stone may originally have been in the alignment. Site plans: Ruggles (1981: 188); TTB. Stone nomenclature follows Ruggles.

Balliscate (Tobermory) (ML4). NM 4996 5413. A 5 m-long three-stone alignment. The end stones are 2.5 m and 1.8 m in height; the central one, which has fallen, is 2.8 m long. Site plans: RCAHMS (1980: Fig. 39); TTB.

Maol Mor (Dervaig A) (ML9). NM 4355 5311. A 10 m-long four-stone alignment. Three stones stand, and are all about 2 m tall; the fourth is prone, and 2.4 m long. Site plans: Thom (1966: Fig. 8); RCAHMS (1980: Fig. 41); TTB.

Dervaig N (Dervaig B) (ML10). NM 4390 5202. Two standing stones 2.5 m and 2.4 m tall, and three prostrate stones all about 2.3 m in length, which appear to have formed an 18 m-long five-stone alignment. The site is now in a clearing in thick forest. Some 250 m to the south-east, and also in the alignment, is an erect stone 1.0 m high × 0.6 m × 0.6 m, which is possibly a standing stone. Site plans: Thom (1966: Fig. 7); RCAHMS (1980: Fig. 42); TTB. Photograph: Fig. 9.1.

---

*   In the site descriptions given below, 'TTB' refers to A. Thom, A.S. Thom & A. Burl, *Stone Rows and Standing Stones*, in preparation and to be published by British Archaeological Reports, Oxford.

Fig. 9.1. The five-stone row at Dervaig N, Mull: View from north-west (Clive Ruggles)

Dervaig S (Dervaig C) (ML11). NM 4385 5163. A 15 m-long three-stone alignment. The present stone heights are between 1.0 m and 1.3 m, although two of them appear to be broken off. The RCAHMS (1972: no. 101(3)) note a fourth stone 1.1 m high which has probably been removed from its original position. Site plans: RCAHMS (1980: Fig. 43); TTB.

Ardnacross (ML12). NM 5422 4915. One standing and five prostrate stones, which appear to have formed two parallel three-stone alignments each about 10m long, in association with three kerb-cairns. The standing stone is 2.4 m tall × 1.0 m × 0.5 m, and the prostrate stones are between about 2 m and 3 m long. Site plans: RCAHMS (1980: Fig. 18); TTB.

The stone rows and aligned pairs of the Kilmartin area

Barbreck (AR3). NM 8315 0641. An alignment of two slabs, one 2.5 m tall × 2.0 m × 0.2 m, the other 1.3 m tall × 1.0 m × 0.2 m, some 3 m apart. An isolated slab 2.0 m tall × 1.5 m × 0.3 m, and oriented roughly parallel to the first two, is situated some 20 m away, perpendicular to the alignment. It is surrounded by three small erect slabs, each about 0.6 m tall × 0.5 m × 0.2 m, forming the central parts of three sides of a rectangle roughly 4.5 m × 3.0 m. Site plans: Patrick (1979: S80); TTB.

Salachary (AR6). NM 8405 0403. Three menhirs which appear to have formed an alignment about 4 m long. Two are about 2.5 m tall, one standing and one leaning at about 20° to the horizontal. The third is prostrate, about 3m long.

Carnasserie (AR12). NM 8345 0080. An alignment of two slabs, respectively about 2.4 m tall × 1.2 m × 0.4 m and 2.3 m tall × 1.5 m × 0.4 m, some 3 m apart. There is a cairn (Campbell & Sandeman 1961: no. 70) some 150 m to the SSW. Site plan: TTB.

Kilmartin (Slockavullin; Temple Wood) (AR13). NR 8282 9760. Some 300 m south-east of the Temple Wood stone circle(s) and 300 m south of the Nether Largie round cairn is a group of five standing stones as planned by Thom (1971: 46). The south-western pair (S_4 & S_5) are aligned slabs, both about 2.5 m tall × 0.8 m × 0.3 m. The central stone (S_1) is a slab 2.5 m tall × 0.9 m × 0.2 m, flanked by four small erect slabs, each some 0.5 m tall × 0.5 m × 0.2 m, forming the central parts of the sides of a rectangle some 3 m × 2 m. Between the south-western pair and the central stone are a similar group of three small slabs forming the central parts of three sides of a rectangle about 5 m × 3 m, but without a central monolith. The north-eastern pair of stones (S_2 & S_3) are an alignment of slabs both about 2.5 m tall × 1.1 m × 0.3 m. Also in this alignment, but some 150 m to the NNW, is a smaller standing stone (S_6) 1.5 m tall × 0.8 m × 0.3 m. The buried stump of a further standing stone (300 m to the west) has been reported by Hawkins (*D & E* 1973: 13). Site plans: Thom (1971: 46); Patrick (1979: S80); TTB. Stone nomenclature follows Thom.

Duncracaig (Ballymeanoch) (AR15). NR 8337 9641. A 15 m-long alignment of four slabs (*a, b, c, d*) up to 4 m high; an adjacent and roughly parallel alignment of two slabs (*e, f*) 4 m apart; and the site of a holed stone, recently excavated (Barber 1978). There are two cairns nearby (Campbell & Sandeman 1961: nos. 64 & 95). Site plans: Thom (1971: 52); Barber (1978: 105); TTB.

Dunamuck I (AR28). NR 8471 9290. A 5 m-long three-stone alignment. The end stones are some 2.5 m tall, and the central one, which has fallen, is some 3.5 m long.

Dunamuck II (AR29). NR 8484 9248. An alignment of two slabs some 3.5 m and 2.5 m tall, situated some 6 m apart. Site plan: TTB.

Table 9.1. *Southern indicated azimuths and declinations at the stone rows of northern Mull and the rows and aligned pairs in the Kilmartin area of Argyll*

Column headings

1	Site reference number
2	Site name (and alignment in question)
3	Nature of the linear feature (R = row; AP = aligned pair; S = slab)
4	Horizon distance category (L = less than 1 km; A = 1 km to 3 km; B = 3 km to 5 km; C = over 5 km)
5	Minimum indicated azimuth
6	Maximum indicated azimuth
7	Mean altitude
8	Rising or setting line
9	Minimum indicated declination
10	Maximum indicated declination
11	Reliability of indicated azimuths on archaeological grounds (OK = reliable; U = less reliable, for reason(s) given in text)
12	Quality of survey (A = reliable; B = less reliable (reason(s) given in text); C = calculated)
13	Date of latest survey

1	2	3	4	5	6	7	8	9	10	11	12	13
ML1	Glengorm	R?	C	160	165	1.0	Rise	-31.0	-30.0	U	A	860511
ML2	Quinish	R?	C	166	170	2.0	Rise	-31.5	-31.0	U	A	760819
ML4	Balliscate	R	A	184	186	5.0	Set	-28.5	-28.5	OK	A	760821
ML9	Maol Mor	R	L	161	163	2.0	Rise	-30.0	-29.5	OK	B	860514
ML10	Dervaig N	R	L	149	151	3.5	Rise	-26.0	-25.0	OK	B	850516
ML11	Dervaig S	R	C	156	158	2.0	Rise	-29.0	-29.0	OK	A	760823
ML12	Ardnacross (NW group)	R	A	195	200	6.5	Set	-26.0	-24.5	U	A	760824
ML12	Ardnacross (SE group)	R	A	206	209	7.0	Set	-23.0	-22.0	U	A	760824
AR3	Barbreck	AP	B	186	188	3.0	Set	-30.5	-30.5	OK	A	810527
AR6	Salachary	R	L	176	178	2.0	Rise	-32.0	-31.5	U	A	850512
AR12	Carnasserie	AP	L	163	171	2.0	Rise	-32.0	-30.0	OK	B	850512
AR13	Kilmartin (to SSW)	R	C	200	207	1.0	Set	-31.0	-29.5	OK	B	810525
AR13	Kilmartin (S_2-S_3)	AP	L	149	150	2.5	Rise	-26.5	-25.5	OK	C	
AR13	Kilmartin (S_1)	S	L	142	144	3.5	Rise	-24.0	-23.0	OK	C	
AR13	Kilmartin (S_4-S_5)	AP	L	141	143	3.0	Rise	-24.0	-23.0	OK	C	
AR15	Duncracaig (*dcba*)	R	B	141	144	2.0	Rise	-26.0	-23.5	OK	B	790630
AR15	Duncracaig (*fe*)	AP	C	153	155	0.5	Rise	-30.0	-29.5	OK	C	
AR28	Dunamuck I	R	A	165	168	2.0	Rise	-31.5	-30.5	OK	B	850513
AR29	Dunamuck II	AP	A	137	141	3.5	Rise	-22.5	-21.0	OK	B	850513
AR31	Achnabreck	AP	L	159	160	2.0	Rise	-30.0	-29.5	U	B	850513

Achnabreck (AR31). NR 8554 9018 & 8563 8993. At the first location is a prostrate stone some 4.5 m long. At the second, about 250 m to the SSE, is a slab some 2.5 m tall × 1.0 m × 0.3 m, roughly oriented upon the prostrate stone.

Southern indications

Survey details, and details of southern indicated azimuths and declinations, have given been given in Ruggles (1984a) for surveys undertaken in years up to and including 1981 (profile diagrams are also included) and in Ruggles (1985) for surveys undertaken in 1985. The following two new surveys were undertaken in 1986.

Glengorm. On the tentative assumption that this site does represent the remains of a stone row of similar form to the others, we can note the orientation of Stone *B* as a possible, though unreliable, indication of the original orientation of the row. This was suggested in Ruggles (1985) but the orientation of Stone *C* (indicated azimuth between 128° and 134°) was quoted in error, giving a declination between -17°.0 and -14°.5. A resurvey in 1986 established the orientation of *B* to be between about 160° and 165°, giving a declination in the range -31°.0 to -30°.0. The profile is towards Cruachan Druim na Croise, some 7 km away.

Maol Mor. The site is surrounded by Forestry Commission trees. The stones are oriented in the south upon the western slopes of a ridge some 150 m away, with distant hills beyond. It is tricky to establish the approximate altitude of the nearby ridge, and hence whether it might have obscured the distant hills as viewed from the stones. An attempt had been made in 1985; another was made in May 1986. We now estimate an altitude of about 2° for the nearby horizon (current ground level), compared with 1°.2 for the distant hills. The difference seems too small for us to be able to say for certain whether the distant view could once have been visible, in view of possible changes in ground level since prehistoric times. For the assumed orientation of the site (161° to 163°) an altitude of 2° gives a declination between -30°.0 and -29°.5; an altitude of 1° yields a correspondingly lower declination.

A summary of the southern indications, incorporating the new survey data, is given in Table 9.1, and the indicated declinations are shown diagrammatically in Fig. 9.2. The interested reader should note that, apart from the inclusion of the new data, these differ from the corresponding tables and figure published in 1985 (Ruggles 1985: Tables 3 & 4, and Fig. 3, lower part) in the following respects:

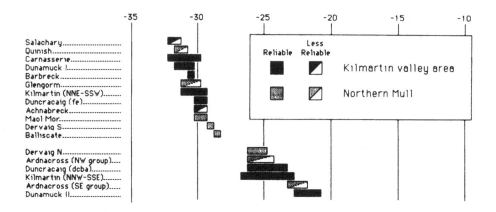

Fig. 9.2. Southern declinations indicated by the stone rows of northern Mull and the stone rows and aligned pairs of the Kilmartin area, Argyll.

(i)   in the table, line $S_6S_2S_3$ at Kilmartin has been replaced by line S_2S_3, and line S_1 (the flat-face indication of the isolated slab) has been added;

(ii) in the figure, lines S_2S_3, S_4S_5 and S_1 at Kilmartin have been replaced by a single NNW-SSE line; and

(iii) Kilmartin valley area sites are now shaded darker than northern Mull sites.

Current interpretations

The sites at Kilmartin, Duncracaig, Dunamuck I and Dunamuck II form a well-known group of standing stone sites consisting of menhirs between 2.5 m and 4 m tall (with the exception of stone S_6 at Kilmartin) and varying in form from the simple three-stone row and aligned pair at the Dunamuck sites, through the parallel row and aligned pair at Duncracaig, to the complex arrangement at Kilmartin. They exist alongside a variety of prehistoric sites of many periods. The aligned pair 3 km north of Kilmartin at Carnasserie, the three-stone row some 3 km north again at Salachary and the site some 2 km north again at Barbreck, which has certain architectural features in common with the Kilmartin site (Patrick 1979) may also be related. Only the Achnabreck site differs noticeably in form: it comprises not a closely-spaced setting or complex of such settings, as do all the others, but two widely spaced menhirs almost 300 m apart. The northernmost is a large prostrate stone some 4.5 m long, and from its vicinity the other, smaller upright appears on the horizon. They may very well never have been directly related, although they do seem to have been aligned.

Amongst this group of sites we find six distinct linear structures all of which indicate a southern declination within 1° of -30°: Barbreck, Carnasserie, Kilmartin (NNE-SSW), Duncracaig (*fe*), Dunamuck I and Achnabreck. The indication is east of south at Barbreck and Kilmartin, and west of south in the other four cases. Only at Salachary, where the orientation of the linear setting appears to be in some doubt owing to its poor state of preservation, was a rather lower declination obtained, between -32° and -31°.5.

The remaining indicated declinations at the stone rows and aligned pairs in the Kilmartin area are as follows. The longer alignment (*dcba*) at Duncracaig yields a southern declination of between -26° and -23°.5, and the aligned pair at Dunamuck II yield a value between -22°.5 and -21°. At the Kilmartin site itself the orientations towards the SSE of the two aligned pairs S_4S_5 and S_2S_3 and of the single slab S_1 yield declinations between -26°.5 and -23°. Two points are worthy of note. Firstly, each of these declinations is within a degree or two of -24°; secondly, each structure indicating a declination near to -24° occurs in close proximity to one indicating a declination in the vicinity of -30°. Thus at Duncracaig row *dcba* (indicating -25°) is situated next to the pair *fe* (indicating -30°); the aligned pair at Dunamuck II (indicating -22°) are situated less than 500 m from the row at Dunamuck I (indicating -31°); and at Kilmartin the orientation of the individual slabs and shorter aligned pairs across the main, longer alignment (which indicates -30°) all indicate around -24°.

This suggests the idea that each of the structures in the Kilmartin area yielding a declination of around -24° is 'secondary' to another yielding a 'primary' declination of around -30°. The other sites indicate the primary direction only. The only fact which seems to run against this interpretation is that at Duncracaig it is the four-stone row which indicates the supposed secondary declination and the aligned pair which indicates -30°.

In northern Mull there are seven sites - Glengorm, Quinish, Balliscate, Maol Mor, Dervaig N, Dervaig S and Ardnacross - each of which is an impressive site originally comprising, it appears, at least three menhirs between about 2 m and 3 m tall (although several have fallen and those at Dervaig S have been broken off). The middle five sites appear to have been rows of at least three standing stones, and Ardnacross to have been two separate three-stone rows flanking three kerb-cairns. We have surmised that the three stones at Glengorm, of which two are known to have been recently re-erected, once also formed a three-stone row.

Those sites where at least two stones stand and the deduced indication seems most reliable are Balliscate, Maol Mor, Dervaig N and Dervaig S. Three yield declinations between -30° and -28°.5; Dervaig N yields a declination around -25°.5. Balliscate is a setting line, the other three rising lines. The row at Quinish and the south-eastern group at Ardnacross only comprise a single stone which still stands, the remainder being prostrate and in most cases partially turf-covered: the intended orientation in these cases is in greater doubt. This might explain the fact that the declination obtained at Quinish, around -31°, is lower than the remainder. The south-eastern group at Ardnacross yields a declination of around -22°.5. The north-western group, which comprises three prostrate stones, yields a value around -25°. The single slab at Glengorm, as surveyed in 1986 and described above, yields a declination in the range -31° to -30°.

Thus the rule observed at the Kilmartin area sites generally appears to extend to the group of sites on northern Mull. The isolated sites - Glegorm, Quinish, Maol Mor and Balliscate - yield 'primary' declinations of around -30°, while the Dervaig N site (indicated declination -25°) is situated only 400 m from Dervaig S (indicated declination -29°). The Ardnacross site, however, appears to be an exception to the rule. Both its three-stone rows appear to have indicated 'secondary' declinations, and it is situated several kilometres from any other known site.

The other stone rows scattered more widely about Argyll seem to fit a more general pattern of orientation between -30° and -19°: thus Duachy, Lorn indicates approximately -21°; Escart, Kintyre indicates -29°.5; Ballochroy, Kintyre indicates approximately -25°; and Clochkeil, Kintyre indicates about -22°.5.

Thus on the basis of present evidence the southern declinations indicated by the stone rows and aligned pairs in two important geographical concentrations - northern Mull and the Kilmartin area of Argyll - seem generally to fit a clear pattern. Either they are oriented within about a degree of -30°, or else they are oriented within a degree or two of -24° and situated close to another row or pair which does indicate -30°. At Dervaig and Dunamuck the primary and secondary indications are a few hundred metres apart. At Duncracaig they are only 40 m apart. The five-stone formation at Kilmartin may have been a variation or even a culmination of this tradition. Here, instead of a simple alignment, we have a long alignment in which both the northernmost and southernmost components, instead of being single stones, have become pairs of slabs oriented across the line. The long alignment indicates the primary direction; the orientation of its components indicates the secondary direction.

The primary orientations appear to present particularly strong evidence of deliberate orientation upon the southern major standstill moon. The secondary structures, if such they were, may have been oriented upon the southern limiting moon at another point in the 18.6-year cycle, but it would then be somewhat surprising that no declinations higher than -21° are obtained. It is also possible that they were oriented upon the midwinter sun. At this stage the two possibilities seem indistinguishable.

Many other points could be made here relating to these results, but would detract from the main purpose of this paper, which is methodological. It should perhaps be mentioned for the benefit of the casual reader, though, that a number of other factors have been taken into account in arriving at these tentative conclusions. A good deal of thought has been given, for example, to factors other than astronomy which might have influenced the placement and orientation of the sites and produced some or all of the effects observed: the lie of the land, for example (see Ruggles 1984a: 243; also *ibid.*: 17-19 for a general discussion). It should perhaps also be mentioned that these results and our interpretation of them should be considered in the light of results from other groups of sites such as the recumbent stone circles of eastern Scotland (Ruggles 1984c; Ruggles & Burl 1985).

It is of some relevance to ask how expected the postulated interest in the rising or setting major standstill moon might be on the basis of ethnographic and archaeological examples world-wide. While knowledge of the lunar month is almost universal, and many seasonal calendars are lunar-based, ethnographic instances of horizon lunar observations are unknown, at least to the present author. This is worrying, and possible explanations must be sought for their importance in prehistoric Scotland. One such might be the particular latitude, where the major standstill moon is seen to scrape along the northern or southern horizon - a rare and spectacular event which could perhaps have assumed great importance. Clearly it would greatly support the case for the reality of the horizon lunar observations in prehistoric Britain if evidence were uncovered of similar practices from communities in similar latitudes.

We would be misguided to launch into unabated speculation about the sites in question, although some cautious interpretations may be justified, especially in the light of cultural evidence. Some ideas have been discussed in Ruggles (1985), and a great many lines of enquiry have opened up for future research.

Discussion - the statistical approach in archaeoastronomy

The most obvious dangers in applying a statistical approach to the study of astronomical alignments relate to the selection of data. Firstly, this must be done fairly, that is without regard for the astronomical possibilities. Paramount attention needs to be paid to the demonstrably fair selection of data: without the assurance that this has been done, the results of any analysis will be at best misleading and at worst meaningless. Secondly, in accumulating site data it is dangerous to rely solely upon the present disposition of the archaeological remains. A site may, for example, have been considerably altered or destroyed, both during its use in prehistoric times or since, and in any case its structure may have been far more complex than that little which remains five millennia later can reveal. There often exists a range of archaeological evidence relating to these matters which must be taken into account.

These problems relate simply to the fair assessment of the available data within the confines of the statistical approach, and can (at least in theory) be minimised given the limitations of the data available. Stage 1 of the project described in this paper attempted to do just this, building upon the work of Thom by identifying inadequacies and undertaking fresh fieldwork in which explicit steps were taken to try to overcome them. At that time the present author took the view, in common with a number of archaeologists (e.g. Thorpe 1983), that the rôle of the archaeoastronomer was to assess, by statistical means, the nature and extent of intentional astronomical alignments in the archaeological record,* accumulating data which could then be considered alongside a variety of other archaeological data in order to interpret astronomical practice in its social context.

Since then the author has learned, mainly through contact with very different ethno- and archaeoastronomical examples in different parts of the world, that it is not easy to separate what we might call 'alignment data' from other types of data, whether one is talking about astronomical practice in particular or more widely. We know of course that astronomical practice can

---

*   One should really think, more generally, in terms of astronomical influences on the design and placement of archaeological structures, rather than just about 'alignments'. The word suggests the mere orientation of structures upon horizon astronomical phenomena, whereas other ways of incorporating astronomical elements into architectural design are also known, such as shadow and light phenomena, 'zenith tubes' and so on.

not be considered in isolation from its cultural context. Numerous examples from a variety of cultures inform us that astronomy is not practised in isolation: astronomical beliefs, observations and rituals invariably form part of a world view which affects many aspects of everyday life. Thus evidence relating to a range of cultural phenomena could be relevant to studies of astronomical influences on the design and placement of prehistoric sites, and should not be dismissed out of hand because it is difficult to quantise.

In studying, for instance, the orientations of the radial line systems etched on to the surface of the Nazca desert in Peru,[†] one would wish to consider (i) the concept of radiality in Andean cultures, upon which there is evidence from ethnohistoric and ethnographic sources; (ii) the place of astronomy in these cultures and the astronomical phenomena which were significant; and (iii) the range of factors, astronomical and otherwise, which influenced the orientation of radial lines. To fail to admit this evidence when it is available would severely limit the value of any analysis. For example, a small proportion of lines oriented upon a particular astronomical phenomenon, such as the rising sun on its day of zenith passage, might be of great interest in view of a known or suspected interest in the same phenomenon in other Andean cultures, but might not of itself show up as statistically significant.

Should we then abandon any attempt to apply a statistical approach to alignment studies in these cases? Clearly not: for to do so would be to revert to the practice - all too common in the early days of 'megalithic astronomy' - of simply seeking out and laying great emphasis upon alignments which fit a particular theory and ignoring all others which do not. Indeed, there is perhaps a tendency amongst American archaeoastronomers to use the presence of the other lines of evidence available to them - the ethnohistorical, the ethno-graphic, and written evidence - as an excuse not to consider the fair selection of alignment data or the background archaeological evidence relating to it; this is a point made by Ruggles & Saunders (1984) about one interpretation of the pecked cross-circle designs at Teotihuacan in Mexico.

In short, the two approaches, the 'statistical' and the 'interpretative', need to be reconciled. An important way in which astronomical practice may be retrievable from the archaeological record is in the effect of astronomical considerations upon the design and placement of ritual and ceremonial sites. We need some way of assessing 'alignment data' in conjunction with background cultural data. The details of the approach will vary, depending upon the extent of independent evidence available (in prehistoric Britain, of

---

[†]   *See* Aveni, A.F. (ed.), *The Lines of Nazca* (in press), and the paper therein entitled 'A statistical examination of the radial line azimuths at Nazca' by C.L.N. Ruggles.

course, it is very scarce indeed), but only in this way, in general, can we hope to provide reliable evidence relating to astronomical practice in past cultures.

The example discussed in this paper draws attention to precisely the same conflict of two approaches. Although the initial analysis of 300 sites tried to be objective and rigorous in the selection of data for statistical analysis, we can by no means claim that the interpretations given in the *Current Intepretations* section above, or the data as presented in Table 9.1 and Fig. 9.2, can be justified in such a manner. Various adjustments have been made to the data set, even since 1985 (a fact which the reader may verify by comparing this paper with Ruggles (1985)). For example stone S_6 at Kilmartin has been left out of the reckoning. With hindsight we can of course claim that this is because it is the only relatively small standing stone found amongst the group of alignments in the northern part of the Kilmartin valley; however, we could not pretend that the fact that line S_2S_3 yields a declination consistent with our findings at the other sites, whereas $S_6S_2S_3$ yields a declination which is anomalously high, has not influenced us in the decision. Similarly, we could claim that the 1986 resurvey at Glengorm was undertaken purely because an error in the data was discovered; but would this error have been discovered quite so soon if the declination originally obtained had not been anomalous?

Should we then, or should we not, indulge in such admittedly subjective interpretations? In particular, should we present data in which such interpretations are implicit? If we had simply presented the evidence as in this paper, and offered our interpretations, without having made explicit the previous stages in the exercise, the reader would have been in no position to judge it fully. Yet in the author's experience subjective interpretations such as these are made extensively, in almost every archaeological argument, and are implicit within virtually every archaeological data set. Not to allow such interpretations would be seen as unduly restrictive in view of the restricted nature of the material record - indeed, it would bring prehistoric archaeology almost to a halt. On the other hand arguments, especially statistical ones, based on the resulting data would certainly trouble a trained scientist.

Many authors, particularly those trained in the humanities, would conclude (and have concluded) that any rigorous statistical approach, since it necessarily excludes a variety of cultural data, is of very limited value indeed. Others, however, - particularly those trained in the numerate sciences - would say the same of any investigation which fails, through the *lack* of such an approach, to attempt to distinguish between deliberate design features and chance occurrences. Ironically, both camps end up by arguing that not to accept their viewpoint would be to open the floodgates to unabated speculation.

We have thus come face to face with some far more insidious problems relating to the statistical approach, problems which stem from the inherent drawbacks and limitations of applying a statistical approach to cultural data. They may be summarised as follows.

(1)   A statistical approach only gives us the power to spot overall trends amongst a large body of data. Human variation, on the other hand, will ensure that any such trends are only of a very general nature. Superficially similar sites may have had complex, differing and changing functions of which astronomy, if it played a part at all, may have entered in various ways, differing from site to site. By sticking to too rigorous a statistical approach we are excluding any possibility of considering this variation and detail.

(2)   In a 'classical' statistical approach at least,* hypothesis must precede data, and a single set of data can be used to test only one hypothesis. In view of the limited nature of the archaeological record this is very restrictive indeed.

(3)   A statistical approach, or a 'classical' one at least, gives us no clear way to take into account our background knowledge (or more accurately our background theories or beliefs) relating to the culture involved, whether directly or indirectly.

These problems are of wider relevance than just to archaeoastronomy - or indeed to archaeology as a whole. The fundamental problem is that evidence acceptable to a numerate scientist is of a very different nature from that acceptable to his counterpart trained in the humanities. It is the view of this author that much of the controversy stirred up by archaeoastronomical investigations stems not from sheer prejudice, as is often claimed on both sides, or even from the lack of detailed background knowledge or ability in mathematics, astronomy or archaeology; but merely from this simple fact. In an interdisciplinary area such as archaeoastronomy two very different approaches to the interpretation of evidence meet head-on, and in order for progress to be made they *must* be reconciled so that the available evidence can be considered in its entirety in a satisfactory way.

We hope that the example described in this paper might begin to suggest a way of bringing about this reconciliation, through an approach where

---

*   There are two main schools in statistics, the 'classical' and the Bayesian. They represent two very different, and often mutually exclusive, approaches (*see* e.g., Barnett 2nd. Edn. 1982)

statistical rigour precedes interpretative reasoning, in stages; this allows a gradual transition from the purely statistical to the more interpretive, with the range of admissible evidence becoming gradually wider. This is one possible approach: another is to adopt a Bayesian, rather than a classical, statistical methodology, which may help to overcome the second and third problems mentioned above (Ruggles 1986). Comparisons across archaeoastronomy, where there are differing amounts of background cultural evidence, may also be a great help, and work along these lines is in progress. There is far to go.

Alexander Thom, by opening up 'megalithic astronomy' as a field of enquiry in the 1950s (see Thom (1955) and the accompanying discussion) and by his strong influence on the gradual development of world archaeoastronomy, has opened up the interdisciplinary arena for a fascinating exchange of views across the 'two cultures' which could have methodological consequences far beyond the mere study of megalithic remains. The reassessment of Thom's statistical evidence in megalithic astronomy has brought these issues to the attention of the present author, and has gradually helped him clarify some of his own ideas. Through examples such as the group of sites described above he hopes to cast a little light not only upon astronomical practice in prehistoric Britain but also upon some very much wider methodological questions.

10

A cluster analysis of astronomical orientations

JON PATRICK and PETER FREEMAN

Summary

An information-theoretic clustering method has been applied to Ruggles' dataset of 276 megalithic alignments in western Scotland. While there is strong evidence that the data are clustered rather than forming a single homogeneous set, there are only weak indications that the clustering is related to any kind of astronomical intention. Some relations between direction of indication and geographical area, horizon distance and whether the indication is on-site or inter-site have also been noted.

Introduction

The difficulties of making an objective assessment of Professor Thom's hypotheses about deliberate construction of megalithic monuments so as to indicate astronomical events are by now well-known. The greatest contribution by far has been that of Thom himself in producing data in such profusion, with such accuracy and under such arduous conditions. For this we shall ever remain in his debt.

The work of Ruggles and his colleagues in recent years (1984a; 1984b; 1984c; 1985) has done much to clarify our understanding of the nature of the controversy surrounding Thom's hypotheses. Careful resurveys of sites and objective criteria for the quality of evidence at each site have provided results that differ markedly from those of Thom, but with vestigial support for astronomical alignments in some senses in some areas.

Our interest is in the statistical analysis of alignment data and we start, in common with other authors, from the null hypothesis of no intentional astronomical indications of any kind. Previous tests of the extent to which the data are consistent with this hypothesis may be divided into the specific and the general. The former (Freeman & Elmore 1979) require detailed specification of what astronomical events were indicated and of the inherent accuracy of the constructions while the latter dispense with this and rest entirely on the assumption of a uniform distribution of azimuths between 0° and 360°. It is, however, perfectly possible for a set of monuments to exhibit very non-uniform azimuths without necessarily showing any astronomical intention (see Ruggles (1984a: Section 1.3) for a full discussion of this point). The essential point we wish to test is that the distribution of azimuths and/or declinations should be smooth and slowly-varying, while any lumpiness in these distributions could easily be interpreted as deliberate preference for some directions over others. As with so many other investigators in this area, we find that Thom had exactly the right ideas many years ago, and the curvigrams that appear in his earliest papers (e.g. Thom 1955) are simple but effective attempts to detect lumpiness. As Ruggles (1984a) points out, however, it is easy for the human eye and brain to be deceived into seeing patterns in such diagrams where none exist. So we continue our search for some more objective way of detecting lumpiness.

Our initial desire to try the general methodology of cluster analysis was tempered considerably by knowing that most clustering programs might be very good at finding clusters when they really exist, but are notoriously bad in that they still tend to find clusters when there aren't any. This is exactly what we want to avoid, since we need a method which, when presented with a set of data from a single continuous distribution, will produce the answer that the best description of the data is as a single large cluster, not a collection of smaller ones. An 'astronomical' dataset, on the other hand, should produce clusters centred around each of the particular azimuths or declinations that were most favoured by the builders. One of the few techniques we knew of that would behave in this eminently sensible way, and one which fitted in with our existing interests in the Wallace information measure (Patrick 1978), is a program called *Snob* written by Wallace and available from Monash University, Department of Computer Science.

The cluster analysis program *Snob*

The program *Snob* (so called because of its desire to differentiate between classes) tries to find the minimum length of message needed to encode a dataset (into binary digits, say) and transmit it along an information channel so that a receiver at the other end can decode it and recover exactly the original data. One way would be to encode each character (digit, decimal point, minus sign etc.) into a pre-set pattern of '0's and '1's, concatenate these into one long string and transmit them. This is roughly what happens when data are typed onto a computer keyboard, for example. If the numbers are coming from some smooth but non-uniform distribution, however, some very clever results in coding theory show how the fact that some numbers occur more frequently than others may be exploited to give a shorter message length. Briefly, those numbers that occur most often are given shorter codes and the rarer numbers the longer ones. The familiar Morse code is a practical example of this principle. In general there is, however, a price to be paid for this economy, since the receiver needs to know what distribution the data are assumed to be coming from before he can decode the message. If this is a normal distribution for example, then the message needs to be prefixed by a section that informs the receiver what the assumed mean and standard deviation are. Only if the total length of the message, including this prefix, is shorter than the length of the code using the pre-set pattern will there be any advantage to be gained.

A similar view applies to detecting clusters within a dataset. If clusters really do exist, then the data for any item (here a megalithic alignment) can most briefly be encoded according to their relative frequency amongst items within the cluster to which that item belongs rather than their (inappropriate) relative frequency amongst all items. The more clearly-defined the clusters are, and the more accurately items are assigned to their 'correct' clusters, the shorter will the message length become. The price to be paid again comprises all the extra information that must accompany the data, namely how many clusters there are, the estimated parameter values of the distribution of data within each cluster and, for each item, which cluster it belongs to. All this has to be encoded and added to the front of the encoded data in order to form a message understandable by the receiver. It could well turn out that after forming the message in this way, its total length becomes greater than that of the simple message that doesn't bother assigning items to clusters. It is therefore quite possible for *Snob* to end up by concluding that there are no clusters in the dataset. This is indeed what it does for the data on Scottish brochs previously clustered by Martlew (1982). On the other hand, if it does

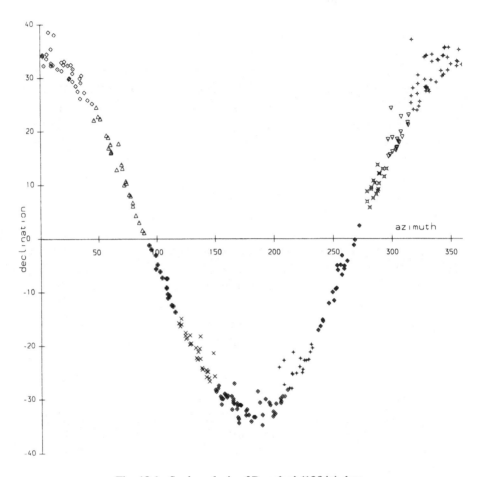

Fig. 10.1. *Snob* analysis of Ruggles' (1984a) data.

find clusters, *Snob* can quantify the strength of evidence for their 'reality' in terms of the reduction in total message length that clustering achieves.

This is, of course, a highly over-simplified account of *Snob* and full details are given in Wallace & Boulton (1968) and in the user manual that accompanies the program.

Snob analysis of Ruggles' data

We chose to try the method on the set of 276 alignments collected and described in Table 11.2 of Ruggles (1984a) as being the most accurate, comprehensive and objectively-chosen set currently available. For each alignment we simply took the central azimuth and declination in order to give *Snob* the chance of detecting clusters in one or both of these. Fig. 10.1 gives a plot of the 276 data points marked by different symbols according to the ten clusters that *Snob* produced. The usual trigonometrical relationship between azimuth and declination is obvious in the diagram (the points deviate from a single smooth curve only because not all horizons are flat at zero altitude). The 'reality' of the clusters is in little doubt, as the minimum message length achieved by clustering is 6325 binary digits, a reduction of 620 bits on that produced by assuming all points belong to just one cluster. In very crude terms (see the *Snob* manual for a full description with caveats), the clustered view of the data provides 2^{620} or about 10 to the power 187 times a better explanation of the data than does the unclustered view.

Because *Snob* did not know that azimuths of 1° and 359° are in fact close together, the two extreme clusters with azimuth ranges 0°-45° and 316°-360° and declinations above 25° should probably be better regarded as a single cluster. This will not be done in the following analysis in case any differences between them are thereby lost.

We shall arbitrarily number the clusters from 1 to 10 in increasing order of azimuth. The numbers of indications in each cluster and the ranges of azimuths and declinations are summarised in Table 10.1. It is readily noticed that Clusters 1, 5, and 10 representing the north-pointing and south-pointing lines are the largest, accounting for nearly half the data. This is, however, only

Table 10.1.

Cluster no.	No. of members	Azimuth range			Declination range		
		Min °	Mean °	Max °	Min °	Mean °	Max °
1	33	1.9	21.2	44.8	25.2	31.4	38.6
2	25	46.2	67.8	88.8	1.2	13.0	24.6
3	17	93.2	105.6	115.4	-13.6	-7.9	-1.2
4	29	118.5	133.4	148.8	-26.6	-21.0	-14.9
5	51	149.4	170.3	210.5	-34.8	-30.8	-27.0
6	19	203.6	219.4	231.6	-28.0	-23.8	-19.8
7	20	236.7	253.5	272.0	-17.0	-7.8	2.4
8	18	278.8	286.5	295.4	5.8	10.0	13.7
9	18	296.8	304.9	314.5	15.3	18.9	24.2
10	46	316.6	336.0	359.7	23.8	30.9	41.0

slightly more than might be expected since they cover a range of 150° of azimuth, so there is no strong evidence for a north-south preference.

Cluster 6 is centred around a declination close to that of the winter solstice but with a range of nearly 30° in azimuth around the south-west direction the cluster is not tightly concentrated enough to be really convincing. Similarly, Cluster 9 varies around the direction of moonset at minor standstill. The range of azimuths is less than 20°, but this covers declinations that vary between 15° and 24°.

It is difficult to compare the results of this analysis with those of Ruggles as the two approaches are so different, but it is perhaps worth commenting on Ruggles' main findings.

(a) *At the lowest level of precision, avoidance of declinations between -15° and +15°*. Fig. 10.1 shows this by the relative scarcity of points in this range. Clusters 3, 7 and 8 cover the range and are all amongst the smallest clusters.

Table 10.2.

Cluster no.	LH	UI	NA	CT	Region ML	LN	AR	JU	IS	KT	Total
1	5	5	0	3	3	1	7	1	2	6	33
2	5	2	0	0	3	0	2	2	4	5	23
3	2	8	0	0	0	0	1	1	2	3	17
4	6	7	1	0	3	1	4	0	3	4	29
5	10	8	0	5	6	0	8	2	5	7	51
6	3	5	0	0	2	0	4	0	0	5	19
7	6	0	0	0	1	1	1	4	4	3	20
8	2	7	0	0	1	0	1	2	1	4	18
9	5	5	0	0	1	0	1	0	4	2	18
10	12	6	2	0	6	1	13	1	1	4	46
Total	56	53	3	8	26	4	42	13	26	43	274

Region codes

LH	Lewis and Harris	LN	Lorn
UI	The Uists	AR	Mid Argyll
NA	North Argyll	JU	Jura
CT	Coll and Tiree	IS	Islay
ML	Mull	KT	Kintyre

(b)   *At the intermediate level, preference for declinations in the range -31°
to -19° and above 27°.* The large Clusters 1, 5 and 10 show a general
north-south tendency but do not distinguish between points on either
side of -31°.

(c)   *At the highest level, marginal evidence of a preference for declinations
close to -30°, -25°, -22°.5, +18°, +27° and +33°.* There is no support
here for any such high-precision indications. The fact that *Snob* did not
find any must warn that they are of no help in explaining the data, as
measured by minimum message length.

It is of interest, in addition to the question of astronomical intention, to look at
the clusters in terms of the other properties of each indication given in
Ruggles' Table 11.2. Table 10.2, for example, shows the geographical
distribution of indications within each cluster. A larger-than-expected number
of the sites in the Uists fall into Clusters 3 and 8, showing a preference for
azimuths in the ranges 90° to 120° and 280° to 296° and an avoidance of those
between 240° and 280°.

The eight indications in Coll and Tiree fall entirely into Clusters 1 and 5,
while nearly half of the 42 indications in Mid Argyll fall into Clusters 1 and
10, with azimuths between 315° and 45°. There is thus a little evidence for
preference for different directions of alignments in different areas, but it is not
particulary strong.

The sub-division of indications into on-site and inter-site is shown in Table
10.3. Just over half of all indications are formed by inter-site alignments, but
the proportion is markedly higher than this in Clusters 5 and 7. In other

Table 10.3.

Cluster no.	On-site	Inter-site	Total
1	15	18	33
2	13	12	25
3	8	9	17
4	16	13	29
5	21	30	51
6	10	9	19
7	7	13	20
8	9	9	18
9	9	9	18
10	22	24	46
Total	130	146	276

Table 10.4.

	On-Site Indications												
Cluster	Class 1		Class 2		Class 3			Class 4		Class 5		Class 6	Total
no.	1a	1b	2a	2c	3a	3b	3c	4a	4b	5a	5b	6	
1	5	1	2	1	2	1	0	0	0	3	0	0	15
2	2	0	1	0	3	1	0	0	1	5	0	0	13
3	0	0	0	0	3	0	0	0	0	5	0	0	8
4	2	1	0	0	1	0	1	3	0	6	1	1	16
5	4	2	4	1	2	0	1	0	0	7	0	0	21
6	1	0	1	1	3	1	0	0	0	3	0	0	10
7	2	0	0	0	1	0	0	1	2	1	0	0	7
8	1	0	0	0	2	1	0	0	1	4	0	0	9
9	0	0	0	0	1	1	0	1	0	4	0	2	9
10	5	3	3	0	0	0	2	2	0	6	0	1	22
Total	22	7	11	3	18	5	4	7	4	44	1	4	130

Table 10.5.

	Inter-Site Indications						
Cluster	Archaeological Status A			Archaeological Status B			Total
no.	Class 1	Class 2	Class 3	Class 1	Class 2	Class 3	
1	2	6	7	1	1	1	18
2	1	3	4	1	3	0	12
3	2	2	3	0	1	1	9
4	5	0	5	0	2	1	13
5	3	7	10	1	7	2	30
6	0	3	5	0	0	1	9
7	1	5	3	1	3	0	13
8	5	2	0	0	1	1	9
9	1	1	5	0	2	0	9
10	2	6	7	1	5	3	24
Total	22	35	49	5	25	10	146

words, there is a preference for indicating directions between south and west by inter-site alignments rather than on-site ones.

Ruggles (1984a) divided the on-site alignments into six classes in terms of their inherent likelihood as astronomical indicators, and further subdivided all but one of these classes. This classification is shown in Table 10.4 for each cluster. The only remarkable feature is that half of the class 1 alignments fall into Clusters 1 and 10. As these cover the range of declinations between 24° and 41°, this could be regarded as some slight evidence for indication of lunar major standstill phenomena. A similar classification of the inter-site indications is given in Table 10.5. Here Cluster 8 stands out, with seven out of its nine members in classes A1 or A2. As they indicate declinations between 6° and 10°, however, they cannot have any astronomical significance.

The distance of the horizon from any indication is one factor that obviously has a possible bearing on its use for astronomical observation. Table 10.6 summarises the data on this and shows quite clearly a preference for larger horizon distances in Clusters 3, 4, 5, and 6, which are those pointing anywhere between east and south-west. Cluster 5, moreover, contains a high proportion of those indications at which the horizon is the sea. These have been excluded from the calculation of mean distances.

Finally, Table 10.7 shows the reliability of the surveyed data. As would be expected, the distribution over the three categories is similar in each cluster. It is, perhaps, worth noting that half the indications in Cluster 1 are calculated rather than surveyed.

Table 10.6.

| Cluster no. | Horizon distance (km) | | | |
	Min	Mean	Max	'Sea'
1	1.0	6.5	24.5	1
2	1.0	10.4	38.5	0
3	1.0	19.4	83.0	0
4	1.0	15.9	120.0	0
5	1.5	16.2	70.0	5
6	1.5	13.3	65.5	2
7	1.0	7.3	25.0	1
8	1.0	4.1	12.0	2
9	1.0	12.4	84.0	1
10	1.0	6.0	47.0	0

Table 10.7.

Cluster no.	Reliably surveyed	Less reliably surveyed	Calculated	Total
1	17	0	16	33
2	16	5	4	25
3	11	1	5	17
4	21	6	2	29
5	29	6	16	51
6	10	2	7	19
7	11	4	5	20
8	13	0	5	18
9	12	1	5	18
10	26	8	12	46
Total	166	33	77	276

Discussion

Although the *Snob* analysis shows there are great advantages, as measured by shorter message length, in thinking of the data as clusters rather than a single homogeneous whole, the clustering is into broad, non-overlapping ranges of either azimuth or declination and there are no clusters closely bunched around any single astronomical direction. Indeed, some clusters cover ranges of no astronomical interest at all. This apparently negative finding really seems to be a 're-discovery' of the universally-agreed proposition that megalithic alignments did not all have an astronomical purpose. It does seem clear, however, that this analysis could be consistent with a broad, low-accuracy astronomical association at some sites, but gives no support at all to the far more specific, high-accuracy theory developed by Thom.

As a by-product of the analysis, we have noticed some rather weak features that lurked unnoticed in Table 11.2 of Ruggles (1984a). There is a preference for some directions rather than others in some areas. The inter-site alignments show a greater preference for directions between south and west than do the on-site alignments. The more inherently plausible astronomical indicators tend to be the north-pointing ones, with azimuths between north-west and north-east. Horizon distances tend to be greatest for south-pointing indicators anywhere between east and south-east.

Acknowledgements

We are very grateful to Clive Ruggles for help with using his data and for many stimulating and informative discussions on this subject.

11

Megalithic observatories in Britain: real or imagined?

RAY NORRIS

Summary

Over the last two decades there has been an accumulation of exciting evidence that appears to show that, as early as 5000 years ago, people in Britain were making precise observations of the sun, moon and stars, and studying small perturbations in the lunar motion. These observations would take place throughout the thousands of megalithic sites in Britain, which are then seen as prehistoric observatories. Within these, data would be accumulated to enable the prediction of celestial phenomena such as eclipses, and allow the construction of a calendar. Recently, however, a small number of rigorous statistical studies of the sites have cast doubt on the astronomical hypotheses, and have posed the question of whether some of the support for these hypotheses has been generated by well-intentioned but over-enthusiastic selections of chance alignments. In this review, the arguments and counter-arguments are presented and examined, and we see what can be salvaged from the astronomical hypotheses after the statistical smoke has cleared.

Introduction

Scattered throughout the wilderness of Britain are thousands of megalithic sites. These exhibit a wide range of forms, from the single standing stones, or menhirs, to carefully laid-out circles of stones, and long stone rows up to 3.6 km long. Most of these were built in the third or second millenium BC, and are frequently associated with cremations or burials. The quantities of manpower expended on these monuments bear testimony to their importance to those who built them, and they must have played a major role in the lives of those who used them, and yet we know little of their purpose or function.

Some hundreds of these have been carefully surveyed, largely by the late Professor A. Thom and his son Dr A.S. Thom, who have constructed from their substantial data a group of hypotheses which propose the following.

(a)   The megalithic sites were laid out using accurate measuring techniques and a unit length (the megalithic yard) which was defined to an accuracy of about 1 mm and whose value was kept within this range throughout Britain.

(b)   The sites were laid out in many cases using elaborate geometrical constructions which involved knowledge of Pythagorean triangles.

(c)   Many of the sites were set up to indicate the rising or setting position of the sun, moon, or a star, at a particularly significant date, and with arc minute accuracy.

The current status of all three of these hypotheses is that they have an ardent following of believers, an equally ardent band of disbelievers, and a rather smaller group of people who consider that Thom's data appear to show a prima facie case, and warrant further investigation. However, this group of agnostics would also insist that more data, and better ways of analysing those data, are necessary before any firm conclusion can be reached. It is perhaps unfortunate that, because nearly all the data have been collected by the Thoms, any evaluation of the astronomical hypothesis is necessarily an examination of their methodology.

There are clearly elements common to all three hypotheses, so that ideally they should be studied together. However, testing the first two hypotheses has proved extremely difficult (e.g. Patrick 1978). The third hypothesis, whilst still presenting a thorny statistical problem, is rather more tractable and is one that has been tackled by a number of workers. It is this astronomical hypothesis with which I will deal in this review.

This review is strictly limited to the sites in Britain, and so omits the culturally unrelated sites of the Americas, on which there is a rapidly growing body of archaeoastronomical evidence, and the megalithic monuments in Brittany. Because of the large volume of literature on British archaeo-astronomy, this review traces the main threads of the archaeastronomical debate, rather than attempting to cover the rich variety of arguments from both sides.

Megalithic sites in Britain

This section shares its title with that of Alexander Thom's first book (Thom 1967), in which he first presented his ideas, previously scattered throughout several journals, as a coherent whole. He lists and describes several hundred sites, of various shapes and sizes, and here I describe some of the types of megalithic monument encompassed within the astronomical hypothesis. The largest group to be described is the stone circles, most of which have diameters between 4 and 30 m. However, there are a few large circles over 100 m in diameter, including the spectacular circle at Avebury (330 m in diameter). The stone circles are extremely common in the unspoilt areas of Britain, often occurring within sight of each other and often constructed in high areas with unobstructed or panoramic views. Burl (1976a) has also suggested that they are preferentially situated near a source of water.

The stone rows are also prolific. Most of those considered by Thom are a few tens of metres long, and consist of standing stones up to 4 m high. They occur as single, isolated rows, as at Swincombe (see below); as single rows associated with burials, as at Ballochroy (see also below), or with circles, as at Eleven Shearers (Thom 1967: 149); and as multiple rows, as at Mid Clyth (*ibid.*: 152). There also appears to be a distinct class of long (up to 3.6 km) rows consisting of short (usually less than 1 m) stones, found mainly in south-west England.

The simplest class of megalithic monument, and that attracting the most scepticism, consists of the single standing stones. A single stone cannot on its own define an azimuth with any great accuracy, even if flattened on one side. There is the additional problem that, on a roughly-hewn rock now weathered and covered with lichens, one man's flat side is another man's cylindrical surface. Nevertheless, Thom's data include several alignments which depend on such flattened surfaces, and which therefore need to be treated with caution.

What does archaeology tell us?

Archaeoastronomers have often been justly criticized for failing to incorporate archaeological data into their hypotheses. This failure, however, may be attributed partly to the paucity of archaeological data on megaliths. I will avoid all archaeological controversy by including all sites under the generic name of megaliths, and the various cultures constructing them will be called the megalith builders.

It is known that the megaliths were erected by various groups at different dates between about 3500 BC and 1500 BC. Most of the structures considered here probably date from around 1800 BC. Of the few sites that have been excavated several show evidence of cremations and burials. From this we can infer that they played some part in funerary rites, and presumably had some religious significance. Beyond this, it has been postulated that they represent temples, meeting places, market places, charnel-houses, or neutral ground where hostile factions could meet. It seems that we actually know very little of the origin of the megaliths. Specifically, the following questions are unanswered.

(a)   What were the megalithic sites used for? Apart from the obvious religious and funerary associations, which do not in themselves provide an answer, we can at the moment only speculate.

(b)   Who built them? It has been argued that these primitive people scratching an existence from the soil could not have had the organizational skills, or continuity of society, to build the larger sites, even if they had the manpower available. This argument leads to the idea that there may have been an élite class of priests or wise men, who ruled and organized the commoners.

(c)   Why are there so many megaliths? Many are within sight of one other, so that some groups probably built more than one megalithic monument. Could they have been status symbols?

(d)   Why did the builders expend so much labour on their construction? For example, Avebury cost an estimated 400 man-years, whilst the related Silbury Hill cost an estimated 7500 man-years. When compared to a total British population of perhaps only a few thousand (Atkinson 1972), it is clear that the construction of these sites was a project of major importance at the time.

The implications of the astronomical hypothesis

Motivation for prehistoric astronomy is not difficult to find. Perhaps the motion of the sun would be observed and studied in order to construct a calendar to help with agriculture. Other motives suggested for the prehistoric Britons are the forecasting of eclipses, and prediction of tides for navigation. In each case, the observations would presumably have been embellished with ritual and religion.

However, if these hypotheses are confirmed, they will have the following serious implications for our present ideas of prehistoric Britain.

(a)   There must have been a stable society, since for the lunar alignments observations would have been necessary over centuries.

(b)   The work force for the construction and operation of the observatories must have been highly organized perhaps by the élite class mentioned above.

(c)   Most controversial of all, there must have been a degree of mathematical and scientific skill that seems totally incompatible with the current view of a primitive, barely agricultural, society.

These implications do seem to be accommodated in a model devised by MacKie (1977a) consisting of a peasant society ruled and dominated by an élite class of highly educated priests. Some support for this view comes from sites like Skara Brae, in the Orkney islands, where a settlement dated to the third millenium BC contains stone furniture (beds, querns, shellfish-tanks, cupboards) totally at odds with the primitive hut-circles usually associated with this date. Examination of the midden at Skara Brae shows that its inhabitants had a superior diet of bread and meat which seems to have been prepared elsewhere (by the servants?). Skara Brae is notable too for its proximity to two fine stone circles (Brodgar and Stenness). Skara Brae in MacKie's model would then represent an ivory tower, occupied by the élite class of priest-astronomers, who were supported and fed by the local commoners.

How to make a megalithic observatory

In order to study the motion of the sun and moon it is necessary to be able to measure their positions. The points on the horizon at which an astronomical body rises and sets depend only on its declination, and are independent of right ascension. It is this measurement upon which most hypotheses of British archaeoastronomy are based.

Over the course of a year the declination of the sun ranges from -23°.5, in the mid-winter, to +23°.5, in the mid-summer. At the solstices the solar declination changes by less than 1 arc minute per day, whilst at the equinoxes it changes by about 24 arc minutes per day. These then are the precisions to which a megalithic calendrical observatory must be built.

In contrast, the moon ranges from a high northerly to a high southerly declination in the course of only a month. Furthermore, this maximum absolute declination ranges from 19° to 29° with an 18.6-year period.

Therefore, a megalithic lunar observatory might be expected to indicate these declinations. In addition, there is a small perturbation of 9 arc minutes, which would need to be studied in order to predict eclipses.

To build a megalithic observatory therefore it is necessary to erect a structure which defines the position on the horizon to arc minute accuracy. The easiest way to do this is to select some distant horizon feature (the foresight), and establish a position (the backsight) at which the observer must stand. This simple recipe has the disadvantage for the astronomer, however, that he and his successors must remember exactly which horizon feature they are supposed to be looking at. (It also has the disadvantage for future archaeoastronomy that the selection of a horizon feature may have more to do with the prejudices of an archaeoastronomer than with the intentions of the megalith builders.) It is therefore advantageous to the builder to erect a structure which not only defines the observing position, but also indicates which horizon feature is to be used (an indicated foresight). Analogously, any statistical study of these sites must also concentrate on those sites with indicated foresights, in order to reduce the personal bias of the investigator.

The actual construction of a megalithic observatory appears straightforward. Suppose, for example, that an indicator of the winter solstice is desired. Having selected their foresight (a distant mountain peak, say), the megalith builders then watch the setting sun each day, and push a stick into the ground at the position from where it is just possible to see the sun setting behind the foresight. Each day the sun sets a little further to the south, and the sticks are pushed in a little further to the right, until one day the sticks get very close together, and then on subsequent days have to be placed to the left of the existing sticks. The solstice has then passed, and a megalith may be erected at the position of the extreme righthand stick. A stone row may also be erected to indicate which mountain top is the correct foresight. We may also conjecture that the whole ceremony would not have been performed with the disinterested objectivity with which modern-day astronomers conduct their research (!), but would have been surrounded with a great deal of religious ritual, perhaps with sacrifices being offered so that the sun would return to drive away the miseries of winter.

In practice, the celebration might well be dampened, for example by heavy rain on the day of the solstice, necessitating either an interpolation between the available data sticks, or a re-observation the next year. In general, we can expect that the setting up of a solar observatory might take a few years. The setting up of a lunar observatory, considering the moon's 18.6-year cycle, might take several generations. A further complication is the variable lunar parallax (Morrison 1980) requiring even longer periods of observation.

The case for prehistoric astronomical observatories

In his first book Thom presented his results of surveys of some 600 sites. In the cases where a site appeared to indicate an alignment he measured its azimuth, converted it to declination, and plotted it on a histogram (Thom 1967: Fig. 8.1). This histogram, which contains over 250 sightlines, appears decidedly non-random, and in fact the peaks appear to cluster around the extreme declinations of the sun, moon, and a few bright stars. In addition, there are some peaks corresponding to the sun's declination on the days half-way between the solstices and equinoxes (Martinmas, Candlemas, May Day, and Lammas), and on the days half-way between these and the solstices and equinoxes (the so-called calendar dates). Thom interpreted this histogram as indicating that not only were the megalith builders marking out the extreme declinations of the sun, moon and stars, which might have indicated some ritual, but were also subdividing the year to make a calendar. This level of the astronomical hypothesis, which maintains simply that there exist astronomical alignments accurate to half a degree or so, is labelled 'Level 1' by Ruggles (1981).

Thom then plotted (1967: 120) the lunar data from the four extremes as a histogram of (observed-calculated) declination. If the alignments indeed had a precision of half a degree or so, then a smooth Gaussian curve would have been obtained. Instead, a bimodal distribution resulted, with a separation between the peaks of about 0°.5 (the apparent lunar diameter). This implies that observations were not made roughly of the centre of the moon, but that rather more accurate observations were made of the lunar limbs. This level of the hypothesis is labelled 'Level 2' by Ruggles.

Subsequent analysis of the lunar lines by Thom (1971) appeared to show an even more significant relationship (Ruggles' Level 3). When only the 'most reliable' lunar lines were plotted on a histogram of (observed-calculated) declination, they seemed to show definite peaks corresponding to the sum and difference of the lunar semi-diameter with the small perturbation term (9.4 arc minutes) of the lunar orbit (Fig. 3a). This implied that not only did the megalith builders construct alignments directed at the limbs of the moon, but they did so to a precision sufficient both for the detection of the small perturbation in the lunar orbit, and for the alignment of the monuments to the moon at the maximum and minimum values of this perturbation.

A further level of sophistication, Level 4, is demonstrated in three further papers (Thom & Thom 1978b; 1980; Thom 1981) in which various small corrections are made to the data, and only the most reliable data are included. The resulting histogram implies a precision in the alignments of better than a 1 arc minute. Because of the variable lunar parallax (Morrison 1980), such a precision could only be achieved by averaging observed positions of the moon over a period of 180 years.

Taken at face value, these data appear very persuasive. However, there is clearly ample room for selection effects to corrupt the data. For example, how do you select the alignments at a site? Given a site of, say, ten standing stones, there are 90 possible alignments, and several of these are likely to hit astronomically significant declinations purely by chance. If all ten are in a straight line, however, we can ignore all except the two alignments along the line, and if these hit an astronomical declination it can be regarded as a significant result. In practice, real sites tend not to be so clear-cut, and it is essential to use objectively-defined selection criteria in order to avoid unintentional bias when selecting the most 'obvious' alignment at a site.

To test these data therefore it is necessary to establish a body of data comparable in size to that of the Thoms, but selected according to rigidly defined criteria. The difficulty of doing so is illustrated by the decades that Thom has spent in collecting his data.

Examples of possible megalithic observatories

Swincombe

This site (Norris 1983) is listed first because it has just one clearly-defined alignment, which unambiguously indicates a single distant notch in the horizon, and because it has not been included in the Thom data. On Thom's hypothesis this single alignment should indicate an astronomical declination. For the same reasons, it is a good illustration of a typical site, and of the techniques and problems associated with archaeoastronomical investigation.

The row of three carefully-tooled stones, each about 1 m high, is above the River Swincombe (SX 631722) on Dartmoor, south-west England. The regularity of the stones, atypical of Dartmoor rows, led to their tentative classification by the Ordnance Survey as tethering posts. However, R. Burnard in 1894 regarded them as true menhirs of considerable antiquity, and they are similar to many authenticated standing stones elsewhere in Britain. Without further data, there seems little reason to doubt that they are associated with the prehistoric hut-circles and small standing stones that surround them.

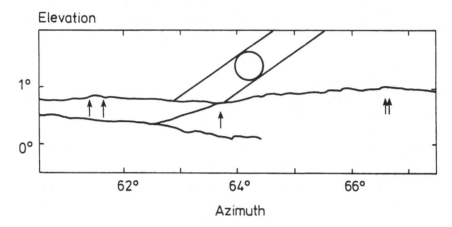

Fig. 11.1. Profile of the indicated horizon at Swincombe (from Norris 1984). Surveyed points are indicated by arrows. Drawn above the horizon is the path of the rising Lammas Day sun in 1800 BC.

All three stones have flat faces, and the line along their faces indicates accurately (to ±1°) a notch in the horizon about 10 km away (Fig. 11.1). The declination of the notch, corrected for refraction, is 16° 33′, which is similar to the declination of 16° 32′ of the lower limb of the rising sun on Lammas Day in 1800 BC. Although this result therefore supports the astronomical hypothesis, it should be regarded more as an illustration of the available data than as a positive confirmation of the astronomical hypothesis. Many more such sites need to be surveyed before a definite result can be claimed.

Ballochroy

Ballochroy is a row of three stones in Kintyre, west Scotland, and is remarkable for showing two accurate solar alignments. Thom (1971) showed that a line along the row indicated the setting midwinter sun over Cara Island 12 km away, whilst a line along the central flat stone indicated the setting midsummer sun over Corra Bheinn, Jura, some 30 km distant. This site attracted attention because both alignments were accurate to a few arc minutes, and so it would be difficult to explain such a site as due to chance alone. The site was independently surveyed by Bailey *et al.* (1975), who agreed with Thom's data. However, more recently it has been suggested that the midwinter alignment would have been obscured at the time of construction by an intervening cairn, and that the midsummer alignment is not necessarily the most obvious alignment along the central flat stone.

Kintraw

Of all sites, Kintraw in Argyll (Thom 1967: 141; 1971: 37) probably supports the astronomical hypothesis more than any other. This is because it is here that, in the tradition of the scientific method, Thom's hypothesis was used to make a prediction that could be tested by conventional archaeology. Kintraw consists of an impressive 3.6 m-high menhir, together with a ruined cairn, a small circle, and a small standing stone, all in a line. This line indicates a notch between two mountains 45 km away, at which the midwinter sun makes a brief appearance as it sets, thus providing an accurate solstitial indicator.

The problem with this site was that an intervening hillside actually prevented an observer from seeing the foresight from the backsight. Professor Thom suggested that an observer could climb up the steep hillside behind the site, in order to see over the obstruction, and actually found what appeared to be a sheep track in the required observing position. Thom suggested that the ledge was not a sheep track but was man-made. The ledge was subsequently excavated, and petrofabric analysis indicated that the ledge was indeed man-made (MacKie 1977a, and references therein).

Thus a prediction based on the astronomical hypothesis has been confirmed by independent archaeological investigation. Although there has since been some debate on the significance of this result (e.g. McCreery *et al.* 1982), it probably remains as the strongest independent support for the astronomical hypothesis.

The statistical approach

Selected samples of alignments

Since opportunites are rare for testing a prediction in the manner of the Kintraw investigation, an alternative approach is to look at the statistics of indicated declinations, and see if there are significantly more astronomical ones than would be expected by chance. The main difficulty with this approach is that there are few sets of data which are sufficiently free of selection effects to be amenable to statistical treatment. A previous difficulty, that there existed no rigorous statistical technique for analysing the data, was removed by the construction of such a technique by Freeman & Elmore (1979).

The Freeman/Elmore test was applied by those authors to a selection of data collected by Barber (1973), who had surveyed a number of circles in Cork and Kerry, Ireland, and to a survey (Cooke *et al.* 1977) of the alignments associated with the Callanish group of sites, in the island of Lewis, Outer Hebrides. In each case the test showed no more alignments than might be expected by chance. However, the data of Barber were subsequently found to be faulty, and when the test was repeated on corrected data (Patrick & Freeman 1983) marginal evidence was found for rough (Level 1) alignments. Another set of data was obtained by Norris *et al.* (1982) on the Barbrook

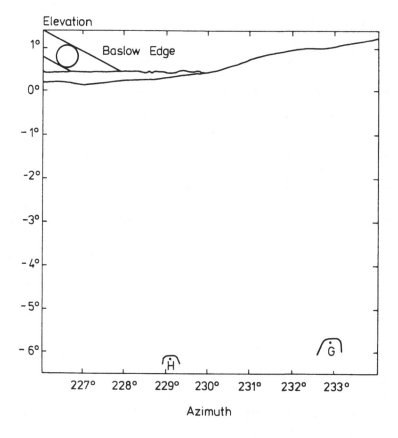

Fig. 11.2. Example of a rough alignment, from the Barbrook I stone circle in Derbyshire (from Norris *et al.* 1982). At the bottom are shown two standing stones, and at the top left is shown the path of the setting midwinter sun in 1800 BC.

group of sites in the Peak District of England. Again the test showed no evidence for accurate alignments, but it did show some marginal evidence for rough alignments. An example of such an alignment is given in Fig. 11.2. These rough alignments would have been of little use for studying the motion of heavenly bodies, or for the construction of a calendar, but could have been used for some ritual associated with the solstices.

The quantities of data used in these tests were small compared to the Thoms' data, and a definitive test may be made only by accumulating a much larger body of data. Such a project, concentrating on the Western Isles and western mainland of Scotland, has been in progress since 1975 (Ruggles 1984a). This study has been executed with particular care being given to the selection of sites using rigorous pre-defined criteria.

The main conclusion to emerge from this study is that a significant number of alignments are roughly (within a degree or so) directed at lunar rising and setting points, particularly the extreme positions. There is also weaker evidence for solar alignments towards the winter solstice, but none towards the summer soltice. An interesting feature of these results is the dominance in the 'positive' results of a group of sites in Mull and Argyll. Ruggles (1985) made this group the subject of a subsequent study which included many sites additional to those in the 1984 study, although still using rigorously pre-defined selection criteria. The results are remarkable in that nearly all the alignments indicate declinations in the range -19° to -31°. Furthermore, many of these are concentrated on declinations within a degree or two of -30° (the minimum lunar declination) or -23° (the minimum solar declination).

These results constitute the first definitive evidence that megalithic man was interested in marking the southern limits of the sun and moon. His interest in the winter solstice, rather than the summer solstice, may indicate either that his summer days were too busy to worry about heavenly phenomena, or that the turning point of the winter sun, attended as it was by the miseries and dangers of that harsh frozen existence, carried more significance than the turning point of the welcome summer sun. His obsession with southern lunar declinations is, however, more difficult to understand, and is only partly explained by the low elevation (and thus high prominence) of the southern moon. In any case, these studies show that his interest in these bodies was of ritual significance (i.e. of low accuracy), rather than an attempt to measure celestial positions as an aid to eclipse prediction or calendar construction.

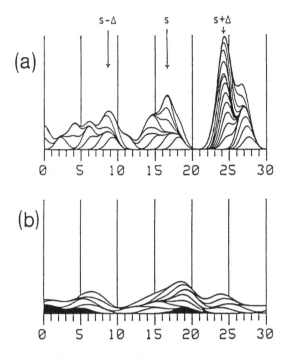

Fig. 11.3. Histograms of the (observed - calculated) declinations of the most reliable lunar alignments, adapted from Ruggles (1982b).

(A)   Thom's data, which shows peaks at the positions corresponding to the sum and difference of the lunar semidiameter s and the lunar orbital perturbation Δ.

(B)   Ruggles' data, which corresponds to Thom's data except for removal of dubious alignments and the use of an explicit selection procedure. The peaks have become insignificant.

The accurate lunar lines

The most controversial aspect of the astronomical hypothesis is the high accuracy with which Thom claims the alignments are established. In order to test this, Ruggles (1982; 1983) investigated the sample of lunar lines in some detail. He visited and resurveyed every site in Thom's sample, and evaluated both the alignments proposed by Thom, and those which seemed at least as feasible. His conclusions are illustrated by Fig. 11.3, which shows a reassessment of Thom's Level 3 data. Ruggles has shown that a few of Thom's alignments are clearly faulty (e.g. foresight invisible from backsight, non-archaeological site) and so they have been rejected. The remaining lines have been resurveyed and have had their uncertainties reassessed. The result, shown in Fig. 3(B), is that there appears to be no evidence for the accurate lunar lines.

The nature of the disagreement

The results discussed above pose the question: how is it that different sets of data produce such different answers? Although occasionally a survey by the Thoms has been found to be faulty (e.g. Ruggles & Norris 1980), such errors are insufficient to explain the discrepancy. The cause of the disagreement for Level 1 alignments appears instead to be the method of selecting the alignments. This is not to say that the Thoms have deliberately chosen to publish only those alignments which fit their theories, but instead emphasizes how easily subjective bias may unintentionally influence data. When surveying a site, the Thoms appear to have been guided by astronomical considerations when choosing their alignments. Only selection criteria, such as those of Cooke *et al.* (1977), which have been rigorously defined prior to the first visit to the site, can ensure freedom from this bias.

The cause for the discrepancy in Levels 2 to 4 is more puzzling, since no-one can estimate positions to arc minute accuracy, and so it is difficult to see how subjective bias can influence the data. Perhaps we should accept that such distributions can occasionally arise purely through chance.

What is left of the astronomy?

The previous section might give the impression that the astronomical hypotheses have disappeared in a puff of statistics. However, it is only on the accurate lunar lines that this disappointing conclusion has been reached. More exciting is the positive conclusion that the megalith builders *were* interested in the motions of the sun and moon, and that they *did* mark the extreme southern solar and lunar positions. The use they made of this knowledge we can only guess at. However, this positive result should give us added impetus to test the remaining archaeoastronomical theories of Thom and others. Such tests, unfortunately, require years of painstaking data collection.

More important than the quantity of data is the quality of data. The work of Ruggles and others has highlighted the real problem of British archaeo-astronomy and shown it to be one of methodology. The nature of the problem is such that different methodologies give different results. It is unlikely that we have yet found the optimum methodology which, although free from subjective bias, yet cannot accidentally discriminate against a type of alignment favoured by the megalith builders.

In conclusion: whilst some of the most extreme claims of archaeoastronomy seem unsupported by recent work reviewed herein, there are areas not yet fully investigated, and other areas (the rough alignments) where the evidence shows signs of accumulating in favour of archaeoastronomy. The case remains open, and the debate continues.

Acknowledgements

I would like to record my appreciation of the late Professor Alexander Thom, who, regardless of the eventual outcome of this debate, has drawn public attention to the rich culture represented by the megaliths. A substantial part of this review was first published in the *Proceedings of the Astronomical Society of Australia*, to whom I am grateful for permission to reprint it here.

12

The stone rows of northern Scotland

LESLIE MYATT

Introduction

Caithness and Sutherland form the most northern parts of the Scottish mainland. Caithness, at the north-east tip, is relatively flat until it borders upon Sutherland where, with its deeply etched valleys, the land rises to over 1000 m. Largely undeveloped, apart from hill farming and the more recent afforestation, this area is rich in archaeological remains, many of which are still unrecorded. There are numerous examples of megalithic remains including isolated standing stones, settings of stone circles and perhaps more intriguing, the settings of stone rows. These are usually made up of a number of rows of small stones radiating in a fan truncated towards the centre from which they radiate. Although there are also examples of parallel stone rows, they are much less common than the fan-shaped settings, more examples of which have been discovered in recent years.

Surveys of some of the stone fans were carried out in the nineteenth century by Sir Henry Dryden, but it was not until the work of Professor Thom that the results of accurate surveys were published. Furthermore, Professor Thom has, through a painstaking analysis, suggested both a geometrical construction and also an astronomical purpose for these settings. It is the work of Thom which has opened up the field for further investigation of both the sites which he surveyed and also new ones which have been discovered recently.

The distribution of stone fans

Although parallel stone rows are found in other parts of Britain, the fan-shaped arrays are only known in the north of Scotland, and in particular in Caithness and the adjoining part of north-east Sutherland. Elsewhere, a fan-shaped setting of four rows is known at Beaghmore, Co. Tyrone, but the rows are more widely spaced than the Scottish examples. Also in Brittany similar sites are well known, although generally on a much grander scale, extending over a greater area and having much larger stones.

The northern Scottish settings are of small stones, often less than 0.5 m above the surface and obscured by the surrounding vegetation. At a number of sites they have been almost completely submerged by the development of peat which makes them difficult to find. It is only by diligent field work that more sites have recently been discovered and undoubtedly yet more will come to light.

In the areas where the fans are found they generally follow the distribution of the Orkney-Cromarty group of chambered tombs. Following easy routes across the country, the Sutherland examples lie above the course of the Helmsdale river following the valley inland from the east coast. They are then again found near the north coast above the valleys of the rivers Naver and Borgie. In Caithness there is a concentration near the east coast centred around the Watenan area and the distribution then follows an easy route through Camster and then across to the course of the Thurso river, and finally near the north coast at Upper Dounreay. That some sites have been destroyed in the more fertile agricultural areas is almost certain, and yet others in the more remote moorland regions probably still await discovery.

There is little reference to the stone rows in the early archaeological papers, although Gunn (1915: 353) does mention two sites in Caithness which are not now known. One is 'between Westfield and the village of Broubster', and the other 'above Borlum in Reay'. An extensive search has failed to reveal either of these sites. Further, the 1:10560 O.S. map ND 06 NW indicates a site at ND 007660 which appears to have been destroyed in the process of land drainage.

Although the best-known site is that at Mid Clyth on the east coast of Caithness, the largest concentration occurs 3 km to the north-east near Loch Watenan which is one of the richest areas locally for antiquities including many chambered tombs. This is a rough moorland area overgrown with

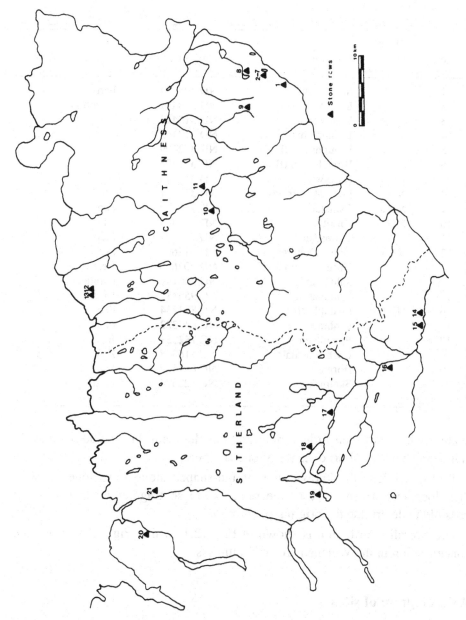

Fig. 12.1. Distribution of stone rows in Caithness and Sutherland

Table 12.1.

Site	Thom no.	Location	Grid reference	Remarks
1	N1/1	Mid Clyth	ND 295384	Fan
2		Loch Watenan	ND 318411	Uncertain
3	N1/9	Watenan	ND 315412	Fan
4		Broughwhin I	ND 313412	
5		Broughwhin II	ND 312409	
6		Broughwhin III	ND 311408	
7		Garrywhin	ND 314413	Fan
8	N1/7	Loch of Yarrows	ND 313441	Fan?
9	N1/14	Camster	ND 260437	Fan
10	N1/17	Dirlot	ND 123485	Fan
11		Tormsdale	ND 148497	Fan
12	N1/3	Upper Dounreay I	ND 012660	Fan
13		Upper Dounreay II	ND 007660	Destroyed
14		Kildonan I	NC 966189	Parallel
15		Kildonan II	NC 955185	Destroyed
16	N2/1	Learable Hill	NC 893234	Fan
17		Kinbrace	NC 827322	Fan
18		Badanloch	NC 782351	Fan
19		Loch Rimsdale	NC 716348	Fan
20		Borgie	NC 661587	Fan?
21		Skelpick	NC 722574	Destroyed

heather which has seen little agricultural development, but includes some six stone fans within a short distance of one another.

Table 12.1 lists the known sites of fan-shaped stone rows together with their locations. In some cases the remains are so fragmented that it is not possible to determine the original pattern.

The overall distribution is shown in Fig. 12.1, whilst Fig. 12.2 shows the concentration in the Watenan area of Caithness.

A description of sites

Until the work of Professor Thom, little had been published on the northern Scottish stone rows. Few plans of the sites existed apart from the results of the less accurate surveys of Sir Henry Dryden which were carried out in the late

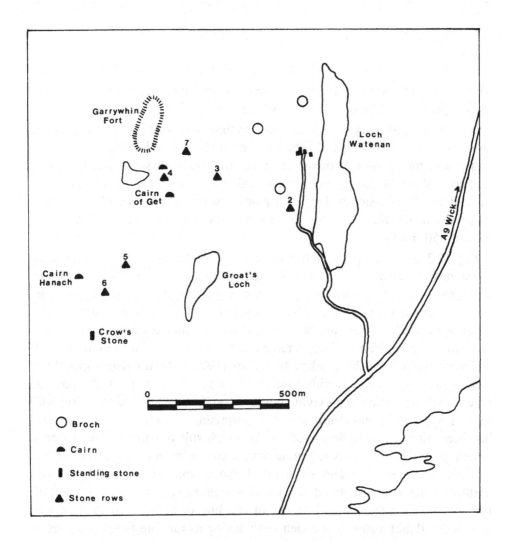

Fig. 12.2. Distribution of stone rows near Watenan, Caithness.

nineteenth century. Even Thom has published plans of only about a quarter of the sites which are currently known to exist, some of which have been discovered since his work was carried out. With more evidence now becoming available further analysis of the stone fans should eventually become possible.

The RCAHMS inventories of Caithness (1911a) and Sutherland (1911b) give descriptions of a number of the sites as they existed at the beginning of this century, but of course do not include the more recent discoveries. The following descriptions include all the currently known sites.

Mid Clyth

This is the best-known and most frequently visited site. Known as the 'hill o' many stanes', it is also one of the most impressive.

The fan-shaped arrays of small stones radiate outwards down the southern slope of a small hill. Whilst the tallest stone is about 1 m in height, 1 m wide and 0.4 m thick, most of them are much smaller. Each stone is set with its long axis along the line of the row, of which there are now 23, containing a total of nearly 180 stones. Dryden's plan drawn in 1871, and which is now preserved in the library of the Society of Antiquaries of Scotland, shows a total of 250 stones.

Fig. 12.3 shows the plan published by Professor Thom (1971: 92), in which the stones are shown to be set out on a grid of radial lines and arcs. There are three fans with the two western ones intersecting and the axis of the main fan is almost due north and south with an azimuth of 358°.5. The centres of the other two fans also lie on this same axis. The main fan has a length of 43.5 m, is 40 m along the base and has a radius of 110 m. Other stones to the east of the main sector have been taken by Thom (1971: 92) to suggest that there were annexes symmetrical with those to the west. Atkinson (1981: 206) on the other hand states that the surviving evidence on the ground gives little support for this hypothesis, and that it would be surprising if such an eastern annexe had been almost totally destroyed, whilst much still remains of the adjacent eastern part of the main rows. Setting out the stones in such a plan would have presented no small problem since the distance from the base to the most northern radial centre is about 125 m. This would place it beyond the top of the hill from where it would be out of the line of sight from most of the stones. No doubt a row of wooden posts acting as ranging poles could have been used to set out the rows from the appropriate centre.

Commenting upon Thom's plan, Burl (1976a: 158) points out that only 29% of the stones actually lie on the grid and that 41% are up to 0.6 m off the line of the row where the rows are only 1.8 m wide. Nevertheless he does accept that some of the stones may have been moved by solifluction. Additionally, Thom & Thom (1978a: 3) have pointed out that other factors may have caused disturbance, including the action of frost. Certainly it has been observed locally, where stones are not anchored to the bedrock below, that owing to the differential thawing of frozen ground in winter on the two sides of stones of this size they do 'float' around. Over a long period of time the movement can be quite appreciable.

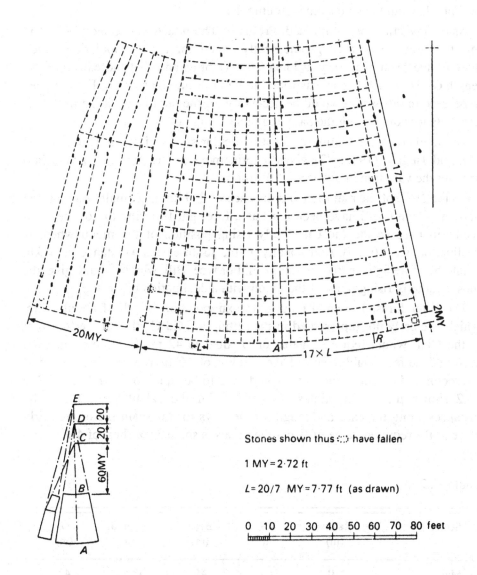

Stones shown thus :::) have fallen

1 MY = 2·72 ft

L = 20/7 MY = 7·77 ft (as drawn)

0 10 20 30 40 50 60 70 80 feet

Fig. 12.3. Mid Clyth stone rows.

In looking for a purpose for the stone rows Thom (1967: 158) suggests that it may have been astronomical. To the north of the Mid Clyth setting is the Hill of Yarrows where between 1900 and 1760 BC Capella would have been seen to set to the west and rise to the east of the hill. The azimuths of the rows in the main sector coincide with this phenomenon over that period. After 1760 BC Capella would have become circumpolar.

As his investigations continued, Professor Thom later suggested (1971: 91) that the stone rows were geometrically suited for the extrapolation which would have been necessary at lunar observatories and that the extrapolation length G (Thom 1971: 85) was found to be built into these sites. For the sites to be used in this way it would be necessary for the fan to have a radius of $4G$, and distances of G along the base and radial length of the rows.

The method of calculating G for a particular site is given in Thom (1971: 85), and Heggie (1972: 47) gives a refinement of the method to take into account the variation in distance of the moon.

To the NNE of the main fan at a distance of 2.9 km is a small notch on the horizon which could have been used as a foresight from the top of the main sector to observe the most northerly rising position of the moon when its declination was $+(\varepsilon + i)$. Additionally, to the south-east at a distance of 80 km is another foresight in the direction of Fordyce Hill which would give the most southerly rising point of the moon with a declination of $-(\varepsilon + i)$.

The latter foresight would require an extrapolation length of about 1 km, which is clearly not represented in the setting at Mid Clyth, but the foresight to the NNE would require a value of G equal to 42 m. This means that the sector of the fan would need to have a radius of $4G$ or 168 m and the base of the sector and length of the rows would need to be equal to G, or 42 m. Table 12.2 shows the actual values compared with the calculated values. The distances along the base and length of the rows for the main sector are fairly close to the required value of G but the radius is some 35% short of $4G$.

Table 12.2.

Sector	Radius (m)	$4G$ (m)	Base (m)	Length (m)	G (m)
Main	110	168	40	43.5	42
SW	126	168	13	26	42

If the fans were used for extrapolation it is surprising that the radius of the main fan is so far short of the value of $4G$. There may have been difficulties in setting it out because of the falling ground at the top of the fan. Nevertheless, despite this shortcoming, the main sector could have been used for extrapolation without too much loss of accuracy.

Although the south-western sector has a somewhat greater radius, it still falls short of the calculated value, and the length and base dimensions would be too short for extrapolation purposes.

Loch Watenan

This is a possible site of stone rows. Above the road, and to the west of Loch Watenan, is a small piece of level ground running north and south below the remains of a broch. At the south end, where the ground begins to fall away, are two small upright stones, about 0.6 m high, firmly set in the ground. To the north of these are a number of much smaller flat stone slabs lying flat and embedded in the surface. They have the appearance of the remains of stone rows but there are insufficient stones to determine a pattern, and the evidence is slight.

Watenan

This is a rather ruinous site, but nevertheless one which enabled Professor Thom (1964: 531) to produce a geometrical plan similar to that of Mid Clyth. This is reproduced in Fig. 12.4. The stones are set on a low ridge of level peat-covered ground running north to south. The remains of four quite distinct rows may be seen with small upright stones still standing. In total, 15 stones are still erect with their flat faces running along the rows and a further 37 are seen to be fallen. Apart from the largest stone, which stands to a height of 0.6 m at the north-eastern corner of the setting, the stones are mostly only ankle high, being submerged in the peat. They are up to 0.6 m in length along the rows.

The plan shows a geometrical grid of radial lines and arcs spaced at intervals of three megalithic yards (2.5 m). The central axis of the fan has an azimuth of 13°.5, with a radius to the base of 280 MY (232 m).

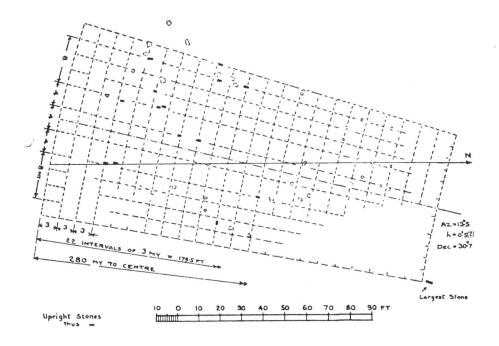

Fig. 12.4. Watenan stone rows.

Broughwhin I

The ruinous remains of a cairn are to be found at this site at the top of a low ridge to the east of the north end of Broughwhin Loch. The remains of stone rows are very slight, but running southwards from the cairn down the slope are three upright small stones with their broad faces running along the line from the cairn. Other stones lie loose on the ground and it is possible that at one time they may have formed part of a setting of stone rows.

Broughwin II

The RCAHMS inventory for Caithness describes the remains of stone rows radiating from an excavated cairn. Unfortunately at the present time the area around the cairn is so thickly overgrown with knee-deep heather that the stones are very difficult to find, but four rows are said to exist including seven stones radiating in a south-western direction from the cairn to a distance of 33 m.

Unless the dense cover of heather can be removed from this site it is difficult to determine the true nature of the remains but the cairn with its central burial cist is easily found and a number of small upright flat slabs not more than 0.5 m above the surface can be located which appear to point towards the cairn.

To the west of the cairn at a distance of 55 m is a single straight row of five small upright small stones running along the top of a ridge. Whether they are connected to, or separate from, the other stones is uncertain.

Broughwhin III

Cairn Hanach, also known as Kenny's cairn, lies to the west of the remains of two rows of small stones. The stones are small and flat, of local sandstone and with their broad faces aligned along the rows. They are situated on level and low-lying ground and run approximately in a SSE to NNW direction.

The row to the west has ten stones over a distance of 34 m whilst the second row to the east has three stones over a distance of 24 m. None of the stones is more than about 0.4 m above the surface.

Garrywhin

Thom (1971: 91) refers to the stone rows at Garrywhin of which a plan appears in Thom (1964: Fig. 2). This plan is undoubtedly that of the Watenan rows, although the grid reference given is that of Garrywhin. It seems unlikely that he would have been able to survey the site at Garrywhin because until about 1980 the heather on this site was almost 1 m high and the small stones were almost impossible to locate. An extensive heath fire then cleared all the vegetation over a large area thus making a survey possible.

Fig. 12.5 shows the results of a survey carried out by Freer and Myatt in 1981, after the fire, when a total of 47 upright stones were located, showing this to be the most extensive setting in the Watenan area.

The first known survey of the site is that of Sir Henry Dryden in 1871 and is reproduced in Anderson (1886: 127) and also in Hadingham (1975: 126). It indicates a total of 46 stones grouped in six rows. In 1911 the RCAHMS

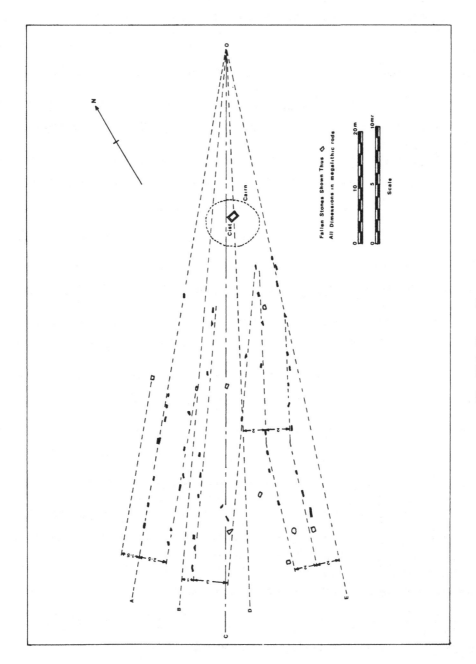

Fig. 12.5. Garrywhin stone rows.

Inventory of Caithness lists eight rows with a total of only 37 stones. Almost certainly at this time there would have been difficulty in locating all the stones because of heather cover. Comparing the Inventory account and the plan given by Dryden, Table 12.3 shows the number of stones in each row, counted from the east.

Table 12.3.

Row	1	2	3	4	5	6	7	8	Total
Stones (Dryden)	11	8	3	9	8	7	-	-	46
Stones (RCAHMS)	10	3	4	1	1	8	5	5	37

The setting of stones is on a peat-covered slope which rises up to the north-east. At the top of the slope are the remains of a cairn in the centre of which is exposed a burial cist. A large capping stone lies displaced at the side and four large slabs of sandstone form the sides of the cist. This was excavated by Dr Joseph Anderson in 1865. In it he found fragments of a beaker of twisted string ornamentation (Clarke 1970: II, 516) amongst which were two human molars. If the cist were associated with the rows, such pottery would give a dating within 2000-1650 BC. From the clay base of the cist were found two oval-shaped pieces of chipped flint (Anderson 1886: 126).

As shown in the plan of Fig. 12.5 the rows of small stones radiate down the slope from the cairn. As in the case of Mid Clyth, it is not possible to view the centre of the fan from the base of the rows since it is beyond the top of the slope.

Superimposed upon the plan is a possible geometrical construction of the setting suggested by Freer (Freer & Myatt 1982: 61). Rows radiate from the centre point O and in addition other rows also run parallel to them separated by distances which can be measured in megalithic rods as shown.

Table 12.4.

Radial line	*OA*	*OB*	*OC*	*OD*	*OE*
Inclination to axis	9° 48′	4° 52′	Axis	2° 25′	12° 16′

The axis is assumed to lie along the line *OC*. The other radial lines are drawn at angles inclined to the axis as shown in Table 12.4. The choice of angle is based upon those previously measured at a similar site of multiple stone rows at Kerlescan in Brittany (Freer & Quinio 1977: 52).

On a clear day it is possible, as at Mid Clyth, to see from the cairn across the Moray Firth to the Hills of Banffshire 80 km to the south. At the present time the heather is begining to recolonize this site and in a few years the stones will be hidden from view again.

Loch of Yarrows

This site, also known as Battle Moss, is close to the east shore of the Loch of Yarrows which is to the north of the stone rows in the Watenan area. The ground is level and the site has probably been partially destroyed in both length and width by cultivation. Both the plan of Dryden (Anderson 1886: 130) and the early Ordnance Survey maps show more stones to the north.

The rows run north and south as shown in Fig. 12.6 which is reproduced from the large scale plan of Thom (1971: 98). The central axis is shown to have an azimuth of 359°.7 and a grid is drawn of lines radiating on a centre 800 ft (244 m) from the base at the north end. Inclined lines are drawn across the radial ones to complete the grid.

For use as an extrapolating sector three possible sites nearby are suggested (Thom 1971: 99).

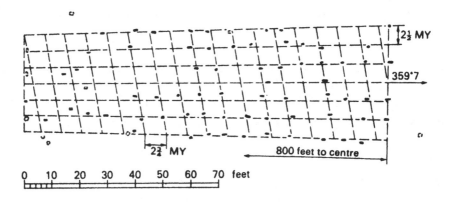

Fig. 12.6. Yarrows stone rows (Thom).

1. An alignment betweeen the most northerly stone of the setting and a cairn at an azimuth of about 191° to give a declination of -($\varepsilon + i$).
2. An alignment between a fallen menhir 1 km to the SSE of the rows and a small break in the horizon to the left of Tannach Hill. This would give a declination of +($\varepsilon + i$).
3. An alignment in the opposite direction, to give a declination of -($\varepsilon + i$) from the horizon feature to the fallen menhir.

A comparison of the present dimensions of the fan according to the plan of Fig. 12.6 and the required values of extrapolation distances for the above alignments is given in Table 12.5.

Table 12.5.

Alignment	Declination	4G (m)	Radius (m)	Length (m)	Base (m)	G (m)
Stone-cairn	-($\varepsilon + i$)	254	244	40	15	64
Stone-hill	+($\varepsilon + i$)	252				63
Hill-stone	-($\varepsilon + i$)	274				69

Depending upon which alignment is taken, the radius of the fan is between 89% and 97% of the required value for 4G. The length and base of the fan are somewhat less than the value of G but of course may have suffered damage by ploughing.

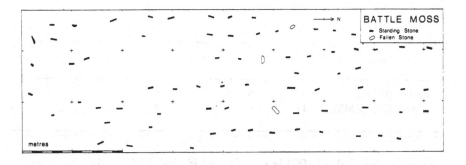

Fig. 12.7. Yarrows stone rows (Ruggles).

Whereas the plan of Thom shows a total of 66 stones standing and a further eleven which are fallen, Ruggles (1981: 193) shows a total of 70 standing and three fallen from a resurvey of the site. The resulting plan is shown in Fig. 12.7 and although he has not carried out a best-fit analysis he suggests that parallel rows are equally consistent with the data.

Camster

Camster is well known for its chambered cairns which are amongst the best preserved on the northern Scottish mainland. It lies inland from the previously mentioned sites of stone rows and may have been on the route northwards across the county to the valley of the River Thurso.

The stone rows are situated about 230 m south of the round cairn in a small valley which rises to the north. The ground in this area is peat covered and for the greater part of the year very wet and boggy. Consequently most of the stones are now buried beneath the surface. Nevertheless it is possible that because of its submergence this could be a well preserved site. Only excavation could show whether this is the case.

The earliest known survey is that of Sir Henry Dryden in 1871. It shows a total of 38 stones erect and above the ground (Hadingham 1975: 126) and a further 34 were thought to exist beneath the surface. In 1911 the RCAHMS Inventory of Caithness listed a total of 34 stones, and Thom (1971: 99) was able to find only 14 definite stones and about a further twelve which were buried. Table 12.6 gives a comparison of the Dryden and RCAHMS surveys with the rows counted from the west.

Table 12.6.

Row	1	2	3	4	5	6	Total
Stones (Dryden)	8	9	6	6	7	2	38
Stones (RCAHMS)	11	9	5	4	3	2	34

Thom's plan (1971: 100) is reproduced in Fig. 12.8. Although there are so few stones shown, he does construct a grid; but, as suggested by Ruggles (1981: 193), to quote a sector radius on the basis of so few stones seems

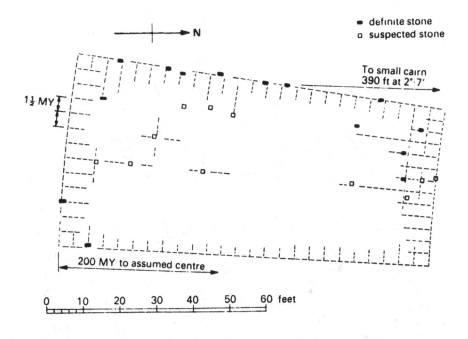

Fig. 12.8. Camster stone rows.

questionable, although Thom himself stated that it was not possible to be perfectly certain of the radius.

As an extrapolating device Thom suggests that the rows were used in conjunction with a sightline from the nearby cairn to a distant boulder 1.8 km away to the SSE giving a declination of $-(\varepsilon + i)$. This would require a value of 202 m for $4G$ as compared with a radius of 166 m for the grid. The length and width at the base of the sector are shown as 31 m and 16 m respectively compared with a value for G of 51 m.

Dirlot

This site, although only a single sector fan, has a number of the features also found at Mid Clyth. It has been set out on the slope of a small hill rising up from the base of the fan, and again it is impossible to view the centre of the fan from the base since it lies beyond the top of the hill. There are over 70 stones which can still be located above the surface but most of which are fallen. They are difficult to locate because of the depth of heather.

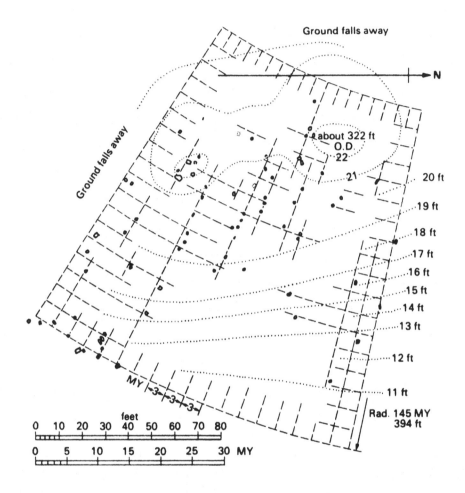

Fig. 12.9. Dirlot stone rows.

Fig. 12.9 shows Thom's plan (1971: 96) which indicates the stones set out on a grid of radial lines and arcs. At the summit of the small hill are two small mounds which do not have the appearance of being natural and could be the remains of burial cairns. The stones are of local sandstone set pointing along the rows. Most are about 0.15 m above the surface except at the base of the fan where they are somewhat larger, and up to 0.6 m in height.

As an extrapolating site Thom (1971: 97) suggests that it could have been used in conjunction with a foresight at Achkeepster, 5.7 km distant and shown on the Ordnance Survey map as the site of three standing stones. This would give a declination of $+(\varepsilon - i)$ for the rising moon at the minor standstill and an extrapolation value for $4G$ of 127 m. This compares with the radius of the fan of 120 m which gives close agreement. The length and base of the fan are shown on the plan to be 42 m and 45 m respectively which compare with a calculated value for G of 32 m.

Although it would have been difficult to see a standing stone as far away as Achkeepster, Thom suggests that the site may have been more easily distinguished perhaps by the erection of a cairn at this position. Such a cairn, if indeed it did exist, may well have been removed for building purposes since at this spot are the remains of a croft house and other buildings in the near vicinity.

From the top of the rows, near the small mounds, Thom (1971: 97) shows the profile of the hills 22.5 km distant to the south-west where the moon would have been seen setting at its major standstill with a declination of $-(\varepsilon + i)$ behind Scaraben. The extrapolation distance of $4G$ for this foresight would have a value of 2432 m which is much greater than the corresponding distance found at the fan. Thom (1971: 98) suggests a row of slabs near to the fan which could have given an extrapolation distance close to this value. Atkinson (1981: 206), however, sees no reason to suppose that these slabs are anything other than the remains of relatively modern stone fence which is quite common in this part of Caithness.

Tormsdale

Down the Thurso river 3.5 km from Dirlot, and on the opposite bank, are the stone rows at Tormsdale. They were first recorded (*D & E* 1982: 49) by the Archaeology Branch of the Ordnance Survey as a result of a survey carried out in this area in 1982. Fig. 12.10 shows the results of a survey carried out by the author and C. Morris (Myatt 1985: 7) during the summer of 1984.

The site is for the most part grass covered, extending over an area of 60 m × 60 m, and bounded on the north and west by an extensive area of peat where cutting has taken place. The ground is almost level, and rises only very slightly to the south-east. The stones in the setting are aligned with their flat surfaces along the rows and, with a few exceptions, are small in height. Many appear as small hummocks where the grass has grown over them.

A total of 103 stones were located of which 43 were upright and the remainder fallen. Taking the co-ordinates of the centre of each stone, a computer analysis was carried out to give a best line fit for each of the rows.

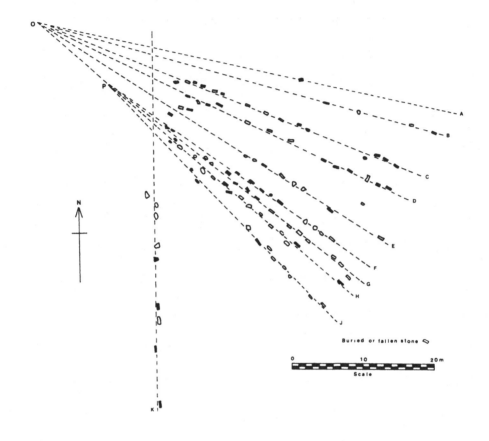

Fig. 12.10. Tormsdale stone rows.

The geometry of the plan is based upon this analysis. It would appear that six rows radiate from point *O* and that a further three rows radiate from row *H* at point *P*. Also, a single row *K*, which crosses the two fans thus formed, runs due north and south. The geometry of the rows thus appears to consist of two conjoined fans together with the separate north-south row. It was not found possible to fit radial arcs joining the stones as indicated on the plans of Professor Thom.

The north-south alignment is also found at Mid Clyth and Loch of Yarrows. At a number of other sites throughout Britain, including Callanish, quite definite indications of the meridian have been found (Thom 1967: 95). It may be significant that the tallest stones in the setting are found at the south end of this row.

A further feature also found at Mid Clyth is that of more than one fan-shaped sector. In the case of Tormsdale they radiate from the two points *O* and *P* which are 13.26 m or 16 MY apart (Thom 1967: 34).

The situation of this site belies the idea that the stone fans are always set out on ground which rises up to the narrow end of the fan - a feature which is found at a number of other sites.

The two sectors appear to have radii of approximately 58.5 m and 43.5 m.

Upper Dounreay I

The Upper Dounreay area of Caithness has a number of antiquities dating from the Neolithic or Bronze Age. On the hill of Cnoc Freiceadain are the remains of two long cairns, one which is known by the name of the hill and the other as Na Tri Shean (Henshall 1963: 267, 282). Five hundred metres to the north are the remains of another cairn having an elliptical setting of eight small stones around it. The stone rows are situated where the ground falls away to a hollow. Also in the vicinity are the short horned cairn of Upper Dounreay (Henshall 1963: 298), a single standing stone 2 m high and a possible prehistoric quarry (Myatt 1977: 46).

The setting of stone rows is in a saddle between two hillocks and the plan of Fig. 12.11 is based upon a survey carried out by R. Freer and the author in 1975. An earlier survey by Sir Henry Dryden in 1871 shows more than twice the number of stones which can be found today and, although his plans are not always accurate, his survey of this site has been relied upon in an attempt to establish the geometrical pattern. The only other survey known of this site is of the west end by Thom (1964: 532).

The suggested geometry shows a number of radial lines from the centre *O* with an assumed axis *OE*. The rows of stones run parallel to these radial lines as found at Garrywhin and the distance between the outer lines of a parallel set is four megalithic rods. In the centre of the site the sets of lines overlap and some of them have been omitted from the plan for clarity. Most of the stones lie within a truncated sector contained within radii of 64 and 100 MY from the centre *O*. A few other stones lie within a sector up to 132 MY from the centre. Table 12.7 shows the inclinations of the radial lines relative to the assumed axis *OE*.

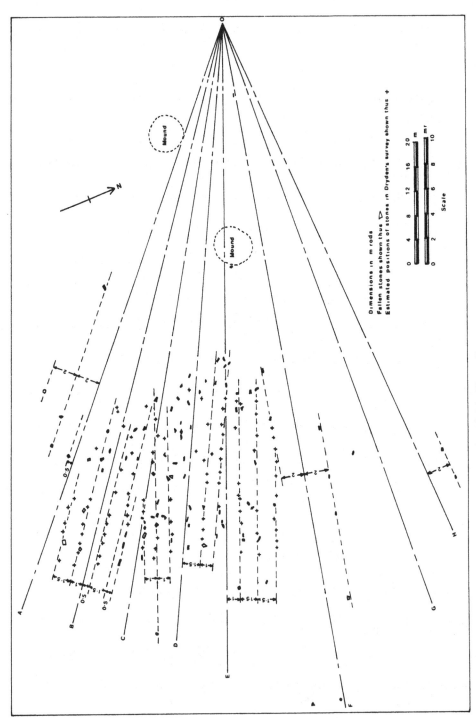

Fig. 12.11. Upper Dounreay stone rows.

Table 12.7.

Radial line	OA	OB	OC	OD	OF	OG	OH
Inclination	19° 36′	14° 44′	9° 48′	4° 52′	9° 48′	19° 36′	24° 19′

Comparing the above table with Table 12.4, there is seen to be a similarity between the construction of the setting at Upper Dounreay and that at Garrywhin where similar angular relationships are found. As in the case of a number of other settings the centre of the fan is not visible from the rows since it is situated over the top of the hillock.

On the higher ground to the west of the rows, but contained within the geometry of the fan, are the remains of two low grass-covered mounds which have the appearance of hut circles or small burial mounds.

If any astronomical observation was associated with this setting there is a possibility that observations could have been made from Cnoc Freiceadain, using St. John's Head some 28 km distant, on the island of Hoy in Orkney, as a foresight. This may have been used to determine the position of the rising moon with a declination of $+(\varepsilon + i)$ at the major standstill. However, such an alignment would require an extrapolation length $4G$ of about 660 m which is considerably greater than the radius of the sector of the fan found at this site.

Upper Dounreay II

The large scale O.S. map indicates the site of stone rows at NGR ND 007660 which is about 500 m to the west of the previous setting. Nothing can now be found of these stone rows, and since the map also shows field drains in this area, it is possible that the rows may have been destroyed at the time the drains were laid. There is no other known record of such a site having existed at this position.

Kildonan I

In Sutherland, the Strath of Kildonan runs inland from the east coast at Helmsdale. This strath shows evidence of habitation from earliest times to the present day, and described in the Sutherland Inventory (RCAHMS 1911b:

132) are stone rows at Torrish Burn. A plan based upon a survey made by the author is shown in Fig. 12.12. Two rows of large boulders of local stone set firmly in the ground rise steeply to the west towards the top of a small hillock. The rows diverge slightly as they rise and on top of the hillock is a low rectangular mound about 8 m × 4 m with a number of small stones around its periphery.

The overall length of each row is under 8 m, the tallest stone is 0.5 m and, because of the size and closeness of the stones, it is not certain that this is the same class of monument as the others which are described. Nevertheless the east-west orientation of the rows should be noted.

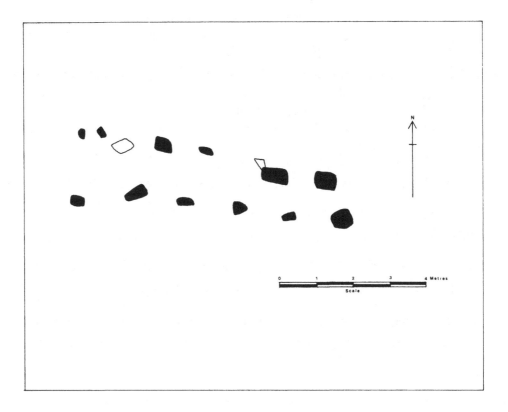

Fig. 12.12. Kildonan stone rows.

Kildonan II

Stone rows are described at Allt Breac in the Sutherland Inventory (RCAHMS 1911b: 133) and are also shown on the large scale O.S. map on the south side of the road along the Strath of Kildonan. Although a plan given in the Inventory indicates the existence of about 14 fan-shaped rows, they can not now be located.

The site is shown as being alongside the road which is single track with passing places. In recent years this road has been realigned and new passing places have been added. It would appear that since there is now a passing place where the site is indicated on the map, the stones may have been either covered over or removed.

Learable Hill

Towards the summit of Learable Hill, west of the River Helmsdale, is the most extensive single site of stone settings known in the north of Scotland. There are stone rows comprising parallel, intersecting and fan-shaped arrays, a stone circle and a single large standing stone, 1.6 m high and inscribed with a plain Christian cross. There are also a number of peat covered mounds, which have the appearance of being man-made. All are contained within an area of 300 m square at around the 160 m contour.

Professor Thom has published a plan of only a small part of this site (Thom 1967: 153). This shows part of the setting of four parallel rows and also the intersecting rows. There are three azimuths indicated by these rows which, in conjunction with foresights on the horizon, give calendar declinations according to the proposed sixteen-month calendar (Thom 1967: 109). The sightings would be on the rising sun and would give indications of the equinoxes and calendar dates in May and August.

The plan of Fig. 12.13 shows the results of a survey carried out by the author and G. Leet during the autumn of 1985 of the fan-shaped array at the south end of the site. A total of 65 upright stones were located with a further 17 which had fallen. The stones are of local rock and tend to be of rounded boulders since the local stone does not split along flat planes like the Caithness flagstone. They appear to conform to a pattern of ten rows radiating from a single point. The inner and outer radii of the fan are 17.5 m and 39 m respectively which makes it much smaller than the Caithness examples.

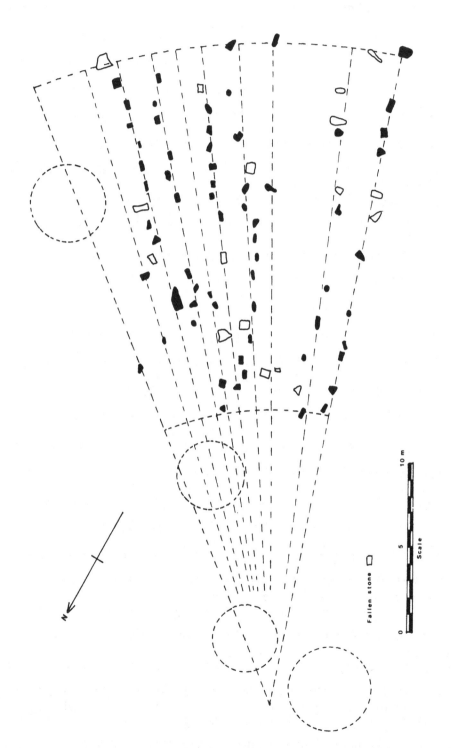

Fig. 12.13. Learable Hill stone rows.

N

Fallen stone ▱

Scale

0   5   10 m

Although the ground rises from the base of the fan to the centre, as in the case of Mid Clyth and Dirlot, it would be quite possible to see the centre of the fan from the base of the rows.

The single large standing stone lies at a distance of 23 m to the north of the centre of the fan. A very noticeable feature of this site is the large number of small mounds, four of which are located on the plan. They do not appear to be related in position to the geometry of the rows either in the fan or the plan of the parallel and intersecting rows shown by Thom but are most commonly distributed in the region between the fan and the large standing stone.

Because of the complex nature of the Learable Hill site, with its numerous stone settings extending over a large area, a further detailed and accurate survey of the complete site is desirable. This would enable an analysis to be made of the interrelation between the individual settings, together with any further possible astronomical significance which the site may have.

If the standing stone was intended as a lunar backsight the most southerly rising moon with a declinaton of $-(\varepsilon + i)$ would appear behind the slope of Beinn Dhorain at a distance of 7.7 km where it forms a notch on the horizon with Creag Dal-Langal. This would give an extrapolation distance, G, of 157 m. This is approximately four times the radius of the fan.

Kinbrace

Along the valley of the River Helmsdale, beyond Kinbrace, another area densely scattered with antiquities is found. This is known locally as Ach-na-h'uaidh (field of the graves) (Sage 1899: 60). There are a number of chambered cairns, hut circles and old field boundaries together with the remains of an old chapel site and burial ground which was still in use until the middle of the last century. Close to the old road, now disused, is a setting of stone rows first reported by the author (Myatt 1975).

The setting is most noticeable because of the stones at the base of the fan which are taller than the rest, being about 0.4 m high. All the stones are of local rounded stone, some of which only just protrude above the surface, and the ground rises gradually from the base of the fan.

Fig. 12.14 shows a plan of the stone rows which is the result of a survey carried out by R. Freer in 1975 (Freer & Myatt 1983: 125). Compared with the Caithness rows, the area covered by the site is very small. The pattern is difficult to determine with certainty, but an assumed geometrical construction has been superimposed on the rows. They lie on lines parallel to the radial lines radiating from point O with spacings between the outer rows of 4 MY

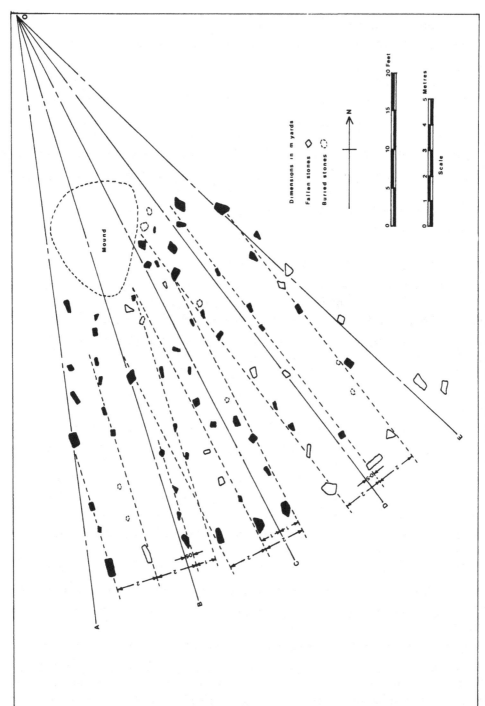

Fig. 12.14. Kinbrace stone rows.

instead of four megalithic rods as found at Upper Dounreay and Garrywhin. Assuming the axis to be along the line *OC*, the other radial lines are drawn at angles inclined to the axis as shown in Table 12.8. These angles should be compared with the corresponding angles found at both Upper Dounreay and Garrywhin.

Table 12.8.

Radial line	*OA*	*OB*	*OD*	*OE*
Inclination	19° 36′	9° 48′	9° 48′	19° 36′

Freer has pointed out (Freer & Myatt 1983: 125) that the alignment *OA* points to a notch on the skyline formed between the slopes of An Cnoc Buidhe and Eldrable Hill at the southern end of the Strath of Kildonan. This locates the rising point of the moon at its minor standstill with a declination of $-(\varepsilon - i)$.

The value of the extrapolation distance, *G*, calculated for the notch is 41 m and the radius of the sector is 22.5 m. This is approximately equal to *G*/2.

In common with many of the other settings of stone rows there is a small peat-covered mound associated with the site. This is shown on the plan at the north-west side at the top of the fan. It does not have the appearance of being natural and has an elevation of about 0.3 m above the natural surface.

Badanloch

This site, which overlooks Loch Badanloch, and is about 0.4 km from the northeast shore, was first recorded (*D & E* 1977: 49) by the Archaeology Division of the Ordnance Survey. It is situated on ground which rises gradually towards the summit of a small hillock, Cnoc Molach. A total of 18 small stones can be located in a fan-shaped array. The largest stones, of which the tallest is 0.62 m, stand at the base of the fan, and the ground rises up towards the centre of the fan.

The result of a survey, carried out by the author in 1982, is shown in the plan of Fig. 12.15. The stones of the setting appear to be set out in eight rows radiating from a single point. Taking row *OE* as the axis, rows *OD* and *OA* are inclined to the axis at angles of 9° 48′ and 19° 36′ respectively which

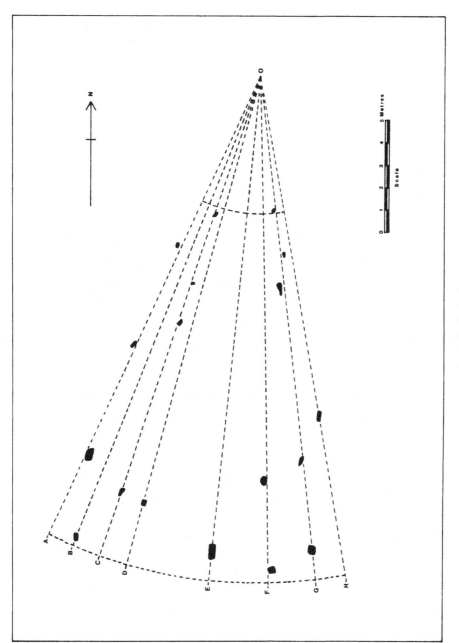

Fig. 12.15. Badanloch stone rows.

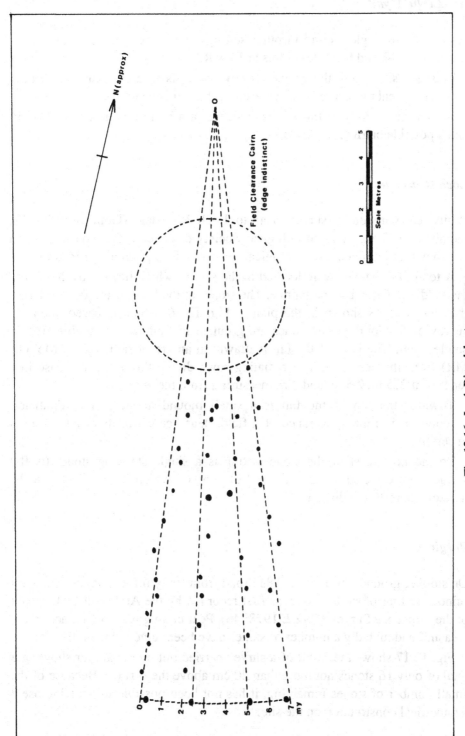

Fig. 12.16. Loch Rimsdale stone rows.

correspond to angles found in other settings. The inner radius of the fan is drawn at 3 MR and the outer radius at 11 MR.

This is a site where the ground is covered in peat, and excavation would probably reveal more stones which are now buried beneath the surface.

Also in the vicinity of this site is evidence of a burnt mound at NC 781356 and a possible cairn at NC 782353.

Loch Rimsdale

Whilst carrying out a survey near to Loch Rimsdale (Gourlay 1975: 7), members of Glasgow Archaeological Society discovered this site in an area which was at imminent archaeological risk from deep ploughing for forestry.

A total of 41 stones were located in the setting which lies almost hidden in the saddle of Cnoc Bad na Fainne. The stones at the base of the fan are larger than the rest. As shown in the plan in Fig. 12.16, which is based upon the survey of R. Gourlay, the stones are grouped in four rows radiating from a single point. The base of the fan is drawn on an arc of radius 27.5 MY (11 MR) from the centre O. The radiating rows divide the arc at the base into lengths of 2.5 m, 2.5 m, and 2 m measuring from the west.

Towards the top of the fan is a small mound about 6 m in diameter. Although it has the appearance of a field clearance cairn, its exact nature is uncertain.

To the north-east of the stone setting is a single standing stone. Its flat surfaces are directed towards the hill of Morven which is prominent on the horizon some 30 km distant.

Borgie

On sloping ground close to the A836 road from Bettyhill to Tongue, this very ruinous setting of stone rows was first recorded by the Archaeology Division of the Ordnance Survey (*D & E* 1978: 46). Peat cutting has taken place in the area and undoubtedly a number of stones have been removed from the site.

Fig. 12.17 shows the result of a survey carried out by the author showing a total of only 16 stones not more than 20 cm above the surface. Because of the small number of stones remaining, it has not been possible to superimpose a geometrical construction on the site.

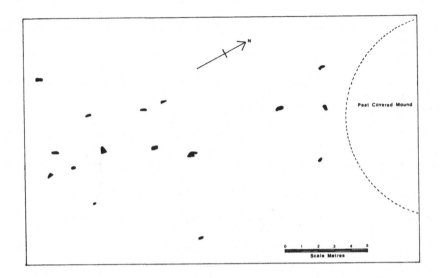

Fig. 12.17. Borgie stone rows.

The ground slopes upwards to the north-east, at which end of the setting is a low peat-covered mound about 10 m in diameter. It has no distinctive features although it does not appear to be natural.

Whilst the geometry of this site is uncertain, it does appear to be aligned roughly in a NNE-SSW direction.

Skelpick

A total of only eleven stones were visible above the surface at this site with the tallest about 20 cm high at the south end. It was first recorded by the author (*D & E* 1975: 55) and was destroyed by land improvement in 1982. It lay not far from the bank of the River Naver, surrounded by higher ground on the other three sides. It is situated in the vicinity of a number of chambered tombs which extend along the Naver Valley.

Fig. 12.18 shows the result of a survey carried out by the author before the site was destroyed; a number of other possible stones beneath the surface were located using a probe. Because of the stony nature of the ground it is not certain that all these stones formed part of the setting.

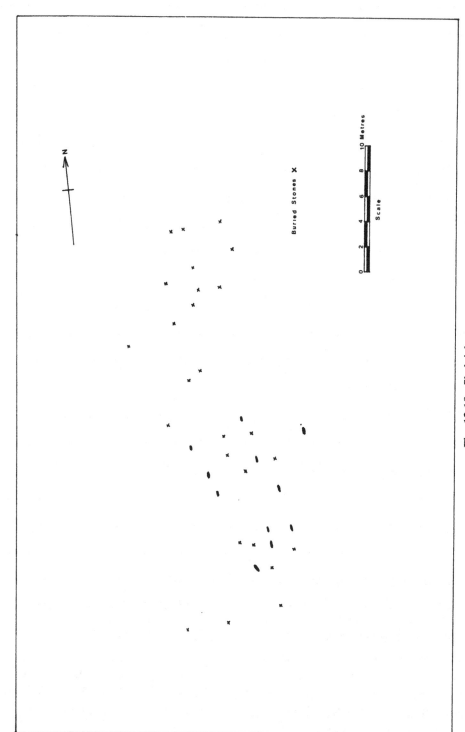

Fig. 12.18. Skelpick stone rows.

The setting is situated on ground which rises slightly to the north and the stones at the southern end are noticeably taller than the rest.

The rows appear to be aligned approximately NNW-SSE but there are insufficient stones definitely associated with the setting to suggest a geometrical construction on the plan.

Comparison of sites

There are a number of features which are common to the sites of stone rows which are found in the north of Scotland, and at the same time features which are unique to certain of the settings.

At all the sites where multiple stone rows are known to exist fan-shaped settings are found. The fans may be single as at Dirlot or multiple as at Mid Clyth. Additionally, parallel rows may be superimposed on the fan as is the case at Upper Dounreay I. It is in these settings particularly that evidence of use of both the megalithic yard and the megalithic rod are found, as also at Garrywhin and again at Kinbrace, although there is evidence in the simple fans also. The Garrywhin site has the addition of a dog leg in the two eastern outer rows.

On Learable Hill is found the most complex site. As well as the fan-shaped rows, parallel rows also exist together with intersecting rows and a stone circle. This is the only known northern site where parallel rows are found.

Stones larger than the rest are not uncommonly found at the base of the fan, almost suggesting that there may be a reason for carefully defining the extremity of the setting. This is often also true with the stone rows on Dartmoor and in Brittany.

It is quite noticeable that at the majority of sites the fans have been set out upon the side of a small hill where the ground rises up towards the centre of the fan. However, this is not a feature which is common to all cases, as is seen at Tormsdale. Furthermore at Mid Clyth, Dirlot and Upper Dounreay I the radiating centre is out of the line of vision from the base of the fan.

Mounds are not uncommonly found at a number of the sites, either within the confines of the setting or in close proximity to it as at Learable Hill. There is definite evidence at Garrywhin that such a mound is associated with a funerary purpose as can be seen from the exposed burial cist. It is tempting to suppose that similar mounds at other sites may have served the same purpose. If this were the case at Learable Hill it would make it one of the largest burial grounds of the period in the north. However, this hypothesis could only be tested by excavation.

Mention was made in the description of some sites of the recurrence of certain angles between the rows of the fans and the axis as noticed by R. Freer. Table 12.9 summarises the results of these findings. Angles close to these values also are found between the rows at other sites.

Table 12.9.

	Garrywhin	Dounreay	Kinbrace	Badanloch	Inclination
		OA	OA	OA	19° 36′
		OB			14° 44′
	OA	OC	OB	OD	9° 48′
	OB	OD			4° 52′
Axis	OC	OE	OE	OE	
	OD				2° 25′
		OF	OD		9° 48′
	OE				12° 16′
		OG	OE		19° 36′
		OH			24° 19′

The other important feature which Professor Thom has found at four of the Caithness sites (Thom: 1971: 104) is that the outer radius of the fan in each case is close to the extrapolation length of 4G. This is required in conjunction with lunar observation since the rising and setting point of the moon moves along the horizon by a considerable amount each day. Such a rapid daily change in azimuth means that it is unusual for it to have maximum declination when on the horizon. For this reason some form of extrapolation would be necessary in order to determine the exact maximum. From this he deduces that there is a characteristic distance, which he calls G, for each site. In using the sectors for extrapolation it would be necessary for the radius of the sector to have a length equal to 4G and length and base of the fan to have maximum values equal to G.

Table 12.10 shows the dimensions found at the four Caithness sites together with the calculated values of G and 4G. The values of G calculated in the above table are slightly greater than those given by Thom (1971: 104) because the corrections suggested by Heggie (1972: 48) have been taken into account.

Table 12.10.

Site	G (m)	Length (m)	Base (m)	4G (m)	Radius (m)
Mid Clyth (main sector)	42	44	40	167	110
Mid Clyth (SW sector)	42	26	15	167	126
Loch of Yarrows	69	40	15	274	244
Camster	50	31	16	202	166
Dirlot	32	44	45	126	120

From the table, the dimensions of Dirlot are very close to the theoretical values needed for extrapolation. At both Camster and Loch of Yarrows the widths of the fans are much too small, but there may be more stones at Camster which is now almost submerged in the peat, and it is most probable that stones have been removed from the width of Loch of Yarrows by ploughing. The radius of the main sector at Mid Clyth is shorter than the calculated value although the dimensions of the length and the base of the fan are about right. Even though the radius is rather short, it has been pointed out by Wood (1978: 128) that the fan could still have been used for extrapolation without too much loss of accuracy. The south-western sector does have a larger radius but the length and base of the fan are both too small.

Possible means of extrapolation have been found by Thom at other sites which are assumed to be lunar observatories. Indeed, if accurate lunar observations were to be made, as is suggested, some method of extrapolation would have been necessary.

At both Learable Hill and Kinbrace, as has already been stated, the dimensions of the fans would appear to be much smaller than those required for extrapolation using the same method as suggested for the Caithness fans. More detailed survey work is yet required on the Sutherland fans to determine possible lunar foresights and their relationships with the fans.

The dimensions of each site, where known, are given in Table 12.11. The azimuths of the fans are measured from the centre looking towards the base. The mean azimuth is that of the centre line and the maximum and minimum azimuths are those of the outer rows. The width of the fan is given in degrees and is the difference between the maximum and minimum azimuth in each case. The two radii given relate to the fan, where the maximum radius is to the base and the minimum radius is to the top of the fan.

Table 12.11.

Site		Location	Azimuth from centre			Width	Radius	
			Min (°)	Mean (°)	Max (°)	(°)	Min (m)	Max (m)
1	(a)	Mid Clyth (Main)	168	178.5	189	21	66	110
	(b)	Mid Clyth NW	193.5	198	202.5	9	-	-
	(c)	Mid Clyth SW	189	192.25	195.5	6.5	100	126
2		Loch Watenan	-	-	-	-	-	-
3		Watenan	190	193.5	197	7	177	232
4		Broughwhin I	-	-	-	-	-	-
5		Broughwhin II	-	-	-	-	-	-
6		Broughwhin III	-	-	-	-	-	-
7		Garrywhin	200.5	211.5	222.5	22	38	94
8		Loch of Yarrows	357.95	359.7	361.45	3.5	204	244
9		Camster	183.25	186	188.75	5.5	135	166
10		Dirlot	102.25	113	123.75	21.5	76	120
11	(a)	Tormsdale N	103	117.25	131.5	28.5	22	56
	(b)	Tormsdale S	125.5	131	136.5	11	11	43
12		Upper Dounreay I	88	109.75	131.5	43.5	56	114
13		Upper Dounreay II	-	-	-	-	-	-
14		Kildonan I	-	-	-	-	-	-
15		Kildonan II	-	-	-	-	-	-
16		Learable Hill	128	144.5	161	33	18	39
17		Kinbrace	133	152.5	172	39	10	23
18		Badanloch	170	187.5	205	35	6	23
19		Rimsdale	160.5	167.75	175	14.5	11	23
20		Borgie	-	-	-	-	-	-
21		Skelpick	-	-	-	-	-	-

The mean azimuths of the fans are shown in Fig. 12.19 where the numbers refer to the sites of Table 12.11. It is interesting to note that, with the exception of Loch of Yarrows, all the azimuths are distributed within an arc of about 102°. They are concentrated between approximately ESE and SSW. For comparison, the extreme southerly rising and setting points of the moon are shown for a level horizon at the major and minor standstills with declination of $-(\varepsilon + i)$ and $-(\varepsilon - i)$. It is perhaps a little surprising to find the azimuths of the rows so concentrated, for if their sole purpose was for extrapolation one might have expected a much more random distribution. The azimuths of the Sutherland fans are much more concentrated than those of Caithness, but of course this may be partly due to the fact that only four sites

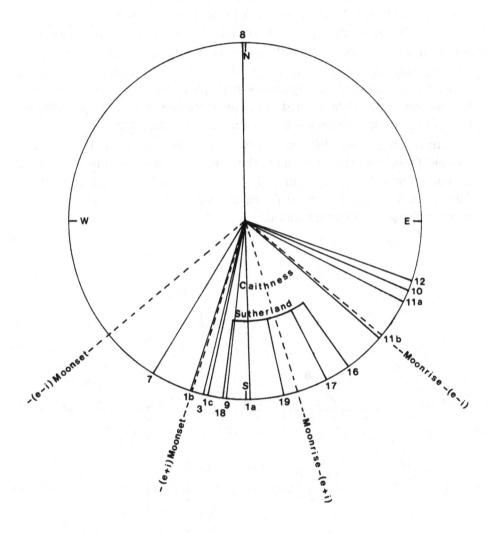

Fig. 12.19. Mean azimuths of fans.

are represented in the diagram. Comparing the Camster passage graves, the orientation of the passages in general range between azimuths of 35° and 152° with a concentration between 65° and 95°. Only three are to the west of north at 223°, 239° and 326°. The great majority appear to face between the lunar risings at the major standstill (A. Burl: *pers. comm.*).

That the azimuth of Loch of Yarrows does not fit into the pattern of the rest is curious. It is aligned almost due north and south but, as has already been suggested, the rows may be parallel and not splayed like the rest.

Fig. 12.20 shows the range of azimuths covered by each of the fans compared with the most southerly limits of moonrise and moonset. It also gives a comparison of the angular widths of the fans. Some fans, such as Camster, may appear to be very narrow perhaps because some of the stones in the outer rows can not be located now, either because they are buried beneath the surface, or else they have been removed during land clearance.

There is a noticeable difference between the dimensions of the Caithness fans and those found in Sutherland. The latter have outer radii to the base of the fan of only between 23 m and 39 m whereas Watenan and Loch of Yarrows both exceed 200 m. At the same time the lengths of the Sutherland rows are also generally much shorter.

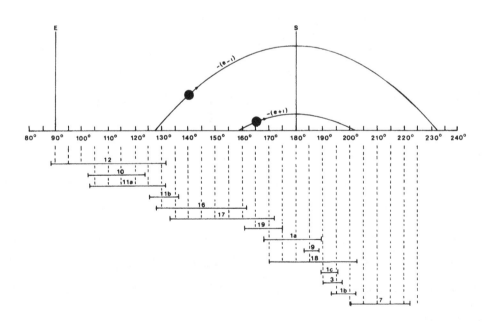

Fig. 12.20. Range of azimuths of stone fans.

Although the dimensions given in Table 12.11 are corrected to the nearest metre, it is interesting to note that at both Badanloch and Loch Rimsdale the plans shown in Figs. 12.15 and 12.16 are each drawn with an outer radius to the base of the fan of 27.5 MY or 11 MR.

Conclusions

Based upon a consideration of his careful surveys of four Caithness settings of stone rows, Thom (1971) has given an interpretation which has far-reaching implications for astronomical understanding by prehistoric society of the period when these alignments were erected. The evidence found at these sites does confirm that they could have been used for extrapolation purposes at lunar observing sites and also adds further strength to the interpretation of other lunar observing sites such as Temple Wood in Argyll (Thom 1971: 45).

Further evidence has been found for the use of both the megalithic yard and the megalithic rod in the setting out of the fans, not only in Caithness but also in Sutherland, both in the sites which he himself has surveyed and others which have been discovered more recently since his work was carried out.

When considering the fans of Sutherland, where they are much smaller than the Caithness examples, it is difficult, with the evidence so far available, to reconcile their use for extrapolation in the same way that the Caithness sites may have been used. For the alignments which have been found, the fans would appear to be of too small a radius. Nevertheless it is still possible that a modification of the method adopted in Caithness was being used.

All the evidence which has been used so far in assessing the geometry and possible use of the stone fans has been derived from what can be seen on the surface. At none of the sites, apart from the cairn at Garrywhin, is there a record of excavation having been carried out. Evidence of how the sectors were set out may exist beneath the surface as was discovered at the stone circle of Temple Wood (Thom, Thom & Burl 1980: 145). Also it is fairly certain that more stones would be found at a number of sites where peat has developed, such as at Camster, and from careful excavation it may be possible to determine more exactly the original positions of the stones where disturbance has taken place.

The other feature of so many sites which can not be assessed at the moment is the presence of mounds which are often situated within, or close to, the fans. Excavation again would reveal whether or not they are associated with burial as at Garrywhin.

Commenting upon the possible astronomical purpose of the stone rows, Burl (1976a: 156) has written 'None of this is incompatible with the belief that the rows were for socio-sepulchral customs'. Certainly the possibility of these monuments having a ritual function should not be overlooked.

In publishing his work on the metrology and astronomical significance of megalithic sites, Professor Thom has inspired the work of others and opened up many new avenues for further investigation. A considerable amount of work can yet be done on the stone rows of northern Scotland. As he has written (Thom 1971: 116): 'It is to be hoped that this monograph does not give the impression that a study of these remains from the astronomical point of view is complete. It is only begun'.

The other comment which he makes concerns the preservation of these monuments. Seven new sites of stone rows have been discovered over the past eleven years and at least one of them has now been destroyed. A number of them do not have scheduled monument status and whereas, in the past, ancient monuments in the north were considered reasonably safe from destruction, this is not now necessarily so with the land developments which are taking place. Stone rows are often very insignificant on the ground, concealed by the heather, and may easily be overlooked.

Acknowledgements

I am most grateful to Dr Aubrey Burl who has encouraged my work on the stone rows and who also read the manuscript for which he offered helpful suggestions on the archaeology of the sites in advance of publication. Mr Robert Freer gave helpful advice in the surveying of sites and also allowed me to use the results of his survey work. Mr Robert Gourlay drew my attention to the site at Loch Rimsdale and provided the results of the survey. Dr Clive Ruggles provided further information on his survey of the site at Loch of Yarrows. Mr Charles Morris, Mr Geoffrey Leet and Mr Elliot Rudie have assisted in survey work. Numerous landowners, including Mr Innes Miller and Sir Jeremy Clay, have willingly given access to sites.

13

Stones in the landscape of Brittany

PIERRE-ROLAND GIOT

For over two hundred and fifty years, wild and lunatic speculations have raged about the standing stones of Brittany; yet despite this, much serious information has been assembled. The essential facts have been brought together in two recent publications (Giot *et al.* 1979; Giot 1983a).

The earliest precise plans, setting some of these stones in the eighteenth century landscape, are the remarkable maps at a scale of 1:14400 surveyed between 1771 and 1785 by the 'Ingénieurs Géographes' along the coasts of Brittany, penetrating inland for two or three kilometres (Service Historique de l'Armée de Terre, Vincennes 1776-83). Unknown to scholars until quite recently (Musée de Brest 1982: 21-9), these maps generally refer to the megaliths as 'pierres' or 'roches', and do not speculate about them.

The first thing to be stressed is that it is quite artificial, and perhaps even criminal, to dissociate the stone rows from the isolated, or apparently isolated, standing stones or menhirs, and from the other archaeological structures that may have been associated with them. In so much tentative speculation, the result has been abstraction from the landscape, and even from the archaeological evidence for the Neolithic period of the monuments themselves.

The next point is that we are dealing only with the ultimate survivors of these monuments, so many of them, as we now realize, having been re-used as a source of stone from Neolithic times onwards. Others have been destroyed by successive religious beliefs, or by superstition. Natural agencies such as weathering have taken their toll, not all stones being equally resistant, and lightning frequently strikes and splits them, or breaks off fragments. If the rate of destruction by lightning observed during the last 50 years is extrapolated over 6000 years or so, many losses can be explained, the broken fragments having been built into field walls and banks. Associated structures and complex monuments have suffered even more when the land on which they stood has been brought under cultivation.

At the time of the first statistics of monuments, in the nineteenth century, there were probably about 1200 isolated menhirs in Brittany, and 100 alignments and megalithic enclosures, the common reckoning being a total of about 4750 standing stones, or more than 5000 with those that have been recognized subsequently. This is, of course, more than in other regions of Western Europe of comparable size, and may be related in part to the availability of outcrops and boulders of granite, quartzite and hard schist; though some large stones have been transported from their source over distances of four to ten kilometres. Recently it has been confirmed (Le Roux 1979; 1981) that for alignments on the continent, and not only for British henges, wooden posts can replace standing stones. It is thus not unexpected that structures, alignments and isolated pillars could have existed in timber, as 'megaxyles' (Giot 1976; 1983b) in areas where convenient stones were rare.

Experimental archaeology and comparative ethno-technology have more or less confirmed current ideas about the extraction, shaping, transport and erection of the stones. Some original ideas have been suggested, which could have been used occasionally. A particular problem concerns the nature of the ropes involved; the stems of wild clematis seem the best material in our climate. To move the very large menhirs, weighing 100 to 300 tonnes, was a task of exceptional difficulty, though to move one of only 50 tonnes would have been an achievement.

The erection of stones of small and medium size, such as most of those in the great alignments, was easier than appears at first sight. If we consider those at Carnac (where 4000 stones survive), there may have been initially about 10000 blocks, if all the lines were ever complete at the same time, with no shifting of stones from one site to another. This represents 20000 to 25000 m³ of granite, or a weight of 50000 to 65000 tonnes at most. A large burial mound such as the Tumulus St. Michel at Carnac represents 60000 m³, or about 150000 tonnes; but of course much smaller stones were involved than in the alignments.

Artefacts and other traces found in the stoneholes of menhirs (mainly in nineteenth century excavations) have provided the basis for dating the monuments to the Neolithic, and at the same time have shown the importance of foundation rituals or ceremonies.

Even though the monuments are crude (though some of granite have been carefully shaped by pounding), their appearance on the skyline gives at a distance the impression of human figures. This anthropomorphic illusion is confirmed by the numerous popular legends about petrified soldiers, dancers or wedding parties; and fertility rites suggest other obvious resemblances.

Menhirs with carvings or decorations, engraved, pecked or in false relief, are not very common. It is possible that originally many more were ornamented, but erosion and weathering will have defaced or obliterated these figures, which are very exposed to the elements; lichen, too, will have hidden them. Granite, in particular, erodes grain by grain, so that many granite menhirs have lost several centimetres from their surfaces. Rain, frost, moss and lichens are the most obvious agencies of destruction, and micro-organisms can cause erosion under overhangs. A few monuments have preserved hafted axes in false relief (St. Denec near Porspoder, Finistère) or 'crooks', also in false relief (Kermarquer near Moustoirac, Morbihan). Incised technique has been used at Er-Lannic, Arzon, and at Le Manio, Carnac (Morbihan), where preservation has been helped by the long burial underground of the decorated parts of those stones.

The giant broken menhir of Locmariaquer shows the very eroded remains of a large hafted axe or 'hache-charrue' on the upper surface of one of its fallen fragments. This example relates of course to the recent demonstration that the capstones of certain passage-graves at Locmariaquer and Gavrinis derive from the breaking up of large menhirs or stelae, whether decorated or not (L'Helgouach 1983; Le Roux 1985).

The very fine menhir of St. Samson-sur-Rance (Côtes-du-Nord), now leaning, has a well-dressed flat face, now on the upper side, completely covered by bands of rectangles, just visible in oblique light towards noon. Between the seven bands of rectangles there are 'crooks' and hafted axes and, more surprising still, small animals (though animals are also visible on the large stone divided between the capstones of the Table des Marchands and Gavrinis passage-graves). The sides of the menhir show the return of the bands of rectangles, and other 'crooks' and hafted axes. Weathering does not help the readability of all this remarkable decoration, and the curvature of the stone makes it difficult to decipher all of it simultaneously.

Cup-marks being of all ages, we will only mention that many menhirs bear them. Some stones could have been painted. The rough ones could have been used as interior skeletons for plaited covers made of vegetable stalks; but there has never been any indication of wattle-and-daub covers. These are preserved only if they have been charred.

Even apparently isolated stones belong to a system integrated into the landscape, often a very complex system. They can be in all kinds of topographical positions, from hill tops (though never right at the summit) through slopes to low-lying ground. Needless to say, standing stones now on the shore (the lowest one at Plouguerneau, Finistère, is close to present mean sea level) or in the middle of brackish marshes must have been erected originally on firm ground. Quite a few are next to a spring or beside a small brook, so that there is a clear association with water. Others are associated with natural rocks, either tors or boulders lying around them, from which they themselves are derived. There is no common feature linking these situations.

A few menhirs, on the axis of one of the sides of a gallery-grave (allée couverte), a few metres away, have been described as 'menhirs-indicateurs'. They are certainly related to the funeral monument.

Sometimes standing stones appear to form small groups with obvious inter-relations. Quite often they seem to be in pairs (if no other stones have been destroyed). These pairs are quite often of the 'Tweedledum-Tweedledee' type, a short stone being associated with a tall one, or a large, broad menhir with a slender one. Usually the stones are a short distance apart, a few metres or tens of metres. Terms such as 'brothers' or 'twins' apply well to them, and the 'talkers' is also a colloquial name.

Triads of menhirs are less easy to be certain of. They rarely form well-spaced triangles, and the stones are at greater distances apart. Long distance relationships are very difficult to prove, and there is the danger of getting entangled in 'ley lines'. On the northern side of the prominent hill of Menez Bré (302 m), one of the 'sacred' mountains of Brittany (Côtes-du-Nord), there are, nearly in a straight line running southeast to northwest, a series of three large menhirs: Kervezennec near Bégard (6.40 m), the Menhir de Pédernec (7.40 m), and Pergat near Louargat (7.50 m); but in this last case there is an accessory stone 2 m high 20 m to the northeast. This is not an alignment, and visibility is masked from one stone to the next; but from all three points there is a good view of Menez Bré. The hill appears to be a kind of centre for a system (now crowned by a chapel dedicated to St. Hervé (Huarnaeuus, protector of horses); around it an important annual fair used to take place, inside the very eroded traces of a hill-fort. If one accepts the approximate alignment of the three menhirs of Porzic and Pasquiou near Vieux-Bourg (Côtes-du-Nord), in a landscape of granite outcrops and boulders, one might try prolonging the line 9 km further east to the menhir of Quintin. It would be no more dangerous to draw a line to one of the many large menhirs on the granite massif of Quintin.

Small alignments, with either only a single row of standing stones, or two or three, crossing each other rather than running parallel, are vulnerable for two reasons. First, most of the stones are quite often only modest in size, and are thus easily broken up. Secondly, as these structures occupy land and impede agriculture, most of their components have been destroyed or dragged aside into banks and hedges. What remains of these monuments has survived the 'Remembrement', or agricultural improvement through the clearing of land in the 1950s and 1960s, the stones being 'obstacles à l'utilisation rationnelle des sols'. Even so, most of them were already so ruined that it was nearly impossible to be sure of their original pattern.

There are a few cases of an association of an alignment with a grinding or polishing stone (Lannoulouarn near Plouguin, Finistère, associated with a fibrolite axe-factory; Lagatjar near Camaret, Finistère).

The large alignments are of course mainly associated with the districts of Brittany in which megaliths of all kinds are densest. The fact that nothing today survives of the large alignment near Penmarc'h, Finistère, where 600 stones could still be counted 150 years ago, is a warning that the absence of a type of archaeological structure from a district where it might be expected is not a conclusive argument. The same can be said of the different kinds of enclosures, rectangular or semi-circular.

In reality, the typology of all these different kinds of monument, of which standing stones are the components, is very diversified. It is probably a very crude over-simplification to consider them all together, undifferentiated; and their significance for those who took the trouble to transport and erect the stones may have been equally diverse, and even for the same monument may have varied during the 2000 years of the megalithic period. We feel more and more cautious before getting entangled in any theoretical speculation.

Some antiquaries were writing of zodiacs or calendars at the very beginning of the nineteenth century. In 1874 H. du Cleuziou, instructed to make an official plan of the Carnac alignments, suggested that they were oriented to solar risings and settings at the solstices and equinoxes, so that these sites have been bedevilled by archaeoastronomy ever since; and in 1894 (and especially in 1904) R. Kerviler made calculations about the units of the megalith builders, which have generated contrary views from scholars ever since. After more than a century of diverse hypotheses and efforts to try to demonstrate them, we do not appear to have made much advance.

It is often difficult to separate more or less objective and scientific theories from the mass of esoteric coincidences and the doctrines about them which flourish in certain schools or sects. Unfortunately these lunacies have contaminated the rest. Nevertheless there are sometimes, in the usually crazy arguments of the devotees of 'alternative archaeology', interesting criticisms of the shortcomings of most orthodox archaeological publications about the significance of the stones. Ignorance, or failure to take account of comparative ethnography, can lead to failings and invite censure. In the huge mass of rubbish accumulated by the 'lunatic fringe' there do exist, if one is to be objective, some interesting observations of facts hitherto overlooked or ignored. The difficulty is to discriminate, to evaluate and exploit them, because of their dubious origin.

On the whole topological, metrological and archaeoastronomical research has suffered from this dubious fringe, much as it has also suffered from the fact that in general it has been divorced from the study of Neolithic societies as such.

Restored to their natural and archaeological environment, and considered as elements of their time, the standing stones of Brittany, like those elsewhere, regain a respectability that they may have lost.

14

The orientation of visibility from the chambered cairns of Eday, Orkney

DAVID FRASER

Background

Megalithic structures are an extant and solid remnant of the Neolithic landscape. Being the most durable human constructions in that landscape, they are our most eloquent informants on the arrangement and meaning of the land to our remote predecessors. In recent decades, a school of scholars, nurtured and directed by Professor Alexander Thom, has postulated that there are megalithic buildings in the human Neolithic landscape which were deliberately located in places from which the sky, and the sun and the moon, could be observed at significant times. The archaeoastronomical notion has been widely tested and there is little doubt now that the inhabitants of Neolithic Britain were serious and educated observers of the sky. Relatively little attention has been given to other possible explanations of the location and orientation of megalithic structures: it is the task of this paper to consider, for one time and place, a wider series of interpretations for the arrangement in a landscape of a class of prehistoric monument. Professor Thom was a leader in the pursuit of rigour and objectivity and I trust that he would have approved of the exercise described here, not least because it involves a close and intimate study of the landscape of a small island off the coast of Scotland.

In considering the location of a megalithic structure in the landscape, there is an endless number of aspects of environment and society to explore. Geology, soils, land use, nearness to water, population density, human irrationality, and many other factors, all combine to produce the complex patterns familiar to all landscape archaeologists. It is possible to isolate individual factors by using a statistical approach: examine a large number of sites and compare them with some hypothetical or actual distribution. This paper takes the single factor of visibility in the landscape; examines the patterns of visibility in three selected areas of the the landscape of Eday; compares them with the patterns of visibility from every chambered cairn in Eday, and every chambered cairn in Orkney; and draws some conclusions on the importance of topography and astronomy in the planning of the Orcadian Neolithic landscape.

Visibility and topography

The observation of astronomical bodies requires clear visibility: from any astronomical viewing point, it must be possible to see a celestial body at some significant time in its path across the sky. Most commonly this significant time is postulated to be the moment the celestial body crosses the horizon. Hence the visible horizon is crucial in any archaeoastronomical argument. But there are other reasons why the visible horizon might be important in the decision to construct a megalithic structure. Borrowing from Ruggles' (1984a: 17) discussion of the orientation of prehistoric structures, there are three possible reasons (including astronomy) why the nature and distance of the visible horizon might have been a factor in the decision to locate a chambered cairn.

i.   *Astronomical reasons.* It may have been desirable to observe and record the motion of celestial bodies. Thus the archaeologist might expect to find evidence of indicated alignments towards the rising and setting points of sun and moon.

ii. *Azimuthal reasons.* Locational decisions may have been taken with the object of securing good visibility (or visibility of some other kind) towards certain points of the compass.

iii. *Ground-based reasons.* It may have been desirable to secure good visibility (or visibility of some other kind) towards individual features in the landscape, or towards general areas. Equally, it may have been desirable to secure good visibility from features or areas in the landscape towards the chambered cairn.

Hypotheses based on postulated astronomical and azimuthal reasons are relatively easy to test and there is a sizeable literature on such hypotheses in which Professor Thom and his family feature prominently. Hypotheses concerning ground-based explanations for observed visibility patterns are much harder to test for two reasons: firstly, there are no universally significant orientations to consider and hence the visibility patterns which conform to any hypothesis are not simple patterns to analyse; and secondly, the testing of any ground-based hypothesis will inevitably require the collection of a large amount of comparative information from points and areas which are of no obvious or immediate importance to archaeologists and hence will be expensive of effort and time. Despite these difficulties, there are at least two published examples of studies which examined the possibility of ground-based visibility patterns as a factor in the location of megalithic structures. The first of these (Davidson 1979) concluded that the chambered cairns of Rousay in Orkney were located in positions from which it is possible to survey the three distinct parts of the island where present day settlement occurs. The second (Barnatt & Pierpoint 1983), as part of an examination of low-precision astronomical hypotheses, concluded that the stone circles of Machrie Moor in Arran were located at prominent points in the landscape and that the impressiveness of these sites may have added to their ritual importance.

The chambered cairns of Orkney and Eday

Orkney has been favoured with much field survey and excavation and the quality of our archaeological information on the Neolithic period is comparable with any small region elsewhere in Britain. There are nearly 80 known chambered cairns, varying greatly in size and design, and yet all diagnostically belonging to the same general cultural tradition and period. A gazeteer and discussion of these cairns may be found in Henshall (1963; 1972) and Fraser (1983). The island of Eday is at the heart of the northern Orcadian archipelago: it is about 13 km long (north-south) and a maximum of 4 km wide (east-west) rising to an altitude of 101 m at the summit of Ward Hill. (As an aside, the place-name 'Ward' comes from a Scandinavian root meaning 'beacon' indicating a later historical preoccupation with visibility in Orkney.) Eday and its outliers Calf of Eday and Faray (Fig. 14.1) contain a great variety of landscape types and are close to being a microcosm of the Orcadian landscape in general. The island-group contains 13 known chambered cairns, no less than eight of which have been the subject of archaeological excavation. Of these the best known (although all are infrequently visited) are the three cairns on Calf of Eday, the impressive Maes Howe-type cairn at the summit of Vinquoy Hill, and the 'double-decker' at Huntersquoy, with two chambers arranged one above the other.

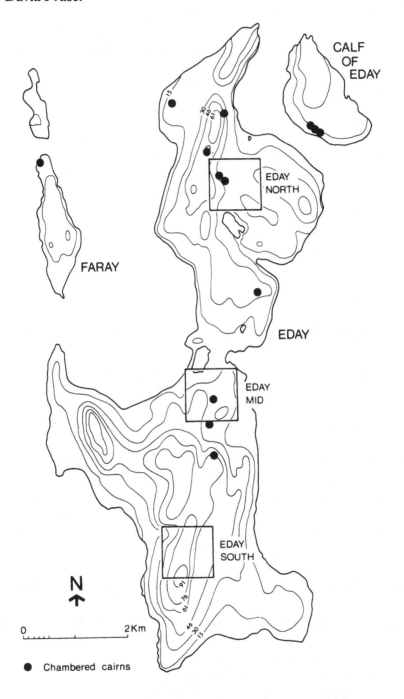

Fig. 14.1. The island of Eday, Orkney, showing the distribution of chambered cairns and the location of the three survey areas.

Method

The hypotheses to be tested depend on the comparison of visibility characteristics of the location of chambered cairns with those of the generality of the landscape of Eday. We thus need some measure of visibility, and some definition of points in the landscape which define firstly each chambered cairn, and secondly the generality of the landscape.

The measure of visibility used is that defined in Fraser (1983: 298). The procedure adopted was to assign each point of the compass (with a resolution of one degree) to one of three visibility classes defined by the distance of the visible horizon. (This measure takes no account of the altitude of the horizon and hence cannot be used to calculate declination information: immediately we are excluding the possibility of testing astronomical hypotheses except at the very lowest of precision levels.) The three visibility classes are defined as follows:

i.   *distant*: visibility exceeding 5 km;

ii.   *intermediate*: visibility between 5 km and 500 m; and

iii.   *restricted*: visibility less than 500 m.

These classes were chosen by experiment during field-work in Caithness and Orkney and were designed to partition the visible horizon into sections which are distinctive from each other and are consistently recognisable throughout Orkney. At this stage, the work of Davidson & Jones (1985: 35) on the existence of tall vegetation in Neolithic Orkney is relevant. They concluded that '...by about 2600 BC, in the Late Neolithic, there was virtually no tree cover remaining ... the major semi-natural components of the Orcadian vegetation seem to have begun to resemble those of the present.' Thus visibility readings taken today are a good approximation to conditions at the time of the cairns.

The operational procedure carried out in the field was to scan the horizon with a prismatic compass. The imaginary circle centred on the observer was divided in plan into sectors of distant, intermediate and restricted visibility, and the direction of the transitions between each pair of neighbouring sectors was noted. All compass readings were converted to degrees true.

Having defined a measure of visibility, we must also define points in the landscape from which measures may be taken. For a point defining a chambered cairn, the observer stands on the ground surface close to the entrance of the passage into the cairn (or on the cairn itself, as close to the entrance as possible, if the mound material obscures the view). By experiment, a difference of a few metres in any direction or in height above the ground surface makes only a negligible difference to the compass reading at this level of precision. For the purposes of this paper, visibility measurements were taken at each of the 13 chambered cairns of Eday, and also at an additional 63 chambered cairns elsewhere in Orkney (the total known population in 1981 at the time of the survey).

The definition of points defining the generality of the landscape of Eday is clearly not as simple a problem. To avoid visiting every point in the island - an infinite number of points - some sampling strategy is required. To avoid large distances between points, a stratified random sample was adopted. Three squares, each one kilometre on the side, were chosen at random from the set of possible squares in the Eday island-group. The three sample squares are shown in Fig. 14.1 and are labelled Eday North, Eday Mid, and Eday South. They are fully described in Fraser (1983: 251-7), and together they reflect the diversity of the terrain of Eday. Each kilometre square was overlain with a 100 metre grid and visibility readings were taken at each of the 100 inter-sections of this grid. Thus 300 readings in total were collected (the operation taking about 20 working hours) and the resultant visibility statistics are assumed to be representative of the generality of the Eday landscape.

It will soon become clear that the argument in this paper depends on the assumption that the distribution of visibility from 300 points in three kilometre squares of Eday is representative of the distribution of visibility in Eday as a whole, and further, is representative of the entire island group of Orkney. This assumption (although admittedly conditioned by the difficulty of data collection) can be justified on the grounds that the three sample areas in Eday contain the same variety of terrain types as can be found in Orkney as a whole.

All results are here presented in graphical form, but the raw numerical data are available from the author. Figs. 14.2 and 14.3 show for each group of points the distribution of distant, intermediate and restricted visibility of orientation. For example, diagram *A* in Fig. 14.2 shows that from the 100 points in kilometre square Eday North, an observer facing 0° (due north) has distant visibility at no points, intermediate visibility at 50 points, and restricted visibility at 50 points. Again, diagram *A* in Fig. 14.3 shows that from the 76 chambered cairns in Orkney, an observer facing 180° (due south) has distant visibilty at 34 cairns (44.7%), intermediate visibility at 29 cairns (38.2%), and restricted visibility at 13 cairns (17.1%).

Orientation of visibility

In these paragraphs we wish to investigate statistically the distribution of orientation of visibility from the chambered cairns of Eday and Orkney (displayed in diagrams *B* and *A* of Fig. 14.3) compared with the distribution of orientation of visibility from the generality of the Eday landscape. Our null hypothesis is not the more common hypothesis of a random distribution (leading to a uniform spread of visibility around the compass), but rather the actual distribution as observed from a selection of 300 points in the Eday landscape, and summarised in diagram *C* of Fig. 14.3.

The statistical test used will be the Kolmogorov-Smirnov one-sample test (Siegal 1956: 47). This is a test of goodness of fit between the distribution of a set of observed values (in our case the orientation of a single visibility class for a group of chambered cairns) and the distribution of a specified actual distribution (the orientation of that visibility class in the three kilometre squares of the Eday landscape). The test determines whether the set of observed values can reasonably be thought to have come from a population with the actual distribution. The test statistic comes from comparing the cumulative frequency distribution of the observed distribution, with the cumulative frequency distribution of the actual distribution, and noting the largest divergence (D) between the two. The probability of divergence depends on the number of observations (N) in the observed distribution, and critical values (C) for rejection of the null hypothesis can thus be determined for comparison with the largest divergence.

Our first test is of the null hypothesis 'the distribution of distant visibility from the chambered cairns of Eday is the same as the distribution of distant visibility from the generality of the Eday landscape'. There are 360 points of the compass and 13 chambered cairns in Eday, making a possible maximum of 4680 observations. Of these, there are 1544 observations (N) of distant visibility. For this number of observations, the critical value (C) for rejection of the null hypothesis at the 0.01 level of significance is 0.041. The test statistic (D) is 0.143. There is less than one chance in a hundred that the null hypothesis is true and it is therefore rejected.

Our second test is of the null hypothesis 'the distribution of restricted visibility from the chambered cairns of Eday is the same as the distribution of restricted visibility from the generality of the Eday landscape'. Of the 4680 possible observations, there are 1387 observations of restricted visibility. For

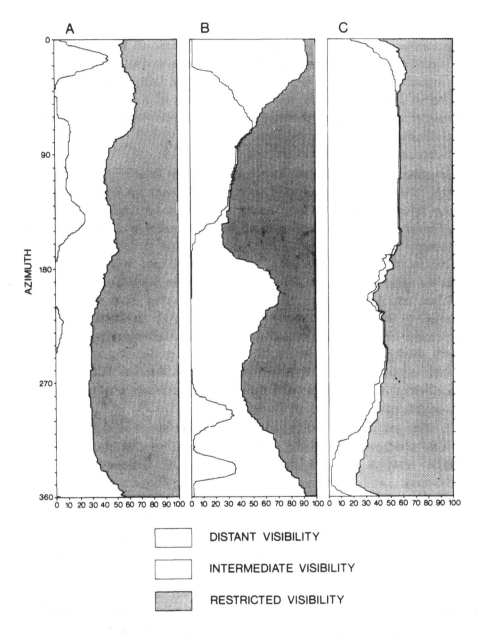

Fig. 14.2. Orientation of visibility from (*A*) Eday North (*n* = 100); (*B*) Eday Mid (*n* = 100); and (*C*) Eday South (*n* = 100).

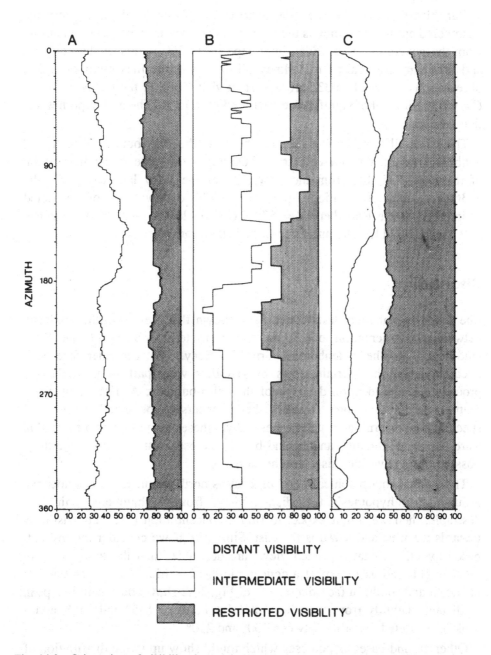

Fig. 14.3. Orientation of visibility from (*A*) the chambered cairns of Orkney (*n* = 76); (*B*) the chambered cairns of Eday (*n* = 13); and (*C*) the generality of the Eday landscape (*n* = 300).

this test, $N = 1387$; $C = 0.044$ for level of significance 0.01; and $D = 0.066$. The null hypothesis is therefore rejected.

Our third null hypothesis is 'the distribution of distant visibility from the chambered cairns of Orkney is the same as the distribution of distant visibility from the generality of the Eday landscape'. The 360 points of the compass and 76 chambered cairns of Orkney allow a possible maximum of 27360 observations, of which 10336 are of distant visibility. For the test $N = 10336$; $C = 0.016$ for level of significance 0.01; and $D = 0.201$. The null hypothesis is thus rejected.

The last null hypothesis examined here is 'the distribution of restricted visibility from the chambered cairns of Orkney is the same as the distribution of restricted visibility from the generality of the Eday landscape'. Of the 27360 possible observations, there are 6724 observations of restricted visibility. For the test, then, $N = 6724$; $C = 0.020$ for level of significance 0.01; and $D = 0.113$. The null hypothesis is therefore rejected.

Discussion

The first two statistical tests above have shown that there is little similarity between the orientation of distant and restricted visibility of the Eday landscape and the chambered cairns of Eday. We can therefore say, unequivocably, that certain types of visibility were part of the selection procedure in the location decision of the cairn-builders. At first sight, there seems to be little more to add - there appears to be equal support for speculation concerning all three types of hypotheses about the visible horizon: astronomical, azimuthal, and ground-based. A closer examination of the three possibilities shows that this is not the case.

The general topographical trend of Eday is north-south. If we consider the ground-based hypothesis that distant views from a chambered cairn are desirable, then we might expect to find a concentration of distant visibility towards the west and towards the east. Similarly, if we consider the ground-based hypothesis that restricted views are desirable, then the topography of Eday might lead us to expect a concentration of restricted visibility towards the north and south. In fact, diagram *B* of Fig. 14.2 shows that there is a peak of distant visibility from the cairns of Eday between 135° and 155°; and a peak of restricted visibility between 200° and 210°.

Other ground-based hypotheses which might show up in the distribution of orientation of visibility include the desire to ensure distant visibility towards a single point - for example Vinquoy Hill or Ward Hill. The summit of Vinquoy

Hill is visible from eleven of the 13 (85%) cairns in Eday (including the cairn situated at the summit); but Vinquoy Hill is very conspicuous from many parts of the northern isles of Orkney, including 209 of the 300 (70%) survey points in Eday North, Mid and South. The summit of Ward Hill can be seen from three of the 13 (23%) Eday cairns and from 107 of the 300 (36%) survey points. Neither of these comparisons is statistically significant and both could have occurred by chance.

After very close examination of the possibilities, we can only conclude that there is no reasonable ground-based hypothesis which may be proved or disproved by further examination of the orientation of visibility from the chambered cairns of Eday.

The remaining statistical tests concern comparisons between the 76 chambered cairns of Orkney and the landscape of Eday as a surrogate measure of the landscape of Orkney as a whole. Again, examination of the orientation of visibility information in diagrams A and C of Fig. 14.3 do not admit of any obvious ground-based hypothesis which may be tested by statistical means. (One conclusion reached in Fraser (1983: 372-6) is perhaps worth repeating here: the distribution of distant visibility from the chambered cairns of Orkney is statistically distinct from the distribution which would be expected had the cairns been located at random in a landscape with a constant proportion of distant visibility at every point of the compass.) What is also worth noting is that there are two distinct peaks of distant visibility from the 76 cairns: between 135° and 145° (corresponding to that observed for the Eday cairns), and between 240° and 290°. It is very difficult to conceive of any type of ground-based hypothesis which might account for these two peaks.

Having traced a long and tortuous path, we appear to have arrived back not very far from our starting point. We have shown that there is a very distinct patterning to the distribution of orientation of visibility from the chambered cairns of Eday (and Orkney), and that there is very little possibility that this patterning occurred as a result of random processes. We have suggested that there are three possible locational decisions by the cairn-builders which might lead to such patterns, and we have shown that there is very little evidence to support any ground-based hypothesis of location. The only conclusion left is that the observed patterns in diagrams A and B of Fig. 14.3 are the result of some azimuthal or astronomical location procedure. The methods of this exercise are not sufficiently detailed to explore such procedures because they involve measurements of orientation and alignments at a higher level of precision.

That is perhaps an appropriate note on which to conclude. The builders of megaliths in Neolithic Britain were undoubtedly the possessors of a body of scientific knowledge and skills which is not fully understood by the archaeologists of the twentieth century, supported as we are by easy access to information and all the tools of a technological age. Alexander Thom was among the first to appreciate the complexity inherent in the planning of the prehistoric landscape, and his pioneering work has stimulated a wealth of speculation and knowledge. Without these the world of archaeology would be a place of poverty and tedium.

Acknowledgements

I am very grateful to Clive Ruggles for his invitation to contribute to this volume, and for the opportunity to remember the stimulating contributions of the late Professor Thom to prehistory. I am also grateful to Alison McGhie who, at very short notice, prepared the illustrations.

15

The Ring of Brodgar, Orkney

GRAHAM RITCHIE

In a small hotel in Kintyre in 1968, the writer was introduced to a gentleman described by his host as 'a Professor who is interested in old stones'. Alexander Thom's kindness on that occasion in explaining his researches at Ballochroy and later in sending copies of many papers on the layout and the astronomical significance of standing stones and stone circles is part of the impetus of this note. We met later on several occasions in hotels near the Ring of Brodgar, Orkney, in the course of his survey of the circle, and a phrase used by Thom at Brodgar is the second trigger for this tribute. It reads in published form: 'We know however from the survey made by Thomas in 1849 that since then some of the stones have been re-erected. This unfortunate form of vandalism makes it difficult to be certain about the exact diameter.' (Thom & Thom 1973: 121-2).

The henge monument known as the Ring of Brodgar is perhaps one of the most impressive prehistoric sites in Britain (NGR HY 294133). Set between the Lochs of Stenness and Harray, it is at the centre of one of the best-known archaeological landscapes, with the Stones of Stenness and Maes Howe in view to the south-east. The stone circle is surrounded by a rock-cut ditch, which is interrupted by causeways on the north-western and south-eastern quadrants; it is possible that there was originally an outer bank. The circle has been described on many occasions; indeed it and the adjacent Stones of Stenness were frequently visited by early antiquaries with sketch book, tapes and compass in hand. But, although we have an unusually full record of the stones, Alexander and A.S. Thom's plans are the most accurate representation of their present layout and their relationship with the surrounding monuments. Professor Thom's strictures about the re-erection of several stones suggested to the author that a re-examination of the stones of the circle might be a useful exercise, in order to work out exactly which had been set up in recent times

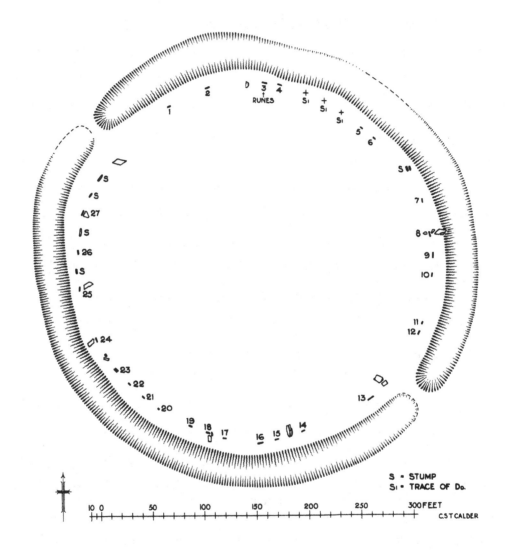

Fig. 15.1. Ring of Brodgar, Orkney: plan by Charles Calder, RCAHMS, 1929.

and to check which stood on the circumference of the ring proposed by Thom, an exercise begun by Thom himself (Thom & Thom 1973: 120-2). Accordingly, during a visit to Orkney in 1985, the stones were re-measured, though not resurveyed. Several numbering sequences of the stones of the circle exist, but that used in the Royal Commission's *Inventory*, based on Charles Calder's careful plane-table survey of 1929, is followed here (RCAHMS 1946: ii, 299-302, no. 875). Calder's plan of the circle cannot compete with the accuracy of the Thoms', but, using four plane-table stations and regular readings for the ditch, his plan shows clearly the positions of the stones within the surrounding earthwork.

The main measurements of the stones, given in the following table, incorporate notes from an Office of Works plan of 1907 (MW/1/1248) and from Calder's original survey (Fig. 15.1). The numbers appearing in brackets are those used in Burke's plan (Fig. 15.7). Discrepancies in height measurements prompted the writer to ask Dr. R.G. Lamb, Orkney archaeologist, to check these; Miss Rachel Johnson and Miss Penny Harris undertook the work using a Suunto optical clinometer/hypsometer in calm conditions in July 1986, and their measurements have been used in the table.

1   2.8 m high, 1.03 m broad and 0.25 m thick; it rises with straight sides to a broken top. (B24.)

2   3.6 m high, 1.5 m broad and 0.4 m thick; it rises with straight sides to a slanting top. (B25.)

The stone indicated between 2 and 3 is recorded on the Office of Works plan with the pencil annotation 'Socket gravelled over', but there is now no trace of the prostrate slab shown on the Commission plan. (B26.)

3   A fragment some 0.9 m high, 1.8 m broad and 0.5 m thick; this bears a cross and a runic inscription (Liestøl 1984). (B27.)

4   A shattered stump protruding 0.47 m, it is 1.1 m broad by 0.35 m thick at the base. (B28.)

There is now no trace of the three stumps between 4 and 5, but they are recorded on the Office of Works plan respectively as a stump, a socket and fragment of stone, and a second socket and stone-fragment.

5   Now 2 m high, the stone is 0.84 m broad and 0.4 m thick, and rises with straight sides to a flat top. (B29.)

6   Now 2.1 m high, it is 0.85 m broad and 0.35 m thick, and rises with straight sides to a slanting top. (B30.)

The shattered stump between 6 and 7 protrudes up to 0.17 m and is 1.05 m broad and 0.23 m thick; there is concrete in the fissures; one stone set at right angles to the slab may be a chocking stone. (B31.)

7   This stone is 3.5 m high, 1.1 m broad and 0.35 m thick; it rises with straight sides to a slanting top with the point to the north. (B32.)

8   The stump of a broken stone protrudes 0.75 m and measures 1.37 m by 0.38 m at the base. The fallen part is about 3.65 m long by 1.4 m broad and up to 0.35 m thick. (B33.)

9   Standing some 2.6 m high, it measures 1.35 m in breadth by 0.25 m in thickness; the southern side is partly broken. (B34.)

10   About 2 m high, 0.9 m broad and 0.28 m thick; it has straight sides and a broken top. (B35.)

11   2.5 m high, 0.8 m broad and 0.42 m thick; it has a distinct shoulder on the north-eastern side above which it slants to a flat top. (B36.)

12   2.7 m high, 1.15 m broad and 0.22 m thick; it rises with straight sides to a slanting top. (B37.)

     To the west of the entrance causeway there are two slabs, one 2.4 m by 1.98 m and at least 0.18 m thick, and the other 1.65 m by 1.12 m and at least 0.2 m thick. (B1.)

13   Some 2.9 m in height, 1.9 m broad and 0.22 m thick; it has straight sides with a shoulder on the north-east and has a top slightly slanting from north-east to south-west. (B2.)

14   About 2.8 m high, 0.9 m broad and 0.5 m thick; it has straight sides (slightly inward slanting) and a slanting top. (B3.)

     Between 14 and 15 there is a fallen slab 3.45 m long by 1.65 m broad and 0.2 m thick; erect it would have had a gabled top. (B4.)

15   3 m high, 1.25 m broad and 0.24 m thick; it has straight sides and a flat top. (B5.)

16   3.1 m high, 1.75 m broad, and 0.25 m thick; it has slanting sides and a pointed top. (B6.)

17   This stone, struck by lightning in the early 1980s, is 2.7 m high, 0.9 m broad and 0.45 m thick; the western side and top have been sheared by the force of the lightning; a chocking stone is visible at the base. Calder recorded the height as 11 ft 6 in (3.5 m) and Burke 12 ft 6 in (3.8 m). (B7.)

18   This stump protrudes 0.35 m, and measures 1.03 m by 0.22 m at the base; the broken fragment measures 1.9 m by at least 1 m. (B8.)

19   2.9 m high, 0.94 m broad and 0.24 m thick; it has straight sides and rises to a slightly gabled top. (B9.)

20   2.6 m high, 0.42 m along the inner side and 0.45 m thick. (B10.)

21   2.6 m high, 0.84 m broad and 0.37 m thick. (B11.)

22   2.5 m high, 0.84 m broad and 0.56 m thick; the south-eastern side is straight, the north-western slightly outward flaring, rising to a pointed top, which slants from south-east to north-west. (B12.)

23   2.5 m high, 1.07 m broad and 0.6 m thick. (B13.)

      A small displaced slab. (B14.)

24   The stump protrudes at least 0.32 m and is 0.87 m by 0.17 m at the base; the broken upper part of the stone is some 2.2 m by 1.1 m and up to 0.22 m thick. (B15.)

25   The stump is still 0.9 m high and 1.04 m by 0.17 m at the base; the broken fragment measures 2.85 m by 1.2 m. (B16.)

      A stump, just protruding, 0.9 m broad and 0.3 m thick. (B17.)

26   About 4.2 m high, 1.36 m broad and 0.32 m thick; it has a straight southern side but a more irregular northern side and rises to a pointed top. (B18.)

      A large stump between 26 and 27 protrudes 0.25 m above ground level and measures 2.1 m by 0.35 m at the base; there are several large chocking stones on the eastern side. (B19.)

27   1.7 m high and 1.6 m by 0.3 m at the base; a large flake has sheared off the northern side. (B20.)

      A stump protruding 0.15 m measures 1 m by 0.1 m at the base, with chocking stones on the north-west and south-east. (B21.)

      A stump protruding 0.2 m and 2.25 m by 0.24 m at the base. (B22.)

      A fallen slab measuring at least 2.65 m by 1.7 m. (B23.)

The different numbering schemes, and differences of opinion as to which stumps or fallen stones are included in the count, have made it difficult to decide how many stones were upright and how many fallen or merely represented by a stump. There are also differences of opinion about the heights of the stones, probably depending on the differing estimates of the ground level; in general our original measurements were more generous than Calder's and closer to those of Burke in 1875.

An examination of some of the earlier illustrations may also complement present day surveys. One of the earliest illustrations of the Ring of Brodgar is that by Richard Pococke, Bishop of Ossory, who visited Orkney in 1760 (Fig. 15.2); Pococke found the ditch very dramatic, but his water-colour gives no impression of the causeways; he shows the stones on the inner lip of the ditch, with an arc of five prostrate slabs on what is probably the southern quadrant of the circle. This illustration (British Library, Add Ms 14257, f. 77v.) formed the basis for a lithograph published in 1887. The expedition of Sir Joseph

Banks to Iceland in 1772 produced some of the most remarkable illustrations and surveys of both the Ring of Brodgar and the Stones of Stenness (set in historical context by Lysaght 1974); the measured drawing of Brodgar (or the Circle of Loda) prepared by Frederick Herm. Walden, a naval architect and surveyor, clearly shows the ditch with the two causeways and also indicates which stones are upright and which are fallen (Fig. 15.3).

The Banks expedition was the first to plot the position of the large number of barrows on the south side of the Ring of Brodgar, careful planning of which is an important part of Professor Thom's own work (Thom & Thom

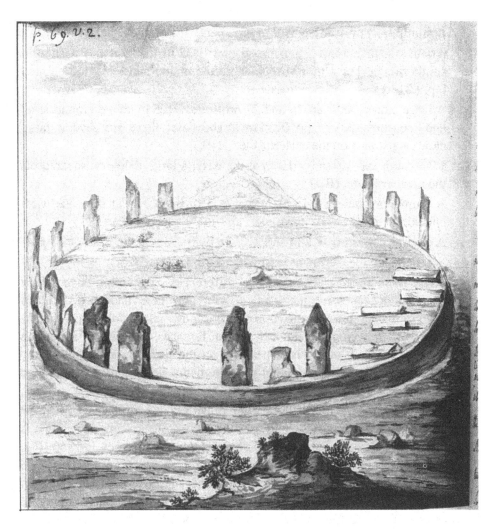

Fig. 15.2. Ring of Brodgar, Orkney: pen and ink wash by Richard Pococke, 1760.

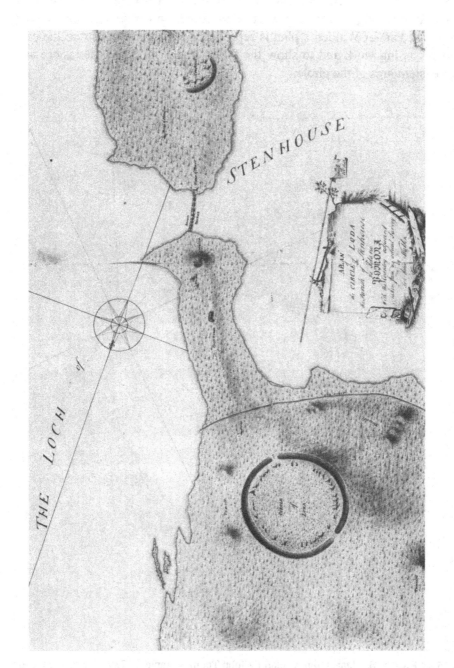

Fig. 15.3. *A Plan of the Circle of Loda in the Parish of Stenhouse ... by Frederick Herm. Walden 1772.*

1973: 114-5; 1975: 102-5). Banks' interests in astronomy suggested to Lysaght that he might even have been a forerunner in the study of prehistoric astronomy. Part of Walden's plan is reproduced here to demonstrate the care of his recording work and to show the fallen stones in the south-eastern and north-eastern arcs of the circle.

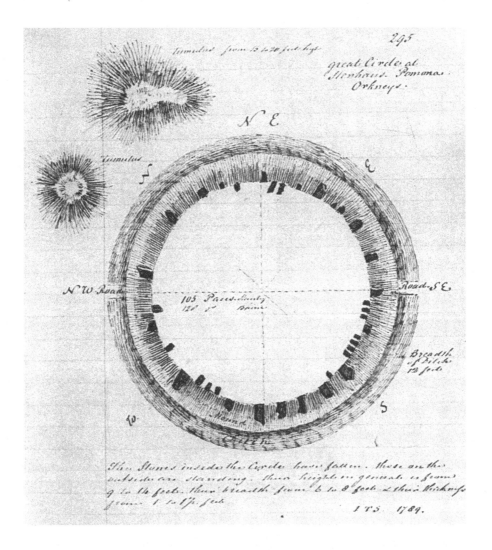

Fig. 15.4. Ring of Brodgar, Orkney: plan by John Thomas Stanley, 1789. The caption reads 'The stones inside the circle have fallen. Those outside are standing ... '.

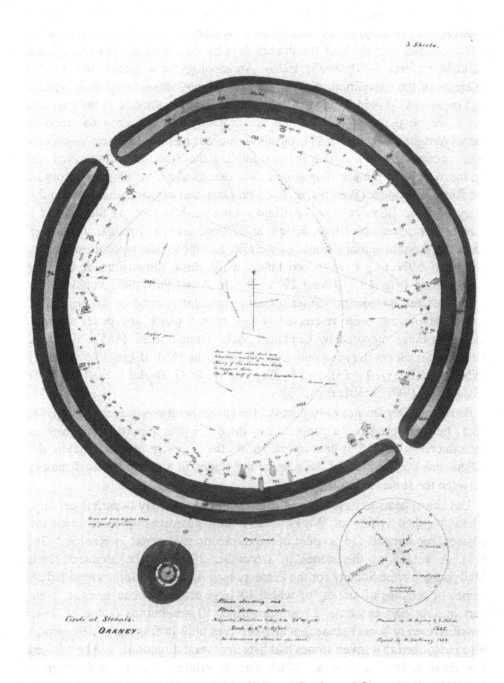

Fig. 15.5. Ring of Brodgar, Orkney: plan by Sir Henry Dryden and George Petrie, 1851.

The expedition of Sir John Thomas Stanley, 6th Baronet, of Alderley Park, Cheshire, to Orkney, Faroe and Iceland in 1789 is not so well known to archaeologists, but three of the diaries kept by the travellers have now been published (West 1970; 1975; 1976). Archaeology was indeed one of the interests of the expedition, and the famous Orkney monuments were visited and measured. Within Brodgar 'there are two Tumuli, the one to the east and the other on the west side, which are probably burying grounds' records James Wright (West 1970: 17), but these are not mentioned elsewhere. John Bain commented on the tradition surrounding the Stone of Odin, speculated on the weight of the standing stones, and remarked on the sloping layout of the Ring of Brodgar (West 1976: 26). The plans and drawings included in the diary of Isaac Benners are of particular interest, however, as the plan is a careful representation of both upright and fallen stones at Brodgar; the stones are located on the circumference of a circle, and those which are illustrated as lying inside the ring have indeed fallen, while those shown outside the ring are standing (Fig. 15.4) (West 1975: 28). It shows the fallen stones to the south-east and the pair on the north-east (particularly noted by Thom), all of which have now been re-erected. More recent plans are better known, including those prepared by Captain F.W.L. Thomas R.N. (1852: 102-5, p. xiii), by Sir Henry Dryden and George Petrie in 1851 (Figs. 15.5 & 15.6) (NMRS ORD 89/10 - 13), and by Captain W. St G. Burke R.E. (Fig. 15.7) (RCAHMS 1946: ii, 301, Fig. 376).

Petrie and Dryden not only planned the stones with considerable care (Fig. 15.5) but also prepared a panorama in three colours, a portion of which is reproduced as Fig. 15.6; the original shows the stones in profile in ochre, the fallen stones in purple and the plan of the stones in a red tint. The distances between the stones are not to scale.

The site appears to have been little altered until the early twentieth century, when the then Office of Works took it into guardianship along with the adjacent monuments (an account of the transactions is given by Ritchie 1976: 6-7). A survey of the stones in December 1905 by the architect Basil Stallybrass for the Society for the Protection of Ancient Buildings reported 38 stones or remains of stones 'of which 13 large stones remain upright, 3 are part upright part prostrate, 11 are prostrate, 10 are stumps (some of these possibly merely socket stones), and 1 survives only in fragments'. His report also mentioned that seven stones had lists in several directions, and he advised that these should be secured. Work was undertaken over three years from 1906 to restore and preserve the circle and to create the monument as we know it today. There is little doubt that careful excavation would have revealed the stone-holes in which the uprights were originally set. There are

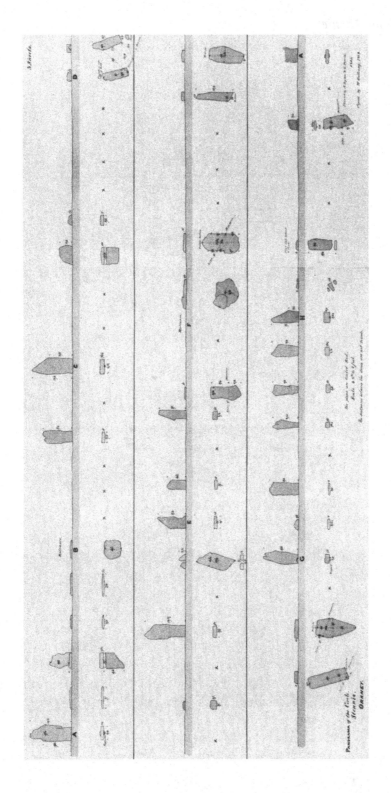

Fig. 15.6. Ring of Brodgar, Orkney: panorama showing the upright and fallen stones by Sir Henry Dryden and George Petrie, 1851.

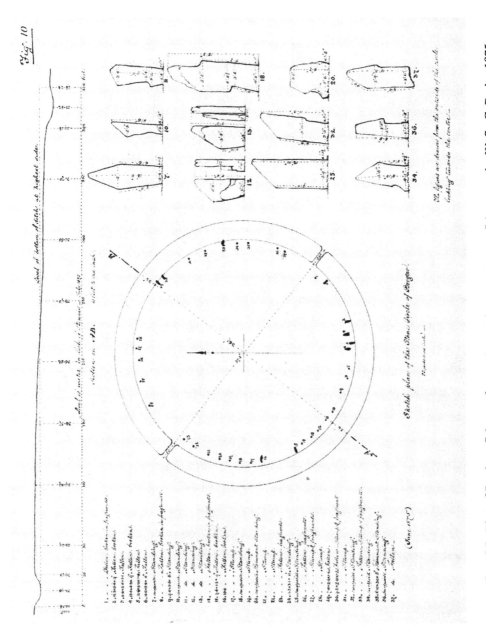

Fig. 15.7. Ring of Brodgar, Orkney: plan, profiles and measurements of the stones by W. St. G. Burke, 1875.

Fig. 15.8. The Ring of Brodgar today.

several interesting photographs of the Stones of Stenness to show the wooden supports that were used in the course of this work. Short and not very satisfactory accounts appear in the *Saga-Book of the Viking Club* (for example 5 (1906): 252-61); in the course of the work at Brodgar one of the stones (no. 3) was found to have a runic inscription, and this seems to have occasioned more excitement than the re-erection of the stones themselves (for example *Orkney and Shetland Miscellany*, 2 (1909): 46-50). This work restored the circle as far as possible to its original state; it is ironic, however, and not often remembered, that one of the most photographed vistas includes four of the re-erected stones (Fig. 15.8).

It is perhaps surprising that no recorded excavation has been undertaken within the circle, though it is also fair to say that survey by fluxgate gradiometer did not reveal any marked anomalies (in contrast to the position at the Stones of Stenness).

Excavations undertaken by Renfrew concentrated on an examination of the ditch, as the main aim of the exercise was the collection of dating evidence; in this they were unsuccessful, but it was discovered that the ditch was cut into solid bedrock and, when measured from ground level, was no less than 10 m wide and 3.4 m deep. It may indeed be that the ditch was the quarry from which the stones of the ring were removed. The possible presence of an outer bank has previously been discussed and is largely unresolved (Renfrew 1979: 42-3; Ritchie 1985: 124).

In 1973 Thom rightly pointed out that several stones had been re-erected since the survey undertaken by Thomas in 1849, marking those that were then prostrate and for which no stump was shown with a *V* (nos. 5, 6, 12, stump, 14 and 15); the present re-assessment of the history of the circle suggests that, rather than the stump adjacent to the causeway having been set up in recent times, stone 13 has been re-erected and stone no. 16 should be added to the tally. The latter seemed unusually out of alignment on the circumference of Thom's ring, and the fact that it has undoubtedly been re-erected probably accounts for this, helping to confirm the accuracy of his survey and corroborating his view that the circle was laid out with great accuracy.

Acknowledgements

This paper is published with the permission of the Commissioners and incorporates material gathered in the course of survey in Orkney in September 1985; the assistance of Mr J. Borland and Dr R.G. Lamb is gratefully acknowledged.

Mr P.J. Ashmore, Mrs L.M. Ferguson, Mr G.S. Maxwell and Dr A. Ritchie have been of considerable assistance in the course of the final preparation of the paper. The help of Miss Rachel Johnson and Miss Penny Harris in verifying the heights of the stones is gratefully acknowledged.

16

The geometry of some megalithic rings

RONALD CURTIS

Introduction

For many years I have had an interest in stone, in stone structures of all kinds and particularly in megaliths. This latent spark was kindled the day I watched Professor Thom on a television programme drawing out his various rings on a sandy beach. Then I read his first books and was immediately convinced of the truth of his findings. Later I had the benefit of kindly and informative discussions by letter and by meeting him in his home.

As a direct result I started serious surveying at the Callanish sites (Fig. 16.1) in 1972, during a family holiday in Harris, and I have continued to do similar fieldwork, mostly in the Outer Hebrides, with the help of many friends. In particular I have had extensive assistance from the late Ian J. Cairns and David J. Cairns on surveys as well as in the analysis of observations, both as regards ground geometry and astronomical alignments. The geometry is largely reported in this paper but the alignments remain to be presented in the fullness of time as all my outdoor and indoor work has to be done in my 'leisure' hours. Being a chartered civil engineer I can understand the vast technical problems involved in establishing sites and moving and erecting stones, and I have developed a healthy respect for the organisational and constructional capabilities of Megalithic Man.

I have a very reliable, though not modern, one-second Tavistock theodolite and basically I continue to use the Professor's methods for instrument checks, referring objects, etc. My procedure for surveying a ring is firstly to examine it as leisurely as possible and to define each corner of each stone with a white pin: then to set up the theodolite at any convenient place inside the ring and record the direction to each pin. Next I measure the distances to the same pins with an accurate tape and allow for ground slope. From these readings the survey is plotted at a scale of 1:100.

Thereafter I have adopted a graphical method (unless otherwise indicated in the text) for finding the best fit. To do this I draw a series of rings on transparent film - ellipse, egg, etc. - each with a radius advancing by 0.5 MY; I overlay each of them on the plan, and find the closest fit and the best orientation by trial and error. At this stage the residual errors can be checked by measuring the distances between the perimeter and the centre of each stone. By careful application it is usually easy to 'see' which type of ring is involved, then to determine which diameter or intermediate size is appropriate. It is quite remarkable how accurately the fit and orientation can be found by this method, a fact that can only be appreciated by actually doing it. It is doubtful if any ring could have been found by any other method to fit the fallen stones at Site H1/8 (Callanish X) and I am convinced that the shape, size, and orientation of the one shown in Fig. 16.6 is correct. It is also easy to see a 'rogue' stone, i.e. one that stands very near the perimeter but which is not part of the geometry of the ring, such as Stone 48 at Site H1/1 (Fig. 16.2) and Stone 11 at The Ring of Stenness (Fig. 16.13). Such 'rogue' stones should not be included automatically in the calculation of a best fit by computer without due consideration.

Problems arise when only a few stones are left standing. It should be remembered that, mathematically, it takes not less than three points to define a circle, four points to define an ellipse, and more points to define a Thomian flattened circle or egg-shape. So the type, size and orientation of any ring can only be determined if sufficient stones still exist around the perimeter. Even with additional points, there are some cases where it is not possible to be certain of the solution. Thus, where all the remaining stones are on one half of the perimeter, the shape of the other half is unknown. Furthermore, because there must always be some imprecision in the recorded position of stones, the correct choice cannot always be made from the presence of a few stones (e.g. between a circle and a near-circular ellipse). At Site H1/2 (Callanish II) and at Site H1/3 (Callanish III) I have given two solutions based on the available surface information. Archaeological excavation can provide further data on the positions of buried stumps or socket holes, and can thereby help to resolve the issue, as was achieved at the Ring of Stenness.

Generally on the plans included in this paper a solid black shape represents a standing stone or the cross-section at ground level of a leaning stone; a full line represents a fallen stone lying on the ground surface; a broken line indicates a buried stone the outline of which has been probed; a '+' represents

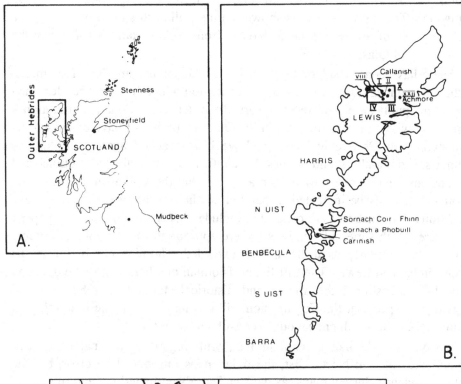

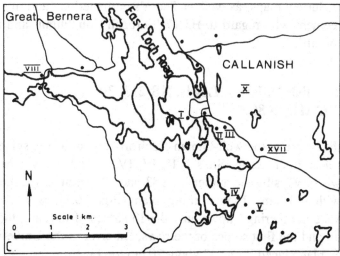

Fig. 16.1. The location of the sites.

the centre of the original socket hole, based on the presence of packing stones, voids, etc. I am much indebted to Stella Ryde for the high quality of these drawings. True north is always shown. At the Callanish sites it is correct to a few minutes of arc; it has been derived from survey stations set up by the University of Glasgow (1978).

There has been considerable profusion of names and numbers for mega-lithic sites under study, with resulting confusion which has not been lessened by some publications of recent years. In order to attempt a unification - primarily with regard to the sites in the Outer Hebrides - I have discussed the situation with Dr Archie Thom and we have reached agreement that his father's site numbers and names be confirmed (after the corrections and extensions which have now been made) and that the Callanish sub-set using Roman numerals be indicated in parallel. As he has asked me to record this additional information I am pleased to include it as Appendix I. In this paper I have used both series of numbers (where appropriate) under each site name and I have similarly used the agreed, extended, Thomian numbering in the areas other than Lewis. The full list of Thomian numbers is now held by the Royal Commission on the Ancient and Historical Monuments of Scotland and it is to be hoped that the Commission will maintain and update it and that all future researchers will endeavour to consult and adopt it.

Lastly, I would like to explain that, with regard to the names of two intervisible sites in North Uist, the older maps introduced an error, but the latest Ordnance Survey maps, as well as Thom's corrected list and this paper are now in agreement with regard to H3/17 'Sornach Coir Fhinn' and H3/18 'Sornach a' Phobuill'.

Callanish (Callanish I), Isle of Lewis. NB 213330
Thom H1/1. RCAHMS 89.*

The ring is at the centre of the well-known cruciform layout on the spine of an elevated ridge near Loch Roag. Sites I, II, III, IV, X and many others are intervisible. Two surveys by the author in 1983 and 1984 confirmed the stone positions to within 5 cm. The resulting plan (Fig. 16.2), which adopts Somerville's stone numbering, shows only the 13 stones of the ring, the large central monolith and the first stones of the rows. An error in the Univeristy of Glasgow survey plan placed Stone 42 about 20 cm too far from the centre.

---

*   All RCAHMS data quoted in this chapter refer to RCAHMS (1928).

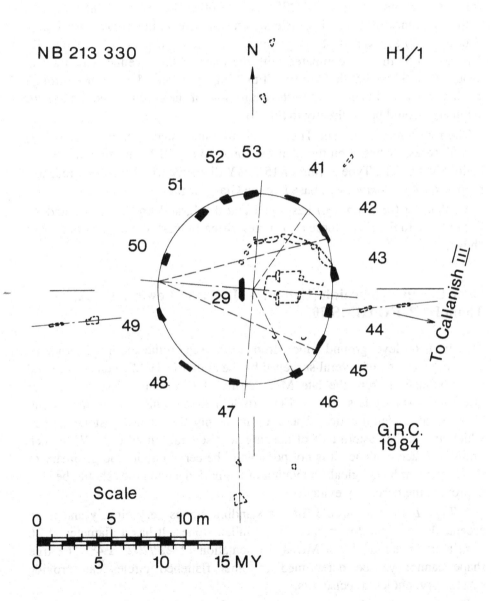

NB 213 330

N

H1/1

52 53

51

41

42

50

43

29

To Callanish III

49

44

48

45

46

47

G.R.C.
1984

Scale

0   5   10 m

0   5   10   15 MY

Fig. 16.2. Callanish I. Flattened Circle Type *A*: major diameter 15.5 MY.

Use of the graphical method produces a Type *A* ring of 15.5 MY diameter, but a closer fit is obtained if the diameter is reduced to about 15.4 MY. The centre is located at a point 1 MY east of the centre of the central stone. It will be noticed that Stone 48 stands about 0.5 m outside the ring. For the best fit the axis of symmetry is at 95°/275° and the other axis is at 5°/185° azimuth with a tolerance of ± 3°. This differs from the orientation I gave previously (Curtis 1979) as a result of the change in the recorded position of Stone 42. The axis at 5° may be compared with the lines of the Avenue which are at about 9° and 11° azimuth. However if the ring were rotated clockwise through 5°, this axis would be parallel to the centre line of the Avenue and the axis of symmetry would point directly to Site III.

The ring which Professor Thom originally suggested (Thom 1967: 122 *ff*) was of necessity based on the plan by Somerville (1912), not then known to be inaccurate. The Type *A* ring of 15.7 MY diameter which he more recently proposed (*pers. comm.*) was based on the Glasgow plan.

M. Ponting (*pers. comm.*) has noted that the 'grain' on Stones 44 and 47 lies radially to the ring - these two stones virtually mark the diameters quoted above.

Cnoc Ceann a Gharaidh (Callanish II), Isle of Lewis. NB 222326 Thom H1/2. RCAHMS 90.

The site is on level ground which drops to the sea on the south-western side. This ring was one of several surveyed by the author in 1972 when he received every assistance from the late Mrs Perrins of Garynahine Estate, and her grieve. The survey is shown in Figs. 16.3(A) & (B) which adopt and extend Thom's stone identification. There are now only five standing stones, all of which are on the eastern half of the ring, and several buried cists. With such limited surface evidence it is not possible to be certain about the geometry of the ring but archaeological excavation of the buried stone holes should be able to provide the necessary evidence.

A Type *B* flattened circle fits the standing stones very closely and is an acceptable fit to the lower ends of the fallen stones. It has a diameter of 26 MY, a perimeter of 30.75 MR and orientation of 22°/202° ± 4°. The true shape cannot yet be determined as other flattened circles can provide satisfactory, but less precise, fits.

The ellipse which I previously proposed (Curtis 1979) fits the stones almost as closely. The diameters are $2a = 26$ MY and $2b = 22$ MY. The ellipse is constructed on a triangle whose sides are therefore 6.93, 11 and 13 MY. It is

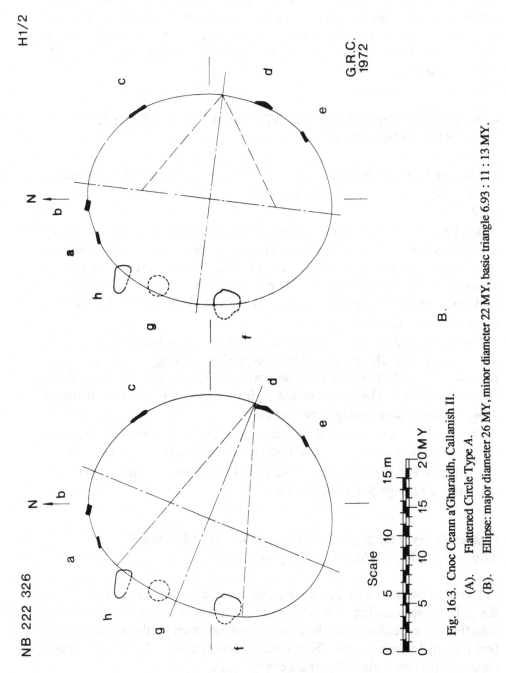

Fig. 16.3. Cnoc Ceann a'Gharaidh, Callanish II.
(A). Flattened Circle Type A.
(B). Ellipse: major diameter 26 MY, minor diameter 22 MY, basic triangle 6.93 : 11 : 13 MY.

one of those triangles which is so close to a true Pythagorean triangle that the difference would not be noticed on the ground. It therefore satisfies the requirements for a megalithic ring. The orientation of the major axis is 7°/187° ± 4°. Thom (1967: 126) suggested the same ellipse with a slightly different orientation.

Cnoc Fillibhir Bheag (Callanish III), Isle of Lewis. NB 225327 Thom H1/3. RCAHMS 91.

The interpretation of this site is complicated by the existence of four standing stones within the ring of nine or more stones, as surveyed by the author in 1972.

Good overall fits are given by either a flattened circle Type *A* (Fig. 16.4(A)) which misses stone *l* or an Egg Type III (Fig. 16.4(B)) which misses stone *k*, adopting and extending Thom's stone identification. In both cases stone *e* does not fit. Obviously excavation is necessary before the true shape can be confirmed. In both cases the axis of symmetry is at 58°.5/238°.5 with a suggested tolerance of ± 2°.

The Type *A* circle has a diameter of 20 MY and a perimeter of 24.47 MR (or 49 half-MRs). The eastern triangle on which the Egg is constructed has sides of 6, 8 and 10 MY and its western triangle was set out with sides of 4.5, 10 and 10.97 MY. The length of the perimeter is 25.29 MR. It is therefore another approximate Pythagorean triangle.

Thom (1967: 126) suggested two concentric ellipses but this was based on his early compass survey which he knew required to be revised (*pers. comm.*).

Ponting & Ponting (1984a: 17) prefer a Type *A* circle with its secondary axis passing through Stones *h*, *p* and *m*.

Ceann Hulavig (Callanish IV), Isle of Lewis. NB 230304 Thom H1/4. RCAHMS 93: 'Garynahine'

This ring stands part way down the north shoulder of a small hill near Loch Roag. If the surrounding peat were removed it would appear as a small ring with five disproportionately tall stones. It was surveyed by the author in 1972 (see Fig. 16.5 which adopts Thom's stone identification). There were almost certainly six stones originally, and a central cairn.

The ellipse which fits neatly through the centres of the stones has diameters $2a = 15$ MY and $2b = 12$ MY. This is a perfect illustration of the use, long

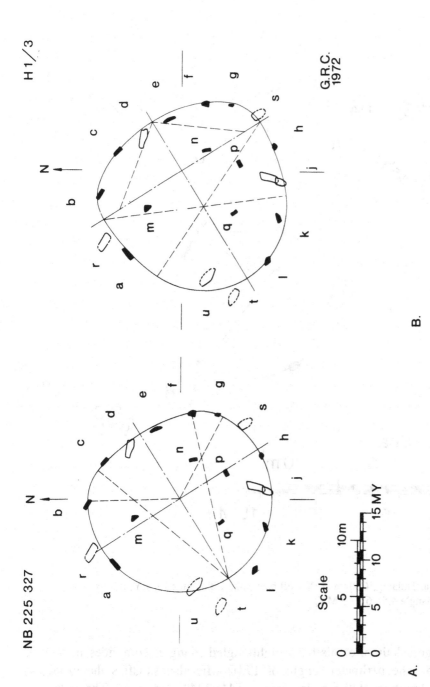

Fig. 16.4. Cnoc Fillibhir Bheag, Callanish III.

(A).   Flattened Circle Type A: major diameter 20 MY.

(B).   Egg Type 3 (Ellipse: major diameter 20 MY, minor diameter 12 MY, basic triangle 6 : 8 : 10 MY; West end: basic triangle 4.5 : 10 : 10.97 MY).

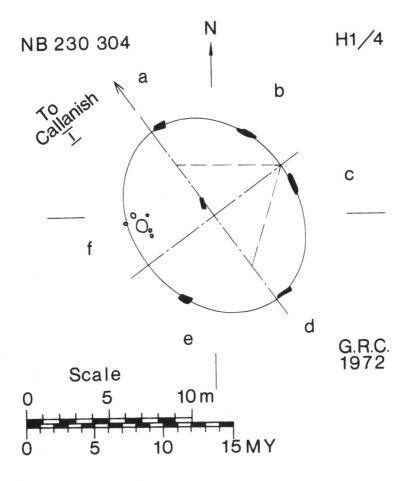

Fig. 16.5. Ceann Hulavig, Callanish IV. Ellipse: major diameter 15 MY, minor diameter 12 MY, basic triangle 4.5 : 6 : 7.5 MY.

before Pythagoras' time, of his basic right-angled triangle, with sides in the ratio of 3:4:5. The perimeter length of 17.02 MR also satisfies the usual criteria. The direction of the major axis is 144°/324° ± 1° or so. The latter direction points directly to the ring at Callanish I.

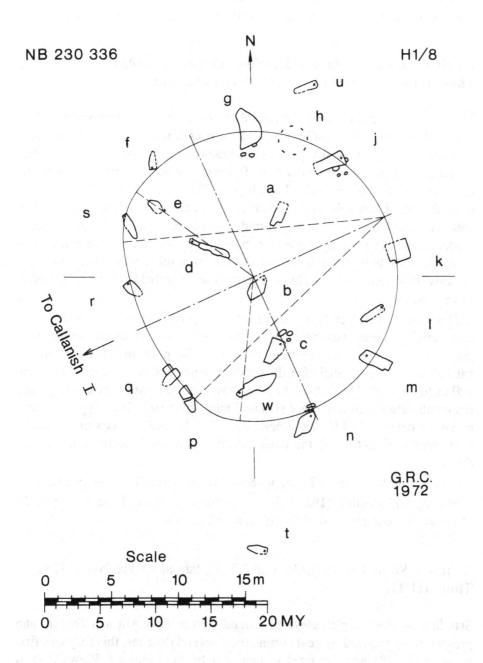

Fig. 16.6. Na Dromannan, Callanish X. Flattened Circle Type *A*: major diameter 26 MY.

Thom (1967: 126) gave an ellipse with diameters of 15.5 MY and 11.5 MY based again tentatively on his compass survey.

Na Dromannan (Callanish X), Isle of Lewis. NB 230336
Thom H1/8. RCAHMS 92: Drum nan Eum (Eun)

This site has fascinated and challenged me ever since my first sight of it in 1972, when I surveyed it (Fig. 16.6), situated as it is on a ridge above a cliff said locally to be one of the quarries used to obtain the standing stones. All its stones are prostrate but the existence of packing stones clustered near one end (marked '+' in the figure) of each monolith and the existence of one excavated pit *h* strongly argue in favour of initial erection, and subsequent collapse. The latter may have occurred as a result of the shallow soil being eroded from the packing stones or possibly by intentional destruction in antiquity. Evidently stone *h* remained standing until it was felled relatively recently. Its fragments have sharp edges and were probably shattered by using fire and water.

There were twelve or 13 stones in the main ring, four or five inside it and two outlying stones nearby, one to the north and one to the south. It is possible that a stone has been omitted from the plan and that a spurious, natural, one has been included. In addition there is a very large stone lying 120 m to the east. The flattened circle shown in the figure very convincingly fits much better than any other standard megalithic ring. It is Type *A* with a major diameter of 26 MY and a perimeter of 31.86 MR. Its axis of symmetry is oriented at 6°/246° ± 3°, the latter direction pointing exactly to the ring of Site I.

Unfortunately Professor Thom never had the opportunity to see this site.

Ponting and Ponting (1984a: 30-1) obtained the same Type *A* ring of 26 MY diameter but gave a slightly different orientation.

Achmore Stone Circle (Callanish XXII), Isle of Lewis. NB 317292
Thom H1/17.

Standing on almost level, elevated ground west of Blar Allt nan Torcan, and progressively exposed by peat cutting over several decades, this ring was first reported in 1981 and surveyed in that year by the Pontings. Views from it include both the Minch and the Atlantic, as well as several other Callanish sites.

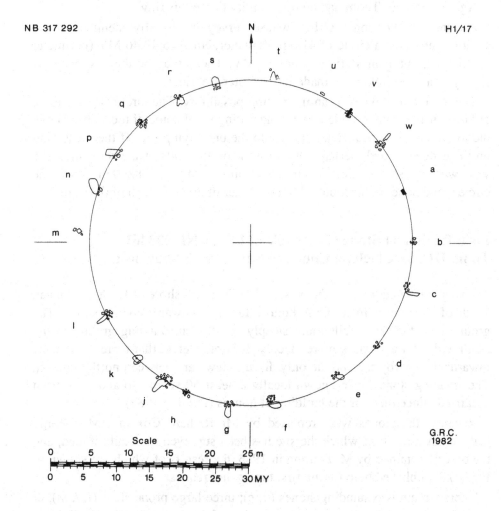

Fig. 16.7. Achmore Stone Circle, Callanish XXII. Circle: diameter 49 MY.

When surveyed by the author in 1982 there were three stones or stumps standing or leaning (*a, c, v*): seven socket holes clearly defined (*b, d, f, g, h, q, w*); seven socket holes less clearly defined (*e, l, n, p, r, s, u*); three stones lying where they fell (*j, k, t*); one position represented by a few packing stones (*m*); and at least one gap which may have had a stone (*l-m*), making a ring of at least 21 stones (Fig. 16.7). The best fit is provided by a circle of 49 MY diameter (perimeter 61.58 MR). Unfortunately it is not possible to fit a circle of 50 MY diameter between the opposing socket holes *h* and *w*.

Again Professor Thom had no opportunity to see this ring.

Ponting and Ponting (1981a), whose survey is virtually identical to the author's, obtained a circle of 41.0 m diameter, equal to 49.46 MY (perimeter 62.15 MR), but again definitely not 50 MY. Theirs may be the first accurate survey of a true circle to be made in the Outer Hebrides.

There remains, however, an interesting possibility. The circle appears to be of late construction as evidenced by the setting up of some of the stones above the lower layers of peat (i.e. not set into the underlying till). If the measuring unit had degenerated, perhaps due to wear on the ends of the rods, until the yard was 2% shorter than Thom's standard (1 MY = 0.8290 m), then the circle could have been set out with a diameter of 50 MY, each of 0.812 m.

Loch Raoinavat Stone Circle, Isle of Lewis. NB 233461
Thom H1/30. 'Clach an Cnoc Laoiran', South Shawbost.

The ring is located towards the west end of the north shore of Loch Raoinavat, south of Shawbost, in a slight hollow falling eastwards to the shore. The ground near the shore falls more steeply and the surrounding ground to the north and west also rises more steeply. It is of interest that there is no view towards the nearby sea and the only distant views are between north and east. The standing stones were known locally at least 50 years ago and have been 'quarried' since then for the building of houses (Ponting 1983).

Recently the stones were reported by Mr Richard Cox and Mr George Mitchell as a result of which the site has been surveyed in detail, plotted, and the best fit obtained by M. Ponting in 1983, the results of which are shown in Fig. 16.8, published here for the first time at her request.

There remain two standing stones (*a, c*); three large prone slabs (*i, j, m*); at least four locations with a recently broken slab or a collection of broken fragments indicating where a stone was trimmed for a lintel before being carried away; and a few other stone holes suggested by features in the ground. With such limited evidence it is not possible to define the ring but the

34.5 MY-diameter circle shown appears to give a good fit. It is probable that archaeological excavation of several stone holes would provide sufficient information for a better determination.

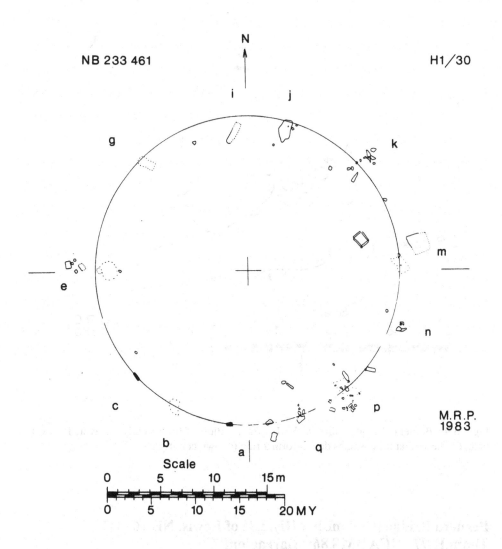

Fig. 16.8. Loch Raoinavat Stone Circle. Circle: diameter 34.5 MY.

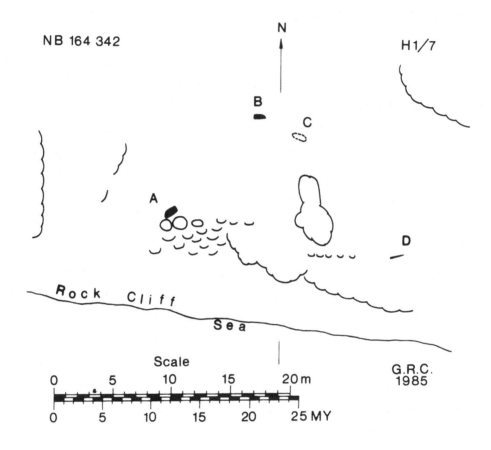

Fig. 16.9. Bernera Bridge, Callanish VIII. The position of the recently excavated socket hole, *C*, shows that these stones do not form a ring (Crown copyright).

Bernera Bridge (Callanish VIII), Isle of Lewis. NB 164342
Thom H1/7. RCAHMS 86: 'Barraglom'

The stones at this impressive site, above a 15 m high rock cliff, stand sentinel across the narrows that separate the island of Great Bernera from Lewis, and now stand overlooking the bridge that serves this island. There are four standing stones; one fell about 1900 and was re-erected in the wrong place in

1985. Thereafter an archaeological excavation was undertaken by M. Ponting, partly on behalf of Count R. de la Lanne Mirrlees, and preparations were made by the author to reinstate this stone in its true socket hole (*C*).

This site has often been described as a 'half circle' (Somerville 1912: Fig. 7) because of the disposition of the three standing stones and because

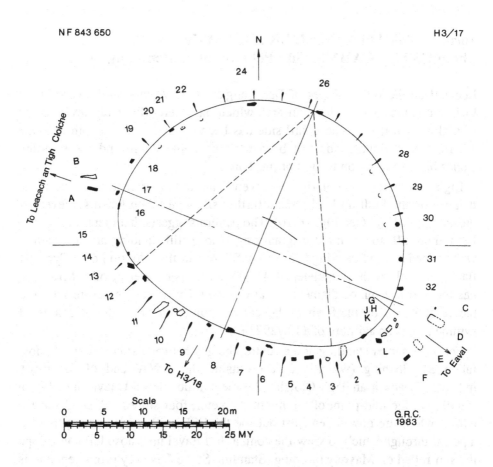

Fig. 16.10. Sornach Coir Fhinn. Flattened Circle Type *B*1: major diameter 44 MY.

there is no space for the southern half owing to the cliff and the sea. However the author's site survey in 1985 shows conclusively that these stones do not form part of a ring (see Fig. 16.9 which adopts Somerville's stone identification).

Somerville (1912: 45-6) reported, *inter alia*, that as viewed from this site a boulder on the horizon indicated Mayday sunrise (Beltane) and a boulder on the southern horizon marks the meridian. The author in collaboration with David J. Cairns can confirm that there are two boulders for the Beltane sunrise and four more boulders accurately monitoring the moon at its major standstill (south).

Sornach Coir Fhinn, North Uist. NF 843650
Thom H3/17. RCAHMS 250: 'Pobuill Fhinn, Ben Langass'.

Located on the sunny slopes of Ben Langass and overlooking a nearby tidal loch, the stones stand within an area which has been effectively levelled out of a sloping hillside. The uphill side has been excavated to a depth of some 1.2 m. The downhill side has been built up some 2.0 m and the standing stones have been set on the top of this bank.

Eight stones remained upright, six were leaning (significantly, mostly those that are on the built up bank) and a further six were fallen when surveyed by the author in 1983 (see Fig. 16.10). The pattern suggests that there might have been about 30 stones in the original ring. Those still standing are sufficiently well spaced around the ring to give confidence in the adoption of a Type *B*1 flattened circle with a diameter of 44 MY and a perimeter of 50.6 MR. This has the shortest minor diameter of any standard type and therefore required the least width of platform to be carved out of the hillside. The axis of symmetry has an azimuth of 112°/292° ± 2°.

For an observer standing at the centre, the gap between stones *A* and *B*, now fallen and leaning (which are just outside the WNW end of the ring), indicated Leacach an Tigh Cloiche, the megalithic site on Uneval hill. When standing at the mid-point of the minor diameter, four stones (*C, D, E* and *F*), two of which are now fallen (just outside the ESE end), would have provided a portal through which to view the conical hill Eaval up whose left hand slope the sun rolled on Mayday morning. Standing Stone *L* is very prominent and is also located outside the ring.

Thom, Thom & Burl (1980: 311-2) illustrate this ring and give its major diameter as 45.59 MY (124 ft).

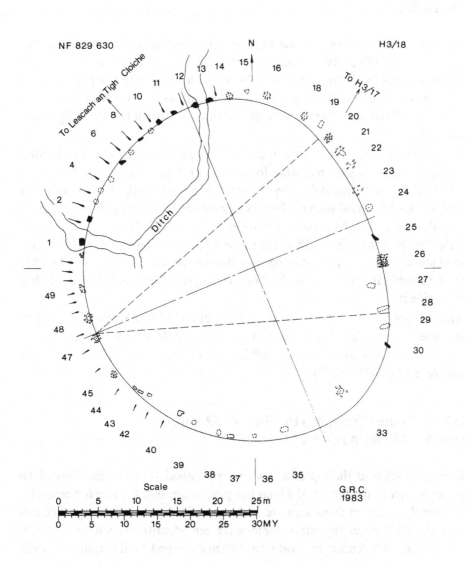

Fig. 16.11. Sornach a Phobuill. Flattened Circle Type *B*: major diameter 52 MY.

M. Ponting (*pers. comm.*) made the interesting suggestion that the ring is a variation of the standard flattened circle and has its major diameter divided into fifths (not thirds). This gives an extremely good fit using three pairs of diametrically-opposed standing stones.

Sornach a Phobuill, North Uist. NF 829630
Thom H3/18. RCAHMS 249: 'Sornach Coir Fhinn, Loch a Phobuill'.

Only some low boulders and the tops of two standing stones mark this ring which lies insignificantly in deep peat near the foot of a gentle slope on the north-western side of Craonaval hill. Over much of the site the peat is 1.0 to 1.5 m deep. Along the lower arc the ground has been artificially built up to form a curved terrace, concentric with the ring, on which the standing stones have been set.

Only ten stones appear to remain erect out of an estimated total of 48. Most of the fallen and buried ones were located by probing and marked before the ground survey was undertaken by the author in 1983 (Fig. 16.11). The ring giving the best fit to all the visible standing stones is a Type *B* flattened circle with a diameter of 52 MY and a perimeter of 64.15 MR. Its axis of symmetry is oriented at 158°/338° with a tolerance of ± 5° or more. However if the buried stones are included it can be seen that the best fit is obtained when this ring is rotated clockwise through about 10 degrees, although Stone 30 then lies outside it.

Thom (1966: 29 & Fig. 14) suggested a circle of 52.0 MY diameter (142 ft) and correctly recorded that Stones 25 and 30, which were set at an angle to the perimeter, indicated other megalithic sites. This is also given in Thom, Thom & Burl (1980: 312-3).

Carinish Stone Circle, North Uist. NF 832602
Thom H3/23. RCAHMS 248.

The ring stands near the top of a low rise with good views of the North Uist and Benbecula hills. The track which has passed through the site for centuries was recently widened from a single lane road to a 6 m carriageway. According to the RCAHMS Inventory there were ten identifiable stones in 1915. Burl (1976a: 147) reports an oval ring. When surveyed by the author in 1985

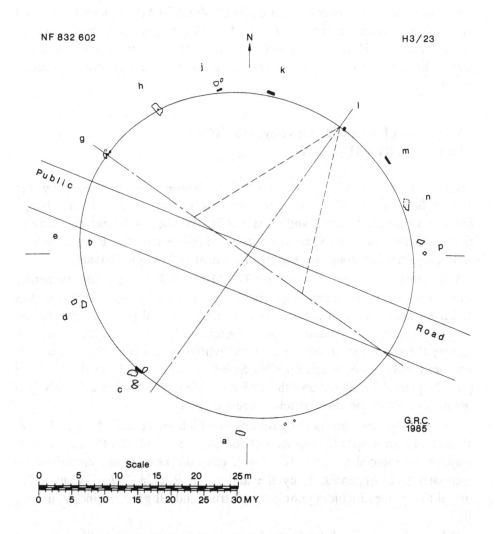

Fig. 16.12. Carinish Stone Circle. Ellipse: major diameter 52 MY, minor diameter 48 MY, basic triangle 10 : 24 : 26 MY.

(Fig. 16.12) two stones were still standing, three were leaning outwards, two had fallen, and several more socket hole positions could be identified.

As standing Stones *c* and *l* are diametrically opposite each other a minor diameter of 48 MY can be accurately established. The best fit is thereafter obtained by an ellipse with a major diameter of 52 MY (lying nearly parallel to the road). This ellipse is based on the true Pythagorean triangle of sides in the ratios of 5:12:13.

Because the long axis of Stone *l* lies across the perimeter instead of parallel to it, it was possibly intended to mark a diameter. Alternatively, if it was not intended to be on the perimeter, a circle of 49 or 50 MY diameter would fit the remaining stones equally well, resulting in a ring almost identical to the one at Achmore. However, a good fit cannot yet be claimed and it is hoped that archaeological excavation of the stone holes could confirm the correct shape.

The Stones of Stenness, Orkney. HY 307125
Thom 01/2. RCAHMS 876.

Located at the centre of a large saucer whose rim is formed by the surrounding Orkney hills, this attractive ring of tall stones within a henge ditch was reduced to two standing stones in the nineteenth century. Early in the twentieth century two fallen stones were re-erected. The Ring of Brogar and many standing stones are visible in the near and middle distance.

When this ring was excavated in 1973-74 by Dr J.N.G. Ritchie the author was invited to survey it and determine its geometry (Fig. 16.13). Stones 2, 3 and the visible stump of Stone 8 were in their original positions; the buried stumps of Stones 4, 10 and 11 were found firmly set among their packing stones; the space for Stone 1 was found within its packers; the pits for the packers for Stones 6, 9 and possibly Stone 12 were found but they did not provide precise locations; and the two re-erected stones were also surveyed but of course their positions cannot be relied upon.

Taking only those stones the positions of which are certain (Nos. 2, 3, 4, 8, 10 and 11), an ellipse having $2a = 39.64$ MY, $2b = 36.97$ MY and with its major axis oriented at about 160°/340°, gave the best fit as determined by computer (see Appendix 8, by the author, in Ritchie 1976). Many other megalithic shapes including circles were tried and all gave distinctly inferior fits.

When it is recognised that Stone 11 does not fit any ring, an ellipse, having $2a = 39$ MY and $2b = 36$ MY, with a perimeter of 47.1 MR and a major axis

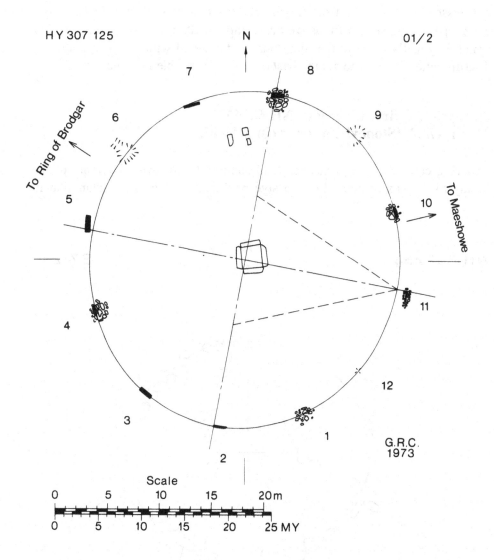

Fig. 16.13. The Stones of Stenness. Ellipse: major diameter 39 MY, minor diameter 36 MY, basic triangle 7.5 : 18 : 19.5 MY.

at 10°.5/190°.5 ± 2°, fits the remaining stones (Nos. 2, 3, 4, 8 and 10) very much better than any other standard shape or size. It is of course based on a multiple of the 5:12:13 Pythagorean triangle. This is shown in Fig. 16.13 which combines Ritchie's excavation results with the author's survey. This ellipse is identical with one of the ellipses at Stanton Drew.

A lesson to be learned from this exercise is that pre-conceived ideas as to what type of ring should fit, as well as computer fits using all stones whether or not they really relate to the ring, must be tempered with understanding and discrimination in order to get the best out of the available information.

Stoneyfield Cairn, Inverness. NH 687454
Thom B7/20. 'Stoneyfield' or 'Achnaclach'.

This ring cairn, lying on flat open land some 0.25 km from the shore of the Moray Firth, was composed of massive boulders more or less rectangular in

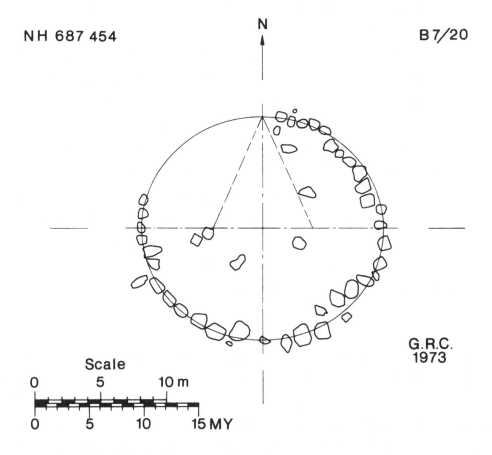

NH 687 454

N

B7/20

Scale

0   5   10 m

0   5   10   15 MY

G.R.C.
1973

Fig. 16.14. Stoneyfield Cairn. Ellipse: major diameter 22 MY, minor diameter 20 MY, basic triangle 4.5 : 10 : 11 MY.

plan, some having fallen inwards or outwards and some having been broken. The remainder of the cairn material was largely absent (Henshall: 1963). There are extensive views of the Ben Wyviss range to the north-west and of nearer hills to the south-east.

Prior to the complete removal of the cairn, which was necessitated by construction of the new A9 Trunk Road, complete archaeological excavation was undertaken by D.D.A. Simpson in the summers of 1972 and 1973 (Simpson 1972; 1973). In April 1973 the author surveyed the ring, as shown in Fig. 16.14, with the assistance of Ian Cairns.

At that time all the boulders had been exposed as a result of the first period of excavation, but they are understood to have been in their undisturbed positions. The best fit to the centre of those boulders which were likely to have been in their original positions is given by an ellipse having diameters of $2a = 22$ MY and $2b = 20$ MY. The perimeter is 52.83 MR. The orientation of the minor axis is due north/south. This ellipse could have been constructed on a triangle with sides of 4.5, 10 and 11 MY - a sufficiently close approximation to a Pythagorean triangle for all practical purposes on the ground.

This ring of boulders was re-erected in about 1975 at NH 687450 and re-named *Raigmore Cairn*.

Mudbeck Ring, North Yorkshire. NY 954077
Thom L6/5

The site, 5.9 km east of Tan Hill, is in a prominent position on a ridge of glacial till at the head of a pass, in upper Arkengarthdale, the valley which leads into Swaledale, North Yorkshire. Situated 680 m west of the confluence of Mud Beck and Arkle Beck, and 300 m north of Arkle Beck itself, it lies at 380 m above sea level in wild, open, lush grass moorland with magnificent views in many directions (Tim Laurie, *pers. comm.*).

The stones were first reported in 1982 by Mr William Stubbs, the local shepherd, via Tim Laurie of Barningham, Richmond, to Archie Thom who made a survey in 1983. The results are first published here in Fig. 16.15 at his request. He has recorded five stones in the ring (standing or leaning slightly), three slabs lying outside the ring, one slab buried on the line of the perimeter, and a small stone just inside the ring. Unfortunately none of the stones which remain standing is in the southern semicircle, which slopes gently to the south. It may be surmised that 'quarrying' has taken place and that possibly the three slabs (Stones 3 and 4) were in the process of being dragged away.

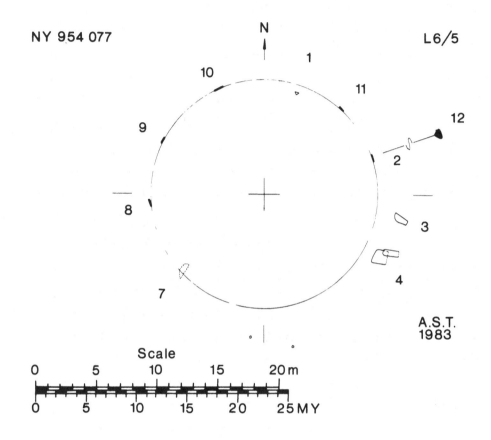

Fig. 16.15. Mudbeck Ring. Circle: diameter 22.53 MY.

One outlying standing stone is located 53.8 m from the centre, towards the ENE.

The best fit, provided by Archie Thom, is a circle of 22.53 MY diameter fitted to the centres of the bases of the 5 standing stones as shown in the figure. The perimeter is 28.32 MR. True north is given on the plan within a tolerance of ± 2°.

Appendix I

Table 16.1 gives a complete list of the sites on the Isle of Lewis which were named and numbered by Professor Thom, with corrections and extensions recently made by Dr A.S. Thom (see introduction). Also shown, in Roman numerals, is the sequence of Callanish site numbers as used by The University of Glasgow (1978), Ponting & Ponting (1981b), and the author. The grid references have been included for full identification purposes.

Table 16.1.

Thom no.	Callanish no.	Grid Ref (NH)	Name
H1/1	C I	213330	Callanish
H1/2	C II	222326	Cnoc Ceann a'Gharaidh
H1/3	C III	225327	Cnoc Fhillibhir Bheag
H1/4	C IV	230304	Ceann Hulavig
H1/5	C V	234299	Airigh nam Bidearan
H1/6	C VI	247303	Cul a' Chleit
H1/7	C VIII	164342	Bernera Bridge (Great Bernera)
H1/7A	C VIIIA	154340	Aird a'Chaolais
H1/8	C X	230336	Na Dromannan
H1/9			Clachan Mora, Steinacleit
H1/10		396540	Steinacleit
H1/11			Steinacleit
H1/12		375538	Clach an Trushel
H1/13		524330	Dursainean, Eye
H1/14		516318	Clach Stein
H1/15		529334	Allt na Muilne, near H1/13
H1/16		204430	Clach an Tursa
H1/17	C XXII	317292	Achmore
H1/18	C VII	232302	Cnoc Dubh
H1/19	C IX	233297	Druim nam Bidearan
H1/21	C XI	222356	Airigh na Beinne Bige
H1/22	C XII	215350	Stonefield
H1/23	C XIII	215341	Sgeir nan Each
H1/24	C XIV	228329	Cnoc Sgeir na h-Uidhe
H1/25	C XV	177346	Airigh Mhaoldonuich
H1/26	C XVI	213338	Cliacabhaidh
H1/27	C XVII	237320	Druim na h-Aon Chloich
H1/28	C XVIII	244292	Loch Crogach
H1/29	C XIX	218331	Buaile Chruaidh
H1/30		233461	Loch Raoinavat
	C XXI		Rubha Tota Sheors
	C XXIII	211355	Cnoc a Phrionnsa

17

Megalithic compound ring geometry

Megalithic rings seem to have a certain style of pattern, and we are indebted to Alexander Thom who took the first steps towards understanding the geometric structure of these monuments (Thom 1967). First steps are always the boldest and most difficult, and Thom deserves considerable praise for taking them. But they are necessarily faltering. As with all profound theoretical accomplishments, however, the errors tend to be less than first perceived, and there is a significant element of truth that remains after the obstacles of criticism have been removed. We will offer a suggestion as to what that element of truth might be. We will then use it to point out a rather surprising and heretofore unknown feature of many of the compound rings.

Megalithic geometry

Thom's answer to the question of geometric construction was mixed. Most of the rings were said to consist of circular arcs with their centres within and on the boundary of the ring. (There were exceptions such as the Avebury ring and the Woodhenge site where the centres for some of the arcs were said to lie outside the ring.) According to Thom, these centres were often located at the apices of Pythagorean triangles. A second kind of construction was proposed for the elliptical ring. Here Thom suggested the common procedure for scribing an ellipse which employed a loop around two anchor points located at the foci. These two ways of describing the geometry of the rings are quite different, and in the interest of parsimony, one should settle on one or the other but certainly not both.

There is an ancillary question here, one that Thom considered only in passing. What procedure was actually used to draw the ring designs? The first proposed scribing scheme chose to follow the main thrust of Thom's geometry and ignore the loop procedure for constructing an ellipse. This method (Cowan 1970) used two anchor points and two pivot points. An anchored rope and its marker approached the pivot then swung around it as the marker traced its path (Fig. 17.1(A)). In the case of the ellipse this procedure was actually closer to Thom's principal geometric plan with its collection of different arcs and their various centers. The resulting figure is an oblate circle, not a true ellipse, but it is indistinguishable from the ellipse in terms of its fit to the 'elliptical' megalithic ring.

A contrasting procedure proposed by Angell (1976) rejected Thom's principal geometric schema and elaborated on the elliptical scribing method for all of the non-circular ring designs. Many of these designs can be drawn by placing a loop around three pivot points, then moving the marker as one would in scribing an ellipse (Fig. 17.1(B)). He later extended this procedure to four pivot points (Angell 1977).

Angell avoided making the claim that the megalithic designers actually employed this technique. Rather he used his proposal as a vehicle for criticism of Thom's geometry. Thom's evidence appears to be restricted to the goodness of fit of his geometry to the field data, but, as Angell demonstrated, a very different method can be used to obtain fits that are at least as good as,

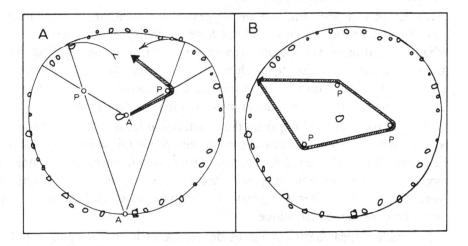

Fig. 17.1. The megalithic ring at Black Marsh (after Thom, Thom & Burl 1980).

(A)   The scribing procedure suggested by the author which closely adheres to Thom's geometric analysis.

(B)   The elliptical scribing procedure suggested by Angell.

and often better than, Thom's geometry. Angell's point in suggesting that the goodness of fit is an insufficient test for Thom's hypothesized geometry is well taken. However, while acceptable goodness of fit is not a sufficient condition for proof it is a necessary condition.

We might ask why Thom selected this particular geometry when others equally fitting could have been used. It would be a fair guess that he wanted support for his 'megalithic yard', an hypothesized prehistoric unit of measure, by showing how the dimensions of his fitted geometry were integral measures of such a unit. In addition, he may have wanted to convince us of the precedence of the Pythagorean triangle which appears in many of his constructions.

Thom's strategy is not without its problems. It sometimes seems that he selected his geometric patterns to fit his purposes rather than the data, e.g. Nine Ladies discussed below. Furthermore, he presents the entire, often complex, design for us to accept without a rationale for it; he offers us whole cloth but has not shown us the threads from which the pattern was woven.

Thom suggested that the existence of a standard unit of measure, and the knowledge of Pythagorean relationships, speak for the mathematical sophistication of the prehistoric designer. Angell dismissed this as overstatement, and in this respect he echoed a similar criticism aimed at the present author by Grossman (1970).

We accept this criticism, and we suggest further that, at the beginning at least, the prehistoric designers did not have a sophisticated pattern in mind. What they did have was a crude way to scribe a ring of unspecified shape, perhaps with the intention to produce a circle. Furthermore, this simplistic method, rather than requiring any mathematical knowledge, actually shows an ignorance of it, particularly if the purpose was to construct a circle.

The ring patterns resulting from the construction technique to be proposed possess some unexpected geometric properties. Since this construction is both necessary and sufficient for the geometric consequences, the empirical support for the consequences will serve as evidence for the scribing procedure. We will offer a suggestion as to what motivated the use of this particular construction technique.

The mathematical basis for the evidence makes it necessary to introduce some elementary concepts about convex figures which will be used throughout this Chapter. Details of these concepts can be found in Yaglom & Boltyanskii (1961).

(1)   A *convex* figure is one in which any two points of the figure can be connected by a straight line that lies completely within the figure. A

square or oval is a convex figure; a cross or a figure with a hole is not. In this paper all convex figures are completely bounded or closed, i.e. all can be encased within some fixed circle.

(2) A *supporting line* is one which passes through one or more boundary points in such a way that the entire figure lies on one side of the line. Circle tangents and a line lying along the edge of a triangle are supporting lines. There are infinitely many supporting lines passing through the corner of a figure such as a square or a triangle. A boundary figure is convex if through each of its boundary points there passes at least one supporting line.

(3) A *diameter* of a convex figure is the least distance between two maximally separated parallel supporting lines. The diagonal connecting two corners of a square is a diameter.

(4) A *width* of a convex figure is the least distance between two minimally separated parallel supporting lines. Any line perpendicular to two opposing edges of a square is a width.

(5) A *figure of equal width* is one in which the width equals the diameter. The most recognizable equal-width figure is the circle.

Let us suppose that the designers of these rings were trying to draw a circle. This was Angell's (1976) assumption, and at the very least it can serve as a heuristic to help us discover the steps taken during the first attempts at laying out a ring.

Depending on how one approaches the problem there are alternative sets of minimum qualifications that must be met when attempting to draw a circle. One alternative includes the following in decreasing order of generality (size of set) and perceptual salience: (i) a circle is a closed convex figure; (ii) it is a figure of equal width; and (iii) it is a figure of equal radius. As for order of generality, not all closed convex figures are equal-width, but all equal-width figures are closed and convex. And, surprisingly, there is an infinite variety of figures of equal width, but there is only one equal-width figure of equal radius - a circle.

How are we to know the order of perceptual salience of these features during the first attempts to reproduce a circle from nature? We begin with a rational guess. The perceptual primacy of (i) above is fairly evident; a child who has no experience with a compass and radial construction will draw a figure that at least meets these criteria when asked to draw a circle. The order of importance of (ii) and (iii), however, is more difficult to judge objectively because of our training in the use of a compass. Yet, for someone who has never seen such an instrument, the equal width of a circle may be perceptually more evident than the characteristic of equal radius; the width is bounded by well defined edges, while in nature the centre end of the radius is rarely marked and must be inferred.

We might assume then that if the megalith builders were trying to construct a circular design the first attempts focused on the diameter and not the radius. That is, they may have used arcs whose radii were diameters of the figure they wanted to draw. This will be called 'diameter construction'. As we shall see, this construction method is consistent with Thom's general principle that megalithic rings were comprised of a collection of arcs drawn from a variety of centres.

Three of the ways a diameter can be used to draw a closed convex figure are relevant for megalithic design and have geometric consequences that are non-intuitive and surprising. Even if megalithic ring designers were not trying to build a circular shape, we can use these unexpected consequences to test the validity of this method of construction.

Construction of cornered rings

Imagine two scribers of a megalithic ring on the opposite ends of a rope trying to walk in a circular path by keeping the rope taut between them. Unfortunately, it would be difficult enough to make a simple closed figure using this procedure let alone a circle. If one remained stationary while the second paced out a short clockwise arc (60° or less), then the second stood still while the first paced out a short clockwise arc, and this was repeated, a cornered, non-circular, closed figure of equal width would eventually be obtained, as Fig. 17.2(A) shows. (A description of the construction procedures discussed in this paper can be found in Rademacher & Töplitz 1957.) If three arcs of 60° were made in this fashion then the resulting shape would be an equal-width triangle called Reauleaux's triangle. This is the best known of all the equal-width figures except the circle. However, as Fig. 17.2(A) indicates, having arcs of equal length is not a necessary condition for cornered figures of equal width. The figure does not even have to be symmetrical.

All equal-width figures, including Reauleaux's triangle, have a number of curious properties, not least amongst them being that all pairs of parallel supporting lines are the same distance apart. If 'rollers' with a Reauleaux triangle cross-section (or any equal-width cross-section for that matter) supported a plank, the plank could roll across them and always remain perfectly parallel to the level ground in spite of the corners on the rollers (see Fig. 17.2(D)).

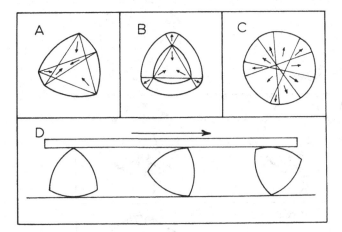

Fig. 17.2. Procedures for producing equal-width figures. The arrows in this and all subsequent figures point to the middle of an arc from the origin of its radius.

(A) The procedure for producing a cornered equal-width figure.

(B) The procedure for producing an equal-width figure with rounded corners.

(C) The general procedure for producing an equal-width figure.

(D) A demonstration of the equal-width property of an equal-width figure. The board will roll evenly across two or more Reauleaux triangles with the same width.

A further consequence is that all shapes drawn using this technique must have an odd number of sides (arcs) and corners (Rademacher & Töplitz 1957). To see this, mark a corner and its opposite arc. Now start moving to the right of the marked corner counting the sides and corners as you pass (do not count the marked corner or its marked side). When the marked side is reached you will have counted n sides and n corners. But each corner you have counted is a pivot for a side on the left of the marked corner. Similarly, the pivotal corner for each side you have counted is to the left of the marked corner. Thus there are n corners and n sides on either side of the marked corner giving $2n$ corners and $2n$ sides. If we now count the marked corner and its marked side we find a total of $2n+1$ corners and $2n+1$ sides for the whole figure. Therefore, the number of sides and the number of corners is odd.

These two results can serve as a rather stringent test for the hypothesis that a diameter construction procedure was used for the megalithic rings. If there exists just one non-circular ring that is a cornered equal-width figure, then a

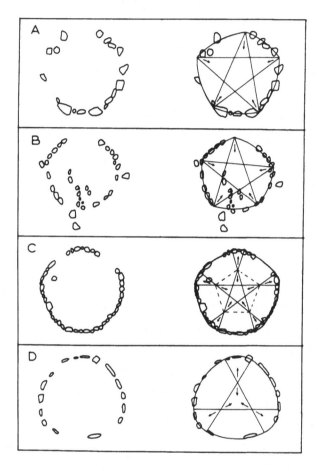

Fig. 17.3. Four megalithic rings and their best-fitting equal-width reconstructions: (A) Moncrieffe House; (B) River Ness; (C) Moel Ty Ucha (all after Thom, Thom & Burl 1980); and (D) Hirnant Cairn (after Thom & Thom 1978).

diameter construction procedure was most certainly used for it, since it is impossible to construct such a figure without it. Furthermore, we can expect such a ring to have an odd number of sides.

Two rings will be reported here, both of which are described by Thom, Thom and Burl (1980) as circular. The first of these is Moncrieffe House (Fig.17.3(A)). If it appears without its 'best fit' circular lines, the cornered character of this ring is immediately evident. Also notice that with its pentagonal shape it meets the criterion of an odd number of sides. An equal-width pentagon provides an excellent fit for this ring.

The other is River Ness (Fig. 17.3(B)). Two of its sides show a clear pentagonal corner. Again, an odd number of sides is called for, and again an equal-width pentagon provides an excellent fit; better, in fact, than a circle.

Thom's misinterpretation of these rings as circular may have come from the fact that the ratio of circumference to diameter is π for all equal-width figures. That is, all figures of equal width h have perimeters equal to the circumference of a circle with diameter h. The reader can easily prove this for Reauleaux's triangle. While the general case (Barbier's theorem) is not terribly difficult, it is too long to include here (but see Yaglom & Bolyanskii 1961: 249).

Rings with rounded corners

If we begin with any cornered equal-width figure and then extend the length of the diameter beyond the corner, a larger equal-width figure with rounded corners can be constructed. The corner becomes a centre for the radius of a tighter arc that rounds out the corner of the larger figure.

For example, a Reauleaux triangle is made by drawing an arc between any two points of an equilateral triangle using the third point as the centre of the arc. This is repeated for the other two corners. To make this figure round-cornered, a point of the triangle is taken as a centre, and a rope (or string) and its marker are extended through and beyond a second point (Fig. 17.2(B)). An arc is drawn by moving away from the second point toward the third. When the third point is reached, the rope is allowed to wrap around it making an arc of a greater bend whose radius is equal to the distance that the rope extends beyond the pivot point it wraps around. This is repeated for the two remaining points. Again, the figures do not have to be regular. The design construction method suggested by the author (Cowan 1970) is well suited for this scribing method (Fig. 17.1(A)).

The compound ring at Moel Ty Ucha is an obvious candidate for this design procedure (Fig. 17.3(C)). Thom's guess as to how this ring was drawn was very close to an equal-width construction, which can be produced from his geometrical lines with only insignificant changes. Thom was probably not aware of the concept of an equal-width figure. If he had known about it, he would have most certainly mentioned it in connection with this ring.

Another ring that fits this sort of construction is Hirnant Cairn (Fig. 17.3(D)). Thom & Thom (1978a) identify this as Type III egg with an elliptical end. Their geometric construction lines make use of three points which appear to be similar to those that Angell would use. The ring at Hirnant

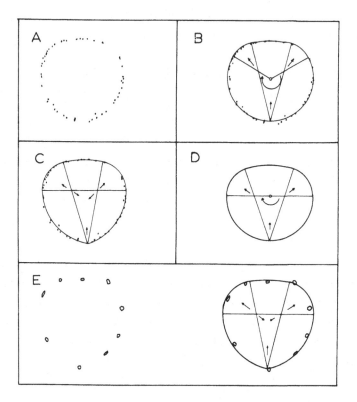

Fig. 17.4. (A) Rough Tor; (B) Rough Tor and Thom's geometric reconstruction; (C) Rough Tor and an equal-width geometric reconstruction; (D) Thom's Type I flattened circle; and (E) Nine Ladies and its best-fitting equal-width reconstruction (all after Thom, Thom & Burl 1980).

Cairn can also be seen as an equal-width figure with three sides, and it is possible that most if not all the Type III eggs can be fit with a rounded Reauleaux figure.

General equal-width construction method

When scribing smoothed cornered figures of equal width the centres of the arcs making up the perimeter are located at the intersections of pairs of crossing width lines (see Fig. 17.2(B)). The same procedure is used with cornered figures except that the point of intersection is on the circumference, and only one arc is drawn; the length of the other is zero (see Fig. 17.2(A)).

Any number of arcs of different lengths (including zero) can be used to make an equal-width figure as long as they are paired off on opposite sides of intersecting width lines; the intersection is their common centre, and their central angles are 60° or less (Fig. 17.2(C)). Thus very irregular equal-width figures of any number of sides can be made from a collection of randomly-placed intersecting width lines that are confined within a closed convex space. One would shift from one intersection to the next drawing pairs of arcs, first one then its opposite member.

Rough Tor is a flattened circle (Type *B* in Thom's system of description) which seems to come to a point, giving it the appearance of a toy top (Fig. 17.4(A)). Thom used the centre to draw a semi-circle through this part, but it is evident that the sides of the monument have less of a bend than the semicircle dictates (Fig. 17.4(B)). An equal-width construction provides a better fit of the points marked by the stones (Fig. 17.4(C)), and the construction is simpler. But what makes the construction interesting is the fact that it makes the geometry of a Type *B* flattened circle ring more like that of Thom's Type *A* flattened circle (Fig. 17.4(D)).

Thom suggested (Thom, Thom & Burl 1980) that Nine Ladies was circular with a diameter and circumference of 13 and 41 MY respectively (Fig. 17.4(E)). However, a circular construction missed the stone at the 'cusp' by such a wide margin that he almost gave it the status of an outlier, calling it simply a 'standing stone.' A general equal-width construction can be fit to this ring without any difficulty.

It is paradoxical that the best example of this design procedure is not an irregular ring but one that is nearly circular in its lay-out. The Lios (Fig. 17.5) is located in County Limerick, Ireland, and no improvement need be made on Burl's description of its construction:

> The Lios is instructive about prehistoric working methods. At the exact centre on the old land surface beneath the clay the excavators found a posthole 13 cm across, too slight for a totem pole but adequate for a focal post from which the 23.8 m radius of the inner bank could be marked out. *This was done not by scribing an entire circle in the turf but by extending the rope to five roughly equidistant places on the eastern circumference and then to five others opposite.* The ten spots were marked with poles... The design looked like one of Thom's compound rings but more haphazardly achieved.

(Burl 1976a: 228; italics added.)

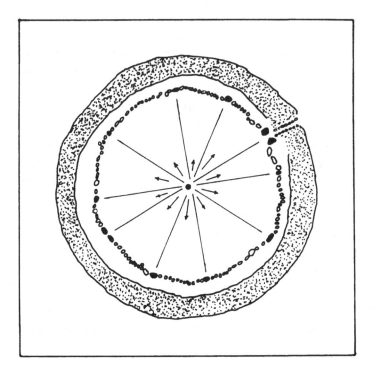

Fig. 17.5. The Lios showing the equal-width plan of its construction. The locations of the perimeter markers are shown in black (after Burl 1976).

It would be a fair guess that, compound appearance notwithstanding, the intent was to construct a circle. Yet the construction is odd. In spite of the fact that the anchor was at the centre, the compass method for drawing a circle was clearly not used; the ring was made by drawing first one arc then the one opposite. If it is assumed that the design of The Lios was scribed following a tradition of equal-width procedures, the mystery of its construction vanishes.

Avebury: a non-equal-width ring

The epitome of compound rings is the Avebury ring and no discussion of such monuments can ignore it. The irregularity of the Avebury ring would seem to demand a general kind of equal-width construction (Fig. 17.6(A)). However, Avebury is not an equal-width monument. Thom offered a spirited argument in favour of a Pythagorean triangle as a rather complex basis for this design (Thom & Thom 1978a). We wish we could share his enthusiasm. Burl's sober reflection cuts to the heart of the matter: 'It must be added that (Thom) was unable to provide any good explanation of why such an involved plan ... should ever have been chosen' (Burl 1979: 180).

Thom and Thom suggest that one of the unique features of Avebury is that its arcs meet at corners rather than smoothly joining one another. As we have seen, this feature is hardly unique to Avebury. There is a particular feature that may indeed be found only at Avebury. For a ring that possesses such smooth arcs, its overall plan might be described as asymmetrical. Burl saw this asymmetry, but is it possible that even this observation is in error?

Fig. 17.6 offers a 'symmetricized' Avebury. The line of symmetry splits the line of the flattest arc, the centre of which lies well outside the ring. Located inside the ring, at a distance equal to the distance that the far centre is outside

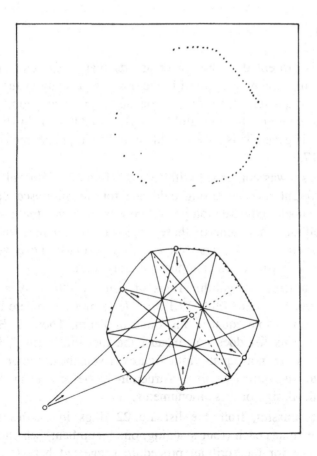

Fig. 17.6. Avebury and a symmetrical reconstruction.

the ring, is the centre for the bottom arc. All the centres for the side arcs are located on the perimeter, and the radii for these arcs are all equal, which gives it a partial equal-width appearance. The diameter on the axis of symmetry and its orthogonal diameter are also equal. The symmetrical outline provides a not unreasonable fit.

However, this construction is contrived and should not be taken seriously. So much of this ring is missing that almost any graphic scheme can be made to fit it as this exercise demonstrates. The possibility that this ring is extremely primitive and crude, or simply a botched job, cannot be discounted. Given the size of this ring, considerable error was not only possible but could be expected.

Avebury remains an enigma.

Conclusions

The hypothesis presented in this paper asserts that a diameter construction procedure and its variants were used in the designs of many of the megalithic rings. Various diameter construction procedures produce geometric shapes that seem to capture the megalithic 'style' mentioned in the opening paragraph of this paper. This style is evident in the simple geometric exercises given in Fig. 17.2.

The hypothesis was concerned with figures of equal width only insofar as such figures might provide strong evidence for the proposed construction method. If a diameter construction procedure was used then there would be an increased likelihood that some of these rings would be equal-width figures. Furthermore, if equal-width rings were found, particularly those with corners, then a diameter construction was almost certainly used.

There are numerous megalithic rings of equal width, only a sample of which were described here. Table 17.1 provides a more complete list of rings of probable and possible equal width found in Thom, Thom & Burl (1980) and elsewhere. This list does not exhaust the megalithic rings, for a large number of them are not equal-width figures. Even these, however, used a partial diameter and equal-width construction if we are to accept Thom's geometry for the design of these monuments.

Furthermore, missing from the list are 22 rings in south-west Ireland containing five stones each (four standing, one recumbent) which might also provide evidence for the scribing procedure suggested here (Burl 1976b). They were not included because acceptable surveys of these rings were not available.

Table 17.1. *Thirty megalithic rings of possible equal width*

Cornered	White Cow Wood
Trowlesworthy	Loanhead Daviot (inner ring)
* Loch Avich	Aquorthies Kingausie (inner ring)
River Ness	Raedykes North
* Tullybeagles (small ring)	* Carnhousie House
Moncrieffe House	* Miltown of Clava (inner and outer ring)
	Castle Dalcross
Rounded	* Tyfos
Tomnaverie	Y-Pigwn North East
Moel Ty Ucha	
Hirnant Cairn (*TT78*)	**General**
Newgrange (*TC70*)	Nine Ladies
	The Lios (*AB76*)
Cornered or Rounded	Blakeley Moss
Odendale	Birkigg Common (inner ring)
Rollright	Rough Tor
Loch Nell (Circle *A*)	Botallack
* Temple Wood	Tyrebagger
* Ardnave	Cambret Moor West

All these rings are listed in Thom, Thom & Burl (1980) unless marked *TT78* (Thom & Thom 1978a) *TC70* (Cowan 1970) or *AB76* (Burl 1976a). Incomplete rings with corners are marked by an asterisk.

A noticeable feature of diameter construction is its crudity; it will not, under ordinary circumstances, succeed in creating a perfect circle. This does not support the belief that the megalithic people were endowed with any special knowledge of geometry particularly if creating a circle was their intent.

Suppose that the equal width property of their designs did not go unnoticed or that their purpose was actually to construct figures of equal width. While they may have discovered the appropriate ways of creating such figures, it does not follow that the megalithic architects knew of the geometry of equal-width construction with the subtlety or finesse of the modern mathematician. It is certainly not our intent to assert that they did. We might, perhaps, guess that they possessed an appreciation for symmetry and strove to achieve it. Beyond this, however, even idle speculation is difficult.

18

The metrology of cup-and-ring carvings

ALAN DAVIS

Introduction

Among the many contributions of Alexander Thom to the study of megalithic
remains, his hypotheses concerning cup-and-ring carvings have been
surprisingly neglected by other researchers. It is clear, however, that Thom
himself considered the subject to be of some importance. In his preface to
Morris (1977) for example, he writes:

> Do they contain a message? Are they the beginning of a form of
> writing? The fact that these questions can be asked shows the enormous
> importance which must be attached to their study and interpretation.

Similar comments may be found in many of Thom's publications and, as in so
many other cases, his own work provides a possible direction for the further
development of such a study. Thom's work on the carvings is to be found in
two papers (Thom 1968; 1969) and was later summarised in Thom & Thom
(1978a) together with a small amount of additional material. Basing his
original analysis on 57 measurements of circular ring diameters, using
rubbings of Scottish carvings made by Morris and Bailey (Thom 1968: 176),
Thom deduces the existence of a unit of measurement, the 'megalithic inch'
(MI) and points out that the value of this unit, 2.0725 cm, is 1/40th of a
megalithic yard (MY), and 1/100th of a megalithic rod (MR).

Considerable space in Thom's papers is devoted to a geometrical analysis
of non-circular rings and spiral carvings in order to demonstrate the use of the
MI in the construction of such shapes, usually involving circular or elliptical
arcs based upon Pythagorean triangles. The quality of fit achieved can be

quite striking in some cases, and does indeed demonstrate that a geometrical interpretation along these lines is perfectly feasible. As evidence for the existence of a megalithic inch, however, such geometrical arguments are notoriously difficult to assess, and critical reaction to them has been mixed. For example, in order to achieve a fit, resort must often be made to lengths which are integer multiples of $\frac{1}{4}$ MI, and this is such a small measure that it offers considerable freedom for adjustment of the geometry to the rubbing. Ultimately it seems that the case for the reality of the MI must rest with the initial statistical analysis of the circular ring diameters.

In his original analysis Thom (1968) employed a method devised by Broadbent (1955), and a nominal probability level of about 4% is quoted. In other words, if we were to analyse similar sets of purely random data repeatedly, we could expect to obtain support for Thom's unit (or 'quantum') as good as, or better than, Thom's on only four occasions out of every 100 tries. This might normally be accepted as a significant result, but in fact it is rather misleading here. To understand why this is so, we need to examine the nature of statistical quantum tests.

The testing of quantum hypotheses

There are two very different circumstances in which we may approach a set of data in order to decide whether it supports a quantum hypothesis, and they require correspondingly different techniques which we shall call Type *A* and Type *B* as follows.

Type A. Here we approach the data with a specific quantum to be tested, whose value has been determined in advance, and which is independent of the data under test.

Type B. In this case we assume no preliminary value for a quantum, but wish to search the data for any quantum over some sensibly determined range of possible values.

The test applied by Thom to the carved ring diameters is in fact a Type *A* test, and the probability level quoted is valid only if the quantum is specified in advance. For the megalithic inch this is clearly not the case, even allowing for the fact that it bears a comparatively simple relation to the megalithic yard. If a Type *B* test of the data is performed (see below), no significant evidence for a quantum survives - a point observed by Heggie (1981).

Although Thom used Broadbent's method in his analysis, more powerful techniques have been developed since then - notably by Kendall (1974) and

Freeman (1976). Throughout this paper, Kendall's test will be used. The method is based upon the test statistic

$$\phi(T) = (2/N)^{1/2} \sum^{i} \cos (2\pi y_i T)$$

where $T = 1/q$ (q being the proposed quantum), the y_i values are the data under test, and N is the number of data points in the sample. Support for a quantum is indicated by a significantly large, positive value for ϕ.

A Type A test is obtained by virtue of the fact that for a given T, ϕ is normally distributed with mean zero and unit standard deviation on the random, non-quantal hypothesis. Applying the test to Thom's data, for example, for the hypothetical quantum of 2.0725 cm, we obtain $\phi = 1.60$. This would correspond to a probability level of about 5.5% provided that the quantum could have been specified in advance. Of course these are the data from which the quantum was originally derived, and so the Type A test is inappropriate - though there is no reason why Thom's hypothesis cannot be tested on independent data using Type A methods, provided his specific quantum of 2.0725 cm is used.

A Type B test is obtained by calculating values of ϕ for all possible quanta over a suitable 'search range' from T_1 to T_2 (T_1 and T_2 representing the lower and upper limits of the T values respectively). The resulting oscillating function $\phi(T)$ is referred to as a 'cosine quantogram' or CQG, and support for the existence of a quantum is implied by the presence of a large peak at some value of T within the search range. Estimation of the significance of such a peak is less straightforward than in the Type A case, and for a detailed discussion the reader is referred to Kendall (1974). It is sufficient here merely to note that the probability level is a function of RMS(y) (the root mean square value of the individual data), and also of the search range T_2 - T_1. A convenient diagram for the estimation of significance levels is given in Davis (1983), but in every CQG chart in this paper the 5% and 1% levels are indicated by dashed horizontal lines (the lower of these corresponding to the 5% level).

By way of example, Fig. 18.1 presents the CQG for Thom's data over the range $T_1 = 0.08$ cm-1 to $T_2 = 0.58$ cm-1, the data having been converted from inches to centimetres to facilitate comparison with the author's own work. As in all subsequent CQG charts here, the vertical lines at the top of the diagram are drawn at values of T corresponding to quanta of 1 MI (furthest right) and its integer multiples up to 5 MI (furthest left). We see here that the 1 MI peak falls well short of the 5% significance level and, further, that a considerably

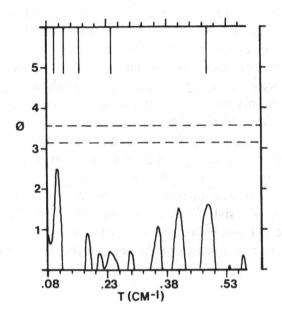

Fig. 18.1.   CQG for 57 ring diameters (after Thom).

larger peak occurs close to $T = 0.1$ - though this peak also has no statistical significance.

Note that in Fig. 18.1, as in all the CQG charts here, only positive values of ϕ are of any relevance; negative values are therefore not shown.

Further testing of the megalithic inch hypothesis

If the question of the reality of the MI is to be resolved, it is clear that the only way forward lies in the further testing of the hypothesis on another, independent body of data. One such test has already been performed, and its results published, by the present author (Davis 1983).

The aims of this first experiment were as follows: (a) to test the MI hypothesis on a small group of cup-and-ring carvings in Yorkshire, near Ilkley; and (b) to determine whether the positions of the cup markings themselves suggested the use of mensuration in their design. Since rubbings are not the most effective way of extracting data from the carvings, their use was rejected in favour of direct measurements made on the rocks using trammels, subsequent to the initial chalking-in of cup centres and rings. Where rings were judged circular, their diameter was estimated by taking two trammel measurements at right angles (between the chalk lines) and accepting the mean of these. For each carving the distances between centres of neighbouring cups were also accepted for analysis, 'neighbouring cups' being defined by Thiessen polygons constructed on an accurate scale plan of the cupmarks drawn up from the measurements made at the site.

Full details of the methods employed are described in Davis (1983), but for the convenience of the reader Fig. 18.2 gives an example of the Thiessen polygon treatment applied to the cupmarks from one of the Ilkley carvings. An additional criterion is sometimes required when dealing with pairs of cups at the outer edges of a design. Consider, for example, cups *C* and *D* in Fig. 18.2; clearly, if the polygon boundaries are extended sufficiently far then *C* and *D* (against all common sense) might be regarded as 'neighbours'. For

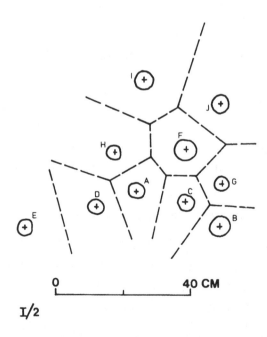

Fig. 18.2. Thiessen polygons.

such cases the arbitrary criterion is invoked that the corresponding vertex of the relevant polygon must lie inside the line joining the cup centres. This excludes C and D from consideration - and, incidentally, also (just) excludes pairs such as J and G. These procedures are necessary merely in order that the selection of distances for analysis be taken out of the hands of the investigator as far as possible, to ensure an unbiased sample of data.

In all, 126 cup centre separations and 39 (assumed circular) ring diameters were presented in the 1983 paper. Analysis of these 165 measurements by Kendall's method yielded a Type B probability level of between 1% and 2% for a quantum of 2.088 cm. That this may be sensibly associated with Thom's MI is demonstrated by a Type A test for the 2.0725 cm quantum, which yields a probability level of about 0.2%. And indeed, the difference between the observed quantum and Thom's was not statistically significant.

Although this earlier work offered substantial support for Thom's hypothesis, it may be criticised in several respects.

(a)   The assessment of circularity of rings was entirely subjective, made at the sites by a purely visual inspection of the chalk marks prior to measurement. Subsequent work, using a direct tracing method, has shown that some of these judgements were probably incorrect. A surprisingly large number of rings show, on careful examination, systematic (and apparently intentional) deviations from true circularity.

(b)   Complete plans of the carvings, showing all discernable details of the designs, were not made - and in one important case (the Panorama Stone) survey coverage of the carved surface was only partial. Thus the data presented in 1983, while forming a satisfactory base for the straightforward testing of the MI hypothesis, are of little use for the further development of the subject.

(c)   The sample size was rather small, the bulk of the data being obtained from only five sites.

Many of these deficiencies have now been remedied, at least in part.

Detailed plans of all the carvings so far studied by the author are now given (albeit at greatly reduced scale) in this paper (Figs. 18.9, 10, 11 & 12), together with cartesian coordinates of the centres of cups (Table 18.3). Any desired distance between cup centres may therefore be calculated. It should be noted that these are the coordinates of the small chalk marks made at the centres of the cups, estimated at the sites by eye before any measurements were made. They are therefore entirely suitable for statistical analysis.

The plans themselves have been made from direct tracings (using felt-tipped pens on large polythene sheets taped over the rock) of the entire chalked design. In addition to this, a number of measurements were made directly on the rock using trammels to enable compensation for stretch and/or slight rock curvature to be made. Such corrections usually proved to be negligible, and never more than about 1% in any case. Where necessary they have already been incorporated into the plans and data presented here. In some cases, weathering of the surface makes the correct interpretation of faint detail uncertain. Some assistance here can be gained by inspecting the carving under cover of a large blanket, allowing light to glance across the rock surface in controlled directions.

In addition to the revision of all sites studied in the 1983 paper, data from a further three Ilkley carvings are now available, together with data from a small number of Northumberland sites. A full site list is given in Table 18.1. Selection of sites for inclusion in this study has been determined by various factors in varying degrees, none of which would bias the sample either for or against a quantum hypothesis; relative ease of access, state of preservation of the carvings, and visually interesting design have all played a part. In most cases the sites selected are simply those which are well-known and documented. A further body of data has been gathered from Baildon Moor in Yorkshire, though at the time of writing the reduction of the raw data has barely begun.

Table 18.1. *Complete Site List*

Reference no.	Name	O.S. Reference
I/1	Weary Hill West	SE 106466
I/2	Weary Hill East	SE 107465
I/3	Panorama Stone	SE 114473
I/4	Barmishaw Stone	SE 112464
I/5	Pancake Ridge 1	SE 131463
I/6	Pancake Ridge 2	SE 131463
I/7	Green Crag	SE 133458
I/8	Backstone Beck	SE 127463
N/1	Roughting Linn	NT 984367
N/2	Weetwood Moor	NU 022282
N/3	Dod Law 1	NU 005317
N/4	Dod Law 2	NU 005317

In the present paper the difficulties concerning the carved rings will be side-stepped. The problems encountered even when dealing with supposed circular rings have already been mentioned, and in many cases of definitely non-circular designs Thom's semi-circular and semi-elliptical geometries do not provide an adequate fit. In the absence of a suitable model for any supposed geometry, therefore, and to eliminate one area of subjective decision, only distances between cup centres will be used for statistical analysis in the discussions which follow. We shall use the term 'neighbouring cups' to refer to a pair of cupmarks whose respective Thiessen polygons share a common side. A 'cup separation' refers to the distance between the centres of neighbouring cups unless stated otherwise. The term 'ringed cup separation' refers to the distance between neighbouring cups where at least one cup of the pair possesses a surrounding ring, or system of rings.

It is important to note at the outset that we have no assurance whatever that cup separations selected in this manner correspond to those putatively measured out by the prehistoric carvers. Even if such measurements were indeed made, our data will undoubtedly contain an unknown number of 'spurious' distances. This does mean that our method of data selection is inevitably, and to an unknown degree, biased against the detection of a real quantum. We can only hope that in a sufficiently large sample (whatever that may mean), a real quantum may still make itself heard above the background noise of 'spurious' distances and measurement errors.

It should be emphasised that the present data are independent of Thom's on two counts. First, Thom's analysis was based upon ring diameters only, whereas we shall limit ourselves to cup separations. Second, Thom's original measurements were obtained solely from Scottish sites (Thom 1968) whereas all the sites discussed here are in England. (Thom did, in fact, incorporate a small number of ring diameters from the Panorama Stone into his later, 1978 analysis, though that does not concern us here.) This independence means that we are at liberty to employ Type *A* tests for the 1 MI quantum, though of course the more searching Type *B* tests are also desirable.

The overall sample size is sufficiently large for the data to be examined in two geographically distinct groups. We shall consider the Ilkley group first.

The Ilkley carvings

Notes on individual sites

Although in principle the definition of neighbouring cups can be achieved objectively by means of Thiessen polygons, in real life the situation is sometimes less clear-cut. There are certain features of some of the carvings which do require decisions to be made by the investigator - and these are not always apparent from the plans alone. Wherever a selection has been made which has a bearing on the acceptance of data for analysis, the reasons for such selection and the nature of it are given below, on a site by site basis.

I/1 (Weary Hill West). In Davis (1983) this site is described as possessing eleven cups. Subsequent visits to the site have led the author to believe that one of these, shallower and more irregular than the others, is probably a natural hollow in the rock surface. It has been omitted here.

I/3 (Panorama Stone). This large slab seems to have suffered damage during its removal from the moor to its present situation in Ilkley town, being cemented together in four sections. It is clearly quite valueless to include in the analysis measurements which span these repaired fractures. Each section has therefore been treated separately, and the frames of reference for the coordinates within each section are independent. The coordinates *cannot* be used to calculate distances which traverse a fracture. There is also an abrupt change of level over part of the rock (indicated in the plan, Fig. 18.10) due to a broken ledge which effectively isolates two ringed cups. These cups appear unlabelled on the plan as traced, but their coordinates are not given; it is not obvious just how their position relative to nearby cups should be measured - if indeed it is sensible to measure them at all.

I/4 (Barmishaw Stone). The surface of this otherwise flat slab slopes sharply away at its south-eastern edge. A single cup is positioned on this sloping section, and its approximate position (unlabelled) is indicated on the plan. No co-ordinates are given for this cup; here again it is hard to see how it should (or could) be related to the others. In addition it should be noted that there is uncertainty as to whether cups *T* and *U* (see Fig. 18.9) once possessed surrounding rings. Repeated visits to the site have not resolved the situation, and sketches made by earlier workers (e.g. Cowling 1940) are not particularly helpful.

I/6 (Pancake Ridge 2). This is a large boulder with a flat, sloping shelf on its northern side. The plan is a complete transcription of the carvings on this shelf, but the many cups scattered over the irregular surface of the boulder elsewhere have been ignored.

I/7 (Green Crag). There is perhaps some doubt as to whether the elongated 'loop' should be regarded (in the context of this investigation) as a 'ring'. The enclosing of more than one cup by a single ring is known to occur elsewhere, however (see I/8 for an example of two cups within a ring), and on this basis the Green Crag 'loop' will be so considered here.

I/8 (Backstone Beck). This complex carving demands that some account be taken of the striking pattern of grooves which appears to divide the design into sections. Neighbouring cup separations have been taken only within each of the nominal cup-filled 'sections' here - but there are ambiguities, and different investigators may well make slightly different decisions. Fortunately the sample obtained from this carving is relatively large, so that the effect of possible selection differences is likely to be small; and however the selection is made, it is of course essential that this be done purely by visual inspection before the analysis begins. In view of the complexity of the design, and the sheer number of cups in such a small area, a more detailed investigation of this carving may be worthwhile at some future date.

On most of the plans, true north is shown. This is based upon a simple magnetic compass bearing, and should be taken only as a rough guide to the orientation of the carving.

Preliminary analysis

Calculating the neighbouring cup separations for each site (subject to the qualifications mentioned in the previous section), and combining all the data from the eight Ilkley sites together, we obtain a total sample of 274 measurements, with $RMS(y) = 21.82$ cm. Sample sizes from each site are given in the second column of Table 18.2.

We begin our analysis in the most obvious way, by lumping all the data together and performing a Type *A* test for the hypothetical quantum of 2.0725 cm. This yields:

$$\phi(1 \text{ MI}) = 1.834 .$$

This is not convincing. Though perhaps formally significant at a probability level of about 3.5%, it does not offer the kind of substantial support for the MI hypothesis that one might have expected (were the hypothesis correct)

Table 18.2. *Site-by-site quantum analysis of the Ilkley carvings (Type A tests)*

Site	N	$\phi(1\ MI)$	Probability level (%)	ϕ_{max}	Optimum value of quantum (cm)	σ (cm)
I/1	16	-1.34	-	-	-	-
I/2	16	1.74	4	3.14	2.115	0.37
I/3	27	2.61	0.5	2.77	2.065	0.46
I/4	43	-1.17	-	-	-	-
I/5	23	0.30	38	1.34	2.098	0.60
I/6	44	0.86	20	0.96	2.082	0.71
I/7	50	1.44	7.5	1.47	2.064	0.64
I/8	55	0.74	23	1.36	2.139	0.69
All	274	1.83	3	1.97	2.082	0.74

Notes. N = number of cup separations.
1 MI is taken as 2.0725 cm.
The optimum value of the quantum is taken as that which maximises $\phi(\phi_{max})$.

with a sample of this size. It is also a result which is at variance with the findings in Davis (1983), despite the fact that a large part of the 1983 data is effectively included here.

In order to perform a Type *B* analysis, we must first decide upon a suitable 'search range'. There is little point in considering possible quanta which are substantially larger than the smallest datum, and this suggests an upper limit to the search of perhaps 10 or 15 cm. Similarly it is pointless to search for quanta so small as to be inevitably drowned in the background noise of measuring errors. The megalithic inch itself must be close to the limit at which a quantum could be detected, given the weathered state of most carvings, and this suggests setting a lower limit of perhaps 1.5 cm to the search. In fact it is convenient to adopt a search range which will be broadly suitable for all the subsets we shall consider, and then stick to it. The range 0.08 cm$^{-1} < T < 0.58$ cm^{-1} seems a reasonable compromise (corresponding to 12.5 cm $> q > 1.72$ cm) and the resulting value of $T_2 - T_1 = 0.5$ is convenient for the estimation of significance levels.

Fig. 18.3(a) gives the CQG over this range for the 274 cup separations. Although the peak corresponding to 1 MI is marginally the largest ($\phi_{max} = 1.966$ at $q = 2.082$ cm) it is by no means prominent, and certainly does not achieve significance on Type *B* criteria. Indeed the data do not appear to be quantal at all.

One might conclude at this point that the quantum hypothesis is not substantiated, and pursue the matter no further. In fact the situation is not so simple. There are two further points to be considered before reaching a conclusion - both of them anticipated in Davis (1983) - and these will now be considered in turn.

(a) We recall that the Thiessen polygon analysis has been imposed upon the designs as a means of obtaining an objectively selected data base. We recall also our awareness that the inclusion of too many 'spurious' cup separations may severely degrade the detectability of a quantum even if it were real. Might it be possible to obtain a sample of data which represents more closely what the prehistoric carvers actually measured (if they did so) while still maintaining our objective selection system? The only thread common to all the carvings is that some cups have rings around them, while others do not. Let us then omit all cup separations where neither cup of each pair possesses a surrounding ring, and re-test the remaining data.

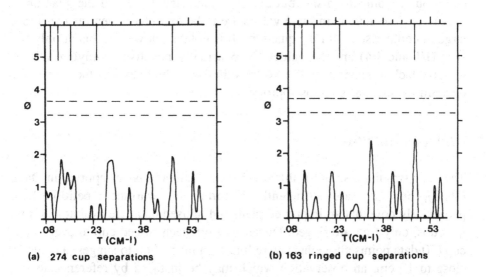

(a) 274 cup separations   (b) 163 ringed cup separations

Fig. 18.3.   CQGs for data from all Ilkley sites combined together.

Performing this exercise leaves us with 163 'ringed' cup separations. Testing these for the hypothetical MI quantum yields the result $\phi(1\ \text{MI}) = 2.315$, which corresponds to a Type *A* probability level of about 1%. The CQG for these data is given in Fig. 18.3(b). The peak near 1 MI is the largest (though by only a narrow margin) and occurs at $q = 2.079$ cm where $\phi_{max} = 2.403$. This still does not achieve Type *B* significance, however, and so offers no convincing reason for rejecting the random, non-quantal hypothesis, despite the apparent improvement in support for the 1 MI quantum.

(b) The second possibility we must consider is that a variety of mensuration practices were in use during the period in which the carvings were made, and that lumping all the data together may be obscuring any evidence for this. Table 18.2 offers a site-by-site Type *A* analysis of the original 274 cup separations. Where some possible support for the 1 MI quantum is indicated (i.e. where $\phi(1\ \text{MI})$ is positive), probability levels are given together with optimised estimates for the putative quantum and the associated maximum ϕ value. The last column gives the value for the standard deviation (σ) at each site, as an indication of the scatter of the individual data about multiples of the hypothetical quantum - this has been estimated using the method suggested by Kendall (1974).

We see from Table 18.2 that statistical support for the quantum varies enormously from site to site, but this is not surprising. Even if the quantum is real, large fluctuations in the ϕ values are to be expected when (as here), σ is large in comparison with the quantum. It is notable, however, that at only two sites (I/1 and I/4) are the values for ϕ actually negative, implying a fairly decisive lack of support for the MI hypothesis. A closer look at these two sites on an individual basis seems appropriate.

I/1: Weary Hill West

The very striking results for this carving have already been reported in Davis (1983), but they are sufficiently important to warrant repetition here, particularly now that an improved plan, and revised data, are available. Of the ten cups, only four are ringed. Distances from each ringed cup to every other cup (30 data points altogether) were found on inspection to suggest a quantum close to 10 cm, an observation which may be justified by reference to Fig. 18.4(b) which presents the data in the form of a kind of probability histogram, or 'curvigram' (see Ruggles 1981 : 156). The superimposed vertical lines are spaced 5 MI apart, and the degree to which these match the major peaks is remarkable. Fig. 18.4(a) presents the CQG for these data, and the dominance of

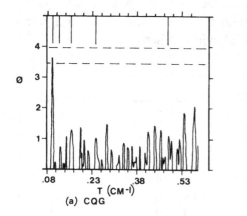
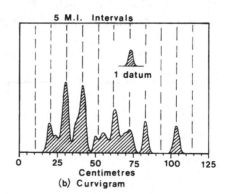

Fig. 18.4. Weary Hill West (I/1): Analysis of 30 distances from four ringed cups.

the peak at $q = 10.45$ cm, where $\phi_{max} = 3.65$, is very evident. With RMS(y) = 56.23 cm, the peak is significant at a probability level of about 2.5%. This is little different from the 1983 result when an extra (but spurious) cup was included.

In fact there are several reasons why this may not be an accurate estimate of probability. The first of these is that in order to achieve a reasonable sample size, *all* distances from the ringed cups (rather than just neighbouring ones) have been used. This does involve a departure from the usual rigorous selection procedure, although since there appears to be no alternative line of approach in this case the point is perhaps not a serious one.

The second point, made to the author by Dr D.C. Heggie (pers. comm.), is that not all the distances are independent owing to the tendency of some of the cups to lie in approximately straight lines. For example (referring here to the plan in Fig. 18.9) if *BC* and *CF* are close to multiples of some quantum, then *BF* will tend also to be close to a multiple. This effect would tend to make the result quoted above seem more significant than it really is.

The third point is that not all the 30 distances are disposable. Measuring from only four cups it is *certain* that we have included in our data distances which could never have been intended to be multiples of the quantum by the prehistoric carvers. This effect will tend to obscure the evidence for a quantum, even if such a quantum is real. Indeed, it is possible that such considerations may counterbalance the arguments of the second point made above.

These points notwithstanding, the startling closeness of the observed quantum to 5 MI (5 × 2.09 cm), the prominence of the peak in the CQG, and the large value of the quantum in relation to the nominal measuring errors conspire to produce an intriguing piece of evidence here.

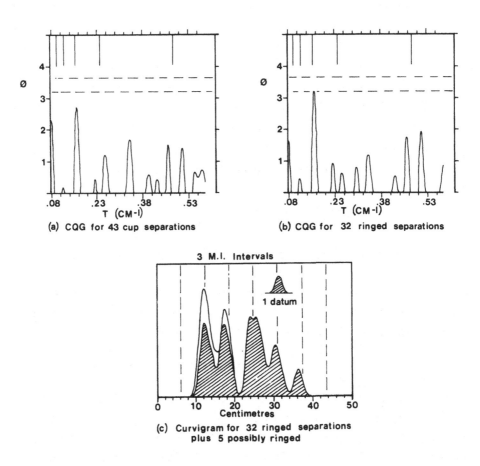

(a) CQG for 43 cup separations

(b) CQG for 32 ringed separations

(c) Curvigram for 32 ringed separations plus 5 possibly ringed

Fig. 18.5. Analysis of data from Barmishaw Stone (I/4).

I/4: Barmishaw Stone

For this carving alone, 43 neighbouring cup separations are available. Testing these data over the usual range produces the CQG given in Fig. 18.5(a). One peak clearly dominates the others (ϕ_{max} = 2.727 at q = 6.116 cm) but does not achieve Type *B* significance. If we restrict the analysis to only ringed cup separations, however, the sample size drops to 32, and the CQG undergoes a significant change (see Fig. 18.5(b)). The height of the aforementioned peak rises to ϕ_{max} = 3.205 at q = 6.148 cm. With RMS(y) at 23.62 cm, this peak is significant at a probability level of 5%. Note how close is this quantum to 3 MI (3 × 2.049 cm). The distribution of the data is given in curvigram form in Fig. 18.5(c), with vertical lines spaced at 3 MI intervals for comparison, and the result is certainly rather striking.

It has been noted earlier that cups *T* and *U* (see Fig. 18.9) may once have possessed surrounding rings, only faint possible traces of which now remain. If these are regarded as ringed cups, and the analysis repeated, we obtain N = 37; ϕ_{max} = 3.449; q = 6.143 cm; and a corresponding Type *B* probability level of 2.5%. These additional data are shown unshaded in Fig. 18.5(c), but it must be emphasised that this extension of the analysis is very uncertain. Once more it is worthy of note that the quantum is quite large in comparison with nominal measuring errors; the detection of such a quantum is therefore quite credible even in a small sample.

Reassessment of the metrology of the Ilkley carvings

It is intriguing that of the eight Ilkley sites, the two carvings which emphatically do not support the MI hypothesis should nevertheless provide evidence, significant on Type *B* criteria, for quanta which are simple multiples of the MI. Further, the values of σ for these two sites are rather large - 2.04 cm for I/1, and 1.32 cm for I/4. If these values are taken as an indication of the precision with which the designs were originally set out, it is by no means surprising that the appreciably smaller 1 MI quantum is undetectable at these sites. Interestingly, none of the other six sites exhibits significant CQG peaks when analysed individually across the usual search range. There seems to be a strong case for omitting both I/1 and I/4 from further consideration, and re-testing the combined data from the remaining six sites.

For these six sites we have 215 neighbouring cup separations. A Type *A* test for 1 MI yields ϕ = 2.96, corresponding to a probability level of 0.2%. This clearly offers considerable support for the MI hypothesis. Fig. 18.6(a)

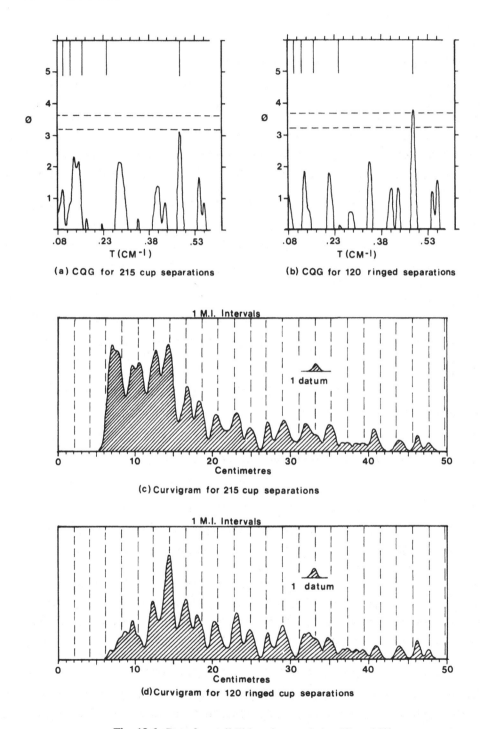

(a) CQG for 215 cup separations

(b) CQG for 120 ringed separations

(c) Curvigram for 215 cup separations

(d) Curvigram for 120 ringed cup separations

Fig. 18.6. Data from all Ilkley sites, omitting I/1 and I/4.

shows the full CQG for these data. The 1 MI peak is certainly dominant, with $\phi_{max} = 3.12$ at $q = 2.081$ cm, and might be described as marginally significant at a probability level of about 6% on Type B criteria.

If next we restrict our attention to only ringed cup separations, the situation changes radically. The sample size becomes 120, with $\phi = 3.71$ for the 1 MI quantum. The CQG for these data, Fig. 18.6(b), is very striking indeed. Here $\phi_{max} = 3.82$ at $q = 2.079$ cm, and with RMS(y) = 24.86 cm this corresponds to a probability level of less than 1% on Type B criteria. If this sample of 120 measurements is acceptable as a suitable data base, then support for the megalithic inch hypothesis is quite strong, and does not depend upon a Type A approach based on Thom's earlier work. The value of σ for this sample is 0.55 cm, which seems broadly comparable in size to the typical accuracy with which most cup centres can be estimated by an investigator today.

Figs. 18.6(c) and 18.6(d) present the two sets of data discussed above in curvigram form. Note that the vertical scale has been adjusted in each case to accommodate the difference in size of the two samples.

The Ilkley carvings: a summary

It may be useful to summarise the above investigations briefly at this point.

(a)   The hypothesis that a unit of 1 MI was universally used in setting out the cupmarks in the Ilkley carvings is not convincingly supported by the present data, though a marginal increase in support is observed if the analysis is restricted to only ringed cup separations.

(b)   Two carvings (I/1 and I/4) exhibit significant evidence for quanta of 5 MI and 3 MI respectively - again when the analysis is restricted to only ringed cups.

(c)   The remaining six sites, combined together, provide strong support for a 1 MI quantum on a Type A test. This apparent support increases considerably when only ringed cups are considered, and achieves statistical significance at the 1% level on a Type B test.

The identification of the two larger quanta in terms of Thom's model as 5 MI and 3 MI may of course be in error, even if the quanta are real. Nevertheless, there does seem to be an appreciable degree of coherence here: the repeated emergence of the significance of ringed cups, and the fact that all putative quanta seem to bear a simple numeric relation to each other do not seem to be coincidental. It is clearly of great interest to see whether any of these tentative findings can be verified by testing a further, independent body of data. This brings us to the Northumberland group.

The Northumberland carvings

Notes on individual sites

The number of sites measured in north-east England is small: only four in all. These are all well-known carvings, and photographs of all of them can be found in Beckensall (1983). The plans made by the author are given in Figs. 18.11 and 18.12, and where coverage is incomplete details are given below.

N/1 (Roughting Linn). The surface of this, the largest and perhaps most impressive carved rock in England, is covered with an enormous variety of carvings which must surely represent the work of many different carvers over an unknown period of time. A complete mapping of the surface was impractical at the time of the author's visit, and one must in any case doubt the value (from the metrological point of view) of attempting to unravel such a complicated series of carvings at this stage. However, on the flat, sloping, northern face of the rock is an impressive large design based around only four small and well-defined cups, each at the centre of a complex ring system. The plan given here is a complete transcription of the carvings on this northern face.

N/3 (Dod Law 1). This is another extensively carved area, and again no attempt was made to map the entire surface. The three well-known groups of cups, however, enclosed by carved grooves of various shapes, seem obvious candidates for statistical analysis and each of these has been planned separately.

N/4 (Dod Law 2). This is a separate small carved slab close to the main outcrop.

Analysis of the Northumberland sites

With so few sites the sample as a whole is unavoidably dominated by data from Dod Law. After construction of Thiessen polygons for each design we obtain a total of 113 cup separations distributed as follows: N/1 (5); N/2 (15); N/3 (87); N/4 (6). There is no trace whatever of a 1 MI quantum in these data.

The CQG for the whole sample is given in Fig. 18.7(a). What is very striking about this is the very large peak at $q = 10.35$ cm, where $\phi_{max} = 4.046$. RMS(y) for the data is 32.62 cm, and if the CQG is taken at face value this peak is

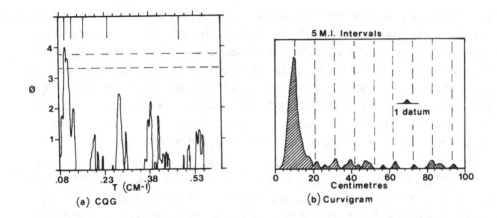

Fig. 18.7. Analysis of 113 cup separations from all Northumberland sites.

significant at a Type *B* probability level of 0.5%. Note that the indicated quantum here is very close indeed to 5 MI (5 × 2.07 cm).

However, if we now examine the distribution of the data, given in curvigram form in Fig. 18.7(b), the interpretation of this result becomes uncertain. The highly significant CQG peak is almost entirely the consequence of an enormous accumulation of data at or near 5 MI, rather than at a variety of multiples. What are we to make of this? Is there any significance in the repeated use of cup separations close to 5 MI? (Of the 87 data from Dod Law 1, 30 lie within 0.5 cm of 5 MI, and over half the data lie within 1 cm.) Or are we observing merely an inevitable tendency for a random set of cups enclosed within a given space to exhibit a mean separation which just happens, by chance, to be about 10.5 cm?

Since the Dod Law 1 site is entirely responsible for the unimodal accumulation at 5 MI in Fig. 18.7(b), it is of interest to determine what contribution to the evidence for a 5 MI quantum comes from the remaining sites. Unfortunately the data from the remaining three sites are very scanty (only 26 measurements altogether), but there is some mild support for the quantum here, with ϕ(5 MI) = 1.84. This is hardly conclusive; however, examination of the three sites individually leads to two results which may be worthy of note.

N/1 (Roughting Linn). There are four cups here, yielding a total of only six possible measurements between them (Fig. 18.11). If these distances are expressed in units of 5 MI we obtain:

$$AB = 7.99; AC = 8.04; AD = 7.08; BC = 6.12; BD = 12.94; CD = 9.05 .$$

It is worth emphasising that the author was unaware of these relationships when the carving was selected for study at the site. Obviously statistical analysis is inappropriate with only six measurements, but it should be noted that the RMS deviation from perfect integer multiples is comparable to the size of the chalk marks made at the cup centres. (This is quite credible - these cups are small and particularly well defined.) How are we to assess the significance of a result like this, obtained from a prominent design on the northern face of the largest known carved site in England? The question remains open.

N/2 (Weetwood Moor). The 14 ringed cup separations from this magnificent carving (Fig. 18.12) produce a CQG whose largest peak has $\phi_{max} = 2.959$ at $q = 6.246$ cm. Is this another manifestation of the 3 MI quantum (3×2.082 cm)? There are too few data to decide the matter, but the result is mentioned here because there are other carvings with multiple rings on Weetwood Moor (not visited by the author) where the quantum might be tested. Details of the carvings in this area are to be found in Beckensall (1983).

Discussion and Conclusion

Clearly any conclusions based on the contents of this paper can be only very tentative. The number of carvings studied is still very small, and in any case our attention has been restricted to English sites. Nevertheless, there are certain indications here which may be of consequence.

The simple hypothesis that a unique quantum of 1 MI (or any other) was used in the design of the carvings receives little if any support from the present data. Indeed, the enormous variety of style of design even within this small sample of sites makes such a conclusion hardly surprising. Unfortunately this very fact presents us with methodological difficulties. If we shuffle the data in various ways, trying alternative hypotheses, we become increasingly in danger of discerning some apparently 'significant' pattern even in random data.

This danger arises in two ways. First, it is possible for the 'shuffling' process to introduce bias into the data unintentionally. In the present case it is difficult to see how such bias could arise - our testing of ringed cup separations, for example, should not in itself give rise to a spurious quantum.

It was simply an obvious experiment to try, given that some cups have rings while others do not. Similarly, the absence of support for one particular quantum (1 MI) should not in itself imply the presence of some other, spurious quantum (such as we find at I/1 and I/4).

The second, perhaps more cogent, problem concerns not the *nature* of the shuffling process, but the mere fact that shuffling has occurred at all. A nominal 5% significance level result from a single hypothesis test may be interesting - but it rapidly becomes less interesting if it is achieved only after several alternative hypotheses have been tried. It is clear that the meaning of the formal significance levels may be considerably degraded by repeated testing of the data. Bearing this in mind, the results of some of the tests described here must in fact be of doubtful statistical significance when viewed purely on their own merits. (The Barmishaw Stone analysis is an obvious example here.) Against this must be balanced the observation that all the putative quanta turn out to be simple MI multiples, invariably associated in some way with ringed cups. It is genuinely difficult to reach a decision in these circumstances, and likely to remain so until more data are available.

Possibly the strongest indications in this paper point towards the use of a quantum close in value to 5 MI at certain sites. We find evidence for its use among the Yorkshire group at Weary Hill West; in Northumberland it may have been used at Roughting Linn and (in rather strange circumstances) at Dod Law.

We find evidence for a quantum close to 3 MI at one site in Yorkshire (Barmishaw Stone), and very tentative indications, based on a very small sample, at Weetwood Moor in Northumberland.

In all these cases (leaving aside the enigmatic Dod Law group) the apparent quantum seems strongly associated with ringed cups. This association is further in evidence when we consider the case for the 1 MI quantum itself, which only emerges to any convincing degree among the Yorkshire sites when ringed cups alone are considered. No confirmatory evidence for this quantum is found among the Northumberland sites, however.

One might feel more confident about these indications if some relation between quantum and carving style could be discerned, but no such relation is apparent. For sites displaying the 5 MI quantum, for example, we have one with single rings (I/1), one with multiple rings (N/1) and another with no rings at all (N/3). This does not, of course, disprove the quantum hypothesis, but neither does it offer additional coherence which might strengthen the case. After completion of the analysis described here, an attempt was made to group the carvings according to whether their cups possessed single or multiple rings, but no obvious patterns emerged from the analysis of these groups. If anything is significant, it appears to be the mere presence of rings rather than their number.

The indications of 1 MI and 5 MI quanta do fit comfortably within the Thom model of prehistoric mensuration in Britain, with $1 \text{ MR} = 2^{1}2 \text{ MY} = 20 \times (5 \text{ MI}) = 100 \text{ MI}$. Indeed, one may point out that two of the six cup separations from Roughting Linn lie within a few mm of 1 MY, and that this kind of precision is consistent with the Thom model. Alternatively, one might justifiably argue that the 5 MI quantum is very close to the mean width of a

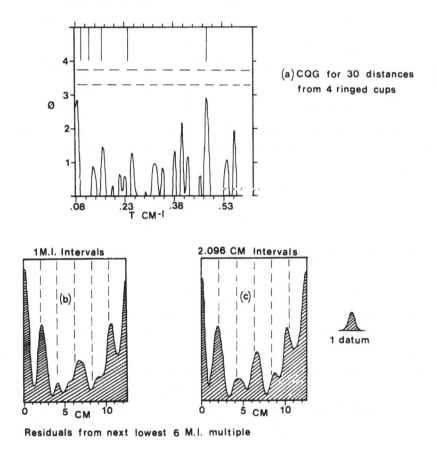

Fig. 18.8. Weary Hill East (I/2): Analysis of 30 distances from four ringed cups.

human hand, and interpret 1 MI as a mean fingerwidth. It would be extremely difficult, if not impossible, to distinguish between these two hypotheses on the basis of present data. Certainly the data from Dod Law would be entirely consistent with the hypothesis that the cups were roughly set out one handwidth apart - and similarly a quantum of 'three fingers' might well be all that is needed to account for Fig. 18.5(c).

Another possibility which may merit further investigation is that we are here seeing (admittedly weak) evidence of differing counting systems based loosely on the 1 MI quantum (however the latter is to be explained) applied with more or less precision at individual sites. Simple number bases of 3 and/or 5 would be perfectly credible, of course. In fact there is one further piece of evidence from one of the Ilkley sites, not discussed previously because of its speculative nature, which may lend support to such a notion.

Comparison of the plans of the two sites Weary Hill West (I/1) and Weary Hill East (I/2) in Fig. 18.9 reveals certain superficial similarities. Both have ten cups, four of which are ringed. The similarities largely end there - the overall scale of I/2 is considerably smaller than that of I/1, as indeed are the cups themselves. However, in view of the results obtained from I/1 by analysing the 30 distances from the ringed cups to all the others, it is of some interest to submit I/2 to the same procedure. The CQG for these 30 data from I/2 is given in Fig. 18.8(a). The 1 MI peak is prominent, and indeed a Type A test yields $\phi(1 \text{ MI}) = 2.18$, corresponding to a probability level of 1.5%. The peak value occurs at $q = 2.096$ cm, where $\phi_{max} = 2.951$. Of further interest, however, is the large peak at the far left of the diagram corresponding to a quantum of about 12 cm, perhaps 6 MI. It was the presence of this second peak which led the author to construct the two curvigrams given in Figs. 18.8(b) and (c). These represent the distribution of the residuals obtained by subtracting the next lowest multiple of 6 MI from each datum in turn. Thus a distance corresponding to (say) 13.6 MI would yield a residual of 13.6 - 12 = 1.6 MI, and it is this latter value (in cm) which would be plotted on the curvigram. In Fig. 18.8(b) the 'standard' value for the MI is used, whereas Fig. 18.8(c) shows the effect of optimising the estimate of the quantum at 2.096 cm, as indicated by the CQG.

The author knows of no rigorous statistical method for assessing the significance of a result of this kind, beyond noting that the presence of two CQG peaks as large as these, where the first quantum is predictable in advance, and the second a simple multiple of the first, is highly unlikely to occur in a single set of random data. The curvigrams are strongly suggestive of the notion that measurements were made from the ringed cups (rather carefully, it seems) corresponding to $6n$, $6n+1$, $6n-1$, and $6n+3$ MI where n is an integer.

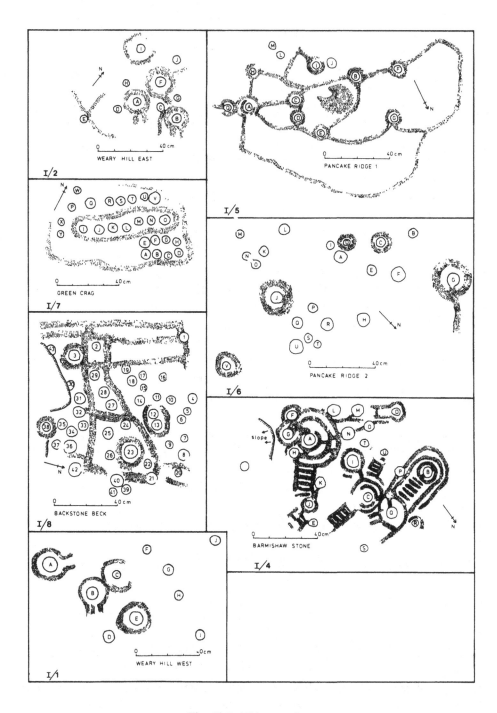

Fig. 18.9. Ilkley carvings.

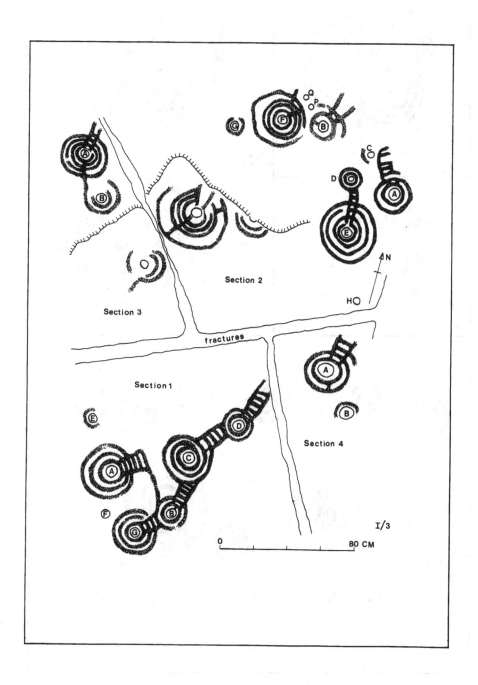

Fig. 18.10. Panorama stone.

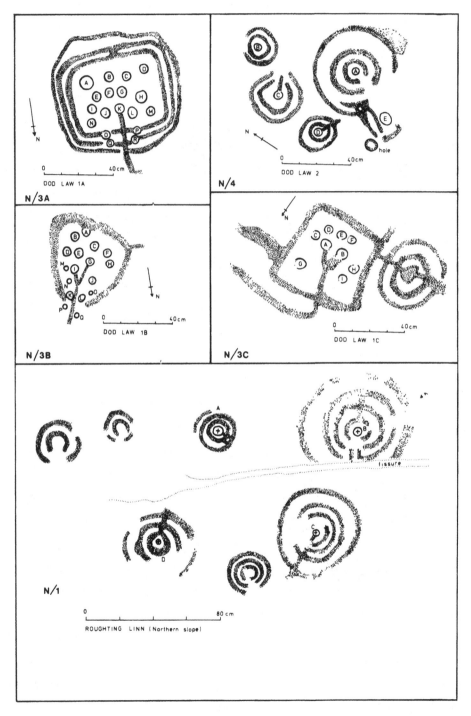

Fig. 18.11. Northumberland carvings.

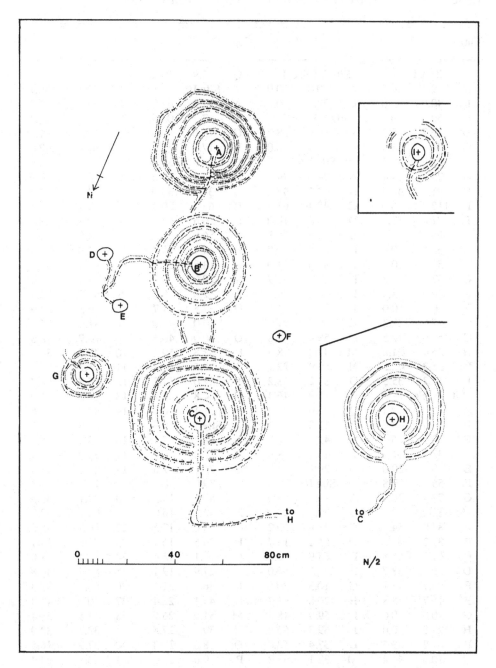

Fig. 18.12. Weetwood Moor.

Table 18.3. *Cup centre coordinates (All measurements in cm)*

Cup	x	y	Cup	x	y	Cup	x	y	Cup	x	y
Site I/1			**Site I/3 Sec 3**			Q	7.9	19.2	Q	20.8	34.0
A	9.2	36.6	A	0.0	0.0	**Site I/6**			R	32.8	36.7
B	40.0	30.0	B	28.1	0.0	A	27.3	62.9	S	39.7	37.9
C	50.3	46.4	**Site I/3 Sec 4**			B	7.7	105.6	T	46.7	39.4
D	59.2	8.7	A	0.0	28.7	C	15.6	86.5	U	54.2	41.4
E	71.4	25.5	B	0.0	0.0	D	17.9	65.6	V	60.5	42.0
F	62.6	67.4	**Site I/4**			E	33.0	82.8	W	11.9	40.8
G	79.4	61.4	A	74.6	24.7	F	34.2	99.6	X	3.5	20.9
H	91.1	48.6	B	2.5	47.6	G	33.3	133.0	Y	4.6	13.8
I	112.2	30.0	C	40.4	61.5	H	63.5	80.8	**Site I/8**		
J	101.0	87.7	D	20.0	10.0	I	20.6	55.8	1	9.3	88.1
Site I/2			E	74.5	75.8	J	56.0	25.9	2	14.0	34.2
A	32.6	10.5	F	84.8	10.0	K	29.1	15.9	3	20.0	20.0
B	57.2	0.0	G	87.6	21.0	L	13.8	27.1	4	46.4	93.1
C	47.2	7.3	H	84.4	32.7	M	20.0	0.0	5	54.4	90.5
D	20.9	5.8	I	48.0	39.1	N	32.8	5.5	6	59.3	86.5
E	0.0	0.0	J	76.1	64.3	O	37.5	11.7	7	70.5	87.9
F	47.5	22.3	K	69.4	51.6	P	60.0	49.0	8	81.0	87.5
G	57.9	12.5	L	58.4	7.8	Q	70.0	40.5	9	74.7	78.6
H	26.4	21.8	M	43.8	8.3	R	68.8	59.2	10	47.6	80.3
I	35.3	42.5	N	50.9	22.1	S	78.5	48.3	11	45.2	71.3
J	57.7	35.5	O	37.2	18.2	T	82.0	54.2	12	55.7	68.7
Site I/3 Sec 1			P	19.9	46.4	U	84.8	41.2	13	62.2	71.0
A	33.7	19.3	Q	26.0	71.6	V	101.1	0.0	14	47.0	60.6
B	69.3	44.6	R	11.8	78.7	**Site I/7**			15	39.3	62.7
C	42.3	66.2	S	43.9	92.7	A	58.4	7.1	16	33.2	75.3
D	34.9	101.0	T	40.4	28.6	B	65.2	8.0	17	31.9	62.4
E	0.0	20.0	U	29.1	34.4	C	72.3	7.6	18	35.1	55.4
F	55.7	7.8	**Site I/5**			D	79.0	10.2	19	28.3	52.3
G	72.2	20.0	A	20.0	20.0	E	57.2	14.3	20	92.1	83.9
Site I/3 Sec 2			B	84.1	43.7	F	63.7	16.0	21	94.8	68.0
A	87.9	54.7	C	48.8	26.2	G	70.0	17.2	22	86.5	65.2
B	31.2	42.8	D	51.1	15.7	H	76.5	16.7	23	78.6	54.7
C	61.0	56.8	E	65.9	7.3	I	18.1	18.4	24	62.1	51.8
D	65.8	38.7	F	110.5	50.0	J	27.9	19.1	25	67.0	40.8
E	90.7	18.1	G	110.3	20.0	K	36.3	21.1	26	80.9	41.3
F	13.7	25.8	H	20.8	41.7	L	45.1	22.4	27	50.0	42.8
G	0.0	0.0	I	59.9	48.8	M	51.9	26.5	28	41.5	38.4
H	127.8	0.0	J	69.3	50.2	N	59.7	27.8	29	32.3	33.0
P	17.2	43.5	L	37.4	53.0	O	68.3	29.3	30	37.9	16.5
Q	10.6	44.4	M	32.1	58.2	P	9.6	30.3	31	46.1	23.4

Table 18.3 (continued).

Cup	x	y	Cup	x	y	Cup	x	y	Cup	x	y
32	54.4	22.9	C	50.0	0.0	M	42.1	21.8	N	14.3	19.2
33	62.8	25.3	D	19.7	71.6	N	7.6	19.6	O	27.0	12.2
34	66.4	18.0	E	23.3	50.3	O	14.9	11.8	P	11.3	4.8
35	61.3	11.9	F	86.6	30.0	P	33.0	12.1	Q	18.3	0.0
36	75.4	16.7	G	6.0	23.9	Q	16.7	7.1	Site N/3 (C)		
37	73.5	7.6	H	167.4	0.0	R	31.2	6.9	A	19.9	38.3
38	63.0	2.5	I	158.6	117.5	Site N/3 (B)			B	30.7	37.7
39	101.6	51.8	Site N/3 (A)			A	24.8	46.9	C	12.2	40.9
40	96.5	46.0	A	7.6	42.8	B	18.3	44.8	D	17.6	46.8
41	102.9	43.8	B	20.6	44.6	C	29.1	39.2	E	26.0	47.1
42	88.8	20.0	C	30.6	43.5	D	13.1	35.9	F	32.4	47.2
43	16.9	5.8	D	41.3	46.4	E	20.3	35.0	G	11.2	22.2
Site N/1			E	12.3	34.4	F	36.5	34.7	H	40.4	33.2
A	38.6	62.4	F	19.9	35.5	G	26.5	30.2	I	37.0	24.7
B	121.2	57.3	G	26.9	34.9	H	38.2	28.9	Site N/4		
C	93.8	0.0	H	37.7	30.7	I	17.5	26.0	A	68.7	45.0
D	0.0	0.0	I	9.0	27.7	J	27.7	19.5	B	12.2	57.8
Site N/2			J	15.9	24.5	K	14.7	11.2	C	24.1	31.1
A	71.0	109.7	K	24.6	25.6	L	21.4	9.7	D	48.0	10.3
B	58.0	62.7	L	31.5	22.3	M	12.8	26.5	E	86.5	18.1

Does this indicate a possible counting base of 6 - or perhaps base 3, with a preference for even multiples? (It may be worthy of note that in each of the two sites (I/4 and N/2) where the 3 MI quantum is suspected, even multiples seem to be preferred.)

All these ideas remain speculative and require further testing. But no one who has worked with these carvings for any length of time can fail to admire the qualities of abstract design exhibited by some of them. Although it may remain uncertain whether measurement was employed in the process, there can be no doubt that the subtlety of shape and arrangement of carvings like the Panorama Stone, or the Barmishaw Stone, was the result of careful deliberation on the part of the men who made them. It seems likely that their meaning will always be elusive, but it has been the remarkable achievement of Alexander Thom to open up a completely new approach to the problem.

The plans and data presented here will enable anyone who so desires to try his or her hand at unravelling the enigma of the carvings, but a considerably larger data base is needed. The books of Morris and Beckensall provide all the information necessary for locating sites in Scotland and Northumberland, and a tremendous amount of raw material is available in the field upon which hypotheses may be tested. The present paper should be seen as no more than a preliminary attempt to build upon the pioneering work of Thom, indicating potentially fruitful directions for future work where possible. If no definite 'conclusion' is forthcoming at this stage, this should hardly surprise us. To quote Thom once more (Thom & Thom 1978a: 181):

> Why should a man spend hours - or rather days - cutting cups in a random fashion on a rock? It would indeed be a breakthrough if someone could crack the code of the cups.

Acknowledgements

Since 1981, when the idea of this investigation was first contemplated, the author has received encouragement, assistance, and advice (often sought, and always freely given) from many quarters. Thanks are due to the following, for their help in various ways, and at various times: Professor R.J.C. Atkinson, Dr H.A.W. Burl, Mrs A. Haigh, Dr D.C. Heggie, Professor D.G. Kendall, Dr E.W. MacKie, Mr R.W.B. Morris, and Dr J.C. Orkney.

Thanks are also due to colleagues and pupils at Lancaster Royal Grammar School for numerous stimulating discussions.

Throughout this period the help and encouragement of Dr Archie Thom have been an invaluable stimulus. And without the insight, inspiration, and example of Professor Alexander Thom, to whose memory this paper is dedicated with profound respect, the investigation would never have begun at all.

19

Megalithic Callanish

MARGARET PONTING

Introduction

Alexander Thom wrote the following to me in a letter dated 21 March 1982.

> For over 60 years I have cruised in the waters of the Hebrides, exploring out-of-the-way places, many of them normally inaccessible by public transport.
>
> On the 1933 cruise in the sailing yacht *Hadassah*, with my son and four friends, we had finished a long day's sail in the North Atlantic. Having left the Sound of Harris that morning, we arrived in East Loch Roag, a beautiful secluded inlet in north western Lewis. I was seeking a quiet anchorage for the night, and navigating with care between rocky islets and promontories, I finally made up my mind to anchor as far up East Loch Roag as my chart allowed me to go with safety.
>
> As we stowed sail after dropping anchor, we looked up, and there, behind the stones of Callanish, was the rising moon. That evening, since I had been concentrating on the navigation as darkness was approaching, I did not know how near we were to the main Callanish site.
>
> After dinner we went ashore to explore. I saw by looking at the Pole Star that there was a north/south line in the complex. This fascinated me, for I knew that when the site was built no star of any magnitude had been at, or near, the pole of the heavens.
>
> Precession had not yet brought Polaris as nearly due north as it now is. I wondered whether the alignment was a chance occurrence or whether it had been deliberately built that way. If it had been deliberate, north/south alignments would probably be found at other megalithic sites.

> I had, of course, known since 1912 about the complex of stones and menhirs at Callanish. Somerville's paper [Somerville (1912)] had intrigued me greatly when I read it.
>
> The Outer Hebrides have a charm of their own, not the least being their remoteness. To realise that megalithic man had lived and worked there as well as on the mainland of Britain aroused my interest in the workings of his mind, and, my interest having been stirred, I have ever since made detailed surveys of all the sites I could find.

My own awareness of an Outer Hebridean network of standing stones developed during long leisurely summer holidays. Although I was intrigued by the stones, I could find little about them in print. Thus I was unaware of Professor Thom's extensive work until a copy of *Megalithic Sites in Britain* (Thom 1967) was given to me as a birthday present soon after moving to live permanently at Callanish. Only then did I realise what a wealth of information had already been gleaned from megalithic remains, and my eyes were opened to further possibilities of research.

Initial approaches from amateurs are not always received kindly, but Professor Thom always replied to letters promptly and courteously, fully answering queries and frequently adding personal touches which made his letters a joy to read.

Living only a mile from the main standing stones at Callanish, and within sight of several other sites, it was inevitable that my interests should focus on the megalithic sites in this area. In the initial stages, I had misgivings about some of Thom's concepts of prehistoric lunar observatories, but as my studies progressed, I came to accept that megalithic man did indeed study the movements of the sun and moon and did construct stone settings such as Callanish as permanent indicators of his studies. Now, as a local resident with more than a decade of Callanish studies behind me and with some preliminary findings already published (Ponting & Ponting 1979; 1981a; 1981b; 1982), I still feel that I have only scratched the surface of possible research at such a profusion of monuments.

Here I propose to outline why I am convinced that the Callanish complex was built and used, *inter alia*, as a lunar observatory specially positioned in the landscape to utilise the horizon circle to maximum dramatic advantage; that a bone artefact found at a nearby habitation site supports Professor Thom's hypothesis of megalithic units of length; and that Callanish is worthy of further study.

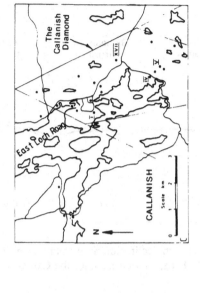

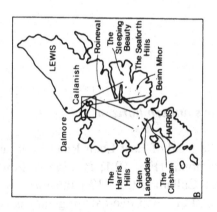

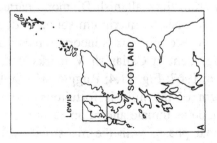

Fig. 19.1. On the Isle of Lewis, the views from the megalithic sites which lie within the 'Callanish Diamond' include the Lewis and Harris hills whose profiles create the dramatic events witnessed only at the moon's maximum southern declination.

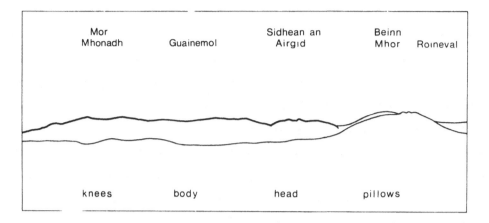

Fig. 19.2. The moon at maximum southern declination rises from Cailleach na Mòinteach - the Sleeping Beauty - in the Seaforth Hills, as seen from all the Callanish Sites within the Callanish Diamond.

Astronomical indications

The Callanish complex and the southern horizon

The sites are concentrated at the head of East Loch Roag in an area that has been called 'the Callanish Diamond' (Fig. 19.1; Ponting & Ponting 1981b: 101). Owing to the latitude of Callanish (58° 12') the moon's path at the maximum southern declination skims two eye-catching ranges of hills. The range to the south-east has the appearance of a supine woman with two pillows (Fig. 19.2). She is known in Gaelic as Cailleach na Mòinteach - literally the old woman of the moors - and in English as 'the Sleeping Beauty' by the local population. Away from the Callanish Diamond parallax distorts her shape. When the moon approaches its maximum southern declination, it rises from the Sleeping Beauty. A record of this is incorporated into several of the sites. For example, the alignment at Callanish V marks this, the moon appearing from her knees (Thom 1967: Fig. 11.4; Ruggles 1984a: 95). It may be argued that the female form of the hills is fortuitous, but widespread legends and traditions of the moon as female and of an earth mother or white goddess, and the linking of lunar phases with the menstrual cycle could be

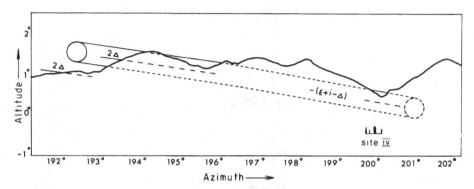

Fig. 19.3. The moon at maximum southern declination setting first into the Clisham, then re-gleaming in the V-notch of Glen Langadale over the circle of Site IV, as seen from the monolith, Site XVII.

said to support the view that the southern lunar extreme and these hills *were* linked in prehistoric people's minds.

The conspicuous range of hills to the south-west includes a dramatic V-shaped valley where the moon reappears shortly after setting. This occurrence would have been seen to a greater or lesser extent from all the sites *except the main one*. For example, from the fallen monolith known as Callanish XVII the moon reappears in Glen Langadale directly over the standing stone circle of Callanish IV (Fig. 19.3).

This reappearance of the moon in the notch would have been a dramatic sight for the populace to witness from any one of the sites and is what makes the Callanish complex special.

The Callanish Main Site

The V-shaped valley is not seen from this site. Nevertheless four astronomical events, three with re-gleams, are indicated by the stones. My theodolite work shows that observations were *not* made primarily from the circle area, but from the north end of the avenue. Firstly, from the latter place, people would have seen the moon at maximum southern declination rising gently from the Sleeping Beauty and skimming across the stones of the east row, then disappearing into the nearby rocky outcrop of Cnoc an Tursa, and finally re-gleaming at the base of the tallest stone, which stands within the circle and at the head of a burial cairn (Fig. 19.4).

Incidentally, Cnoc an Tursa obscures the Clisham hills and Glen Langadale from observers, both at the circle and at the north end of the avenue, and has

Fig. 19.4. The moon at maximum southern declination as seen from the north end of the avenue (Site I). The Clisham profile cannot be seen from this position. The moon rises from the neck of the Sleeping Beauty; skims the tops of the stones of the east row; sets into the nearby horizon of Cnoc an Tursa; and reappears within the circle. This is one of the three astronomical events viewed from, and indicated from, the north end of the avenue.

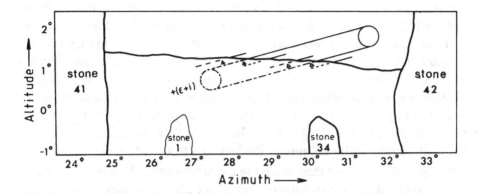

Fig. 19.5. The moon at maximum northern declination, seen across the circle at Site I from stone 9, rising from a point on the horizon indicated by the position and attitude of stone 34. Stones 41 and 42 are part of the circle, and stone 1 is in the avenue.

confused researchers working directly from maps. Although the avenue (and the long axis of the circle) is aligned towards the Clisham range, it cannot have been intended that these hills were to be viewed along it, for by building the avenue a few metres further west, a clear sight line was available. If people had tried to view the path of the moon at its southern maximum declination from the circle area, virtually nothing significant would have been seen due to Cnoc an Tursa.

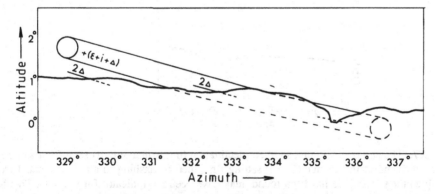

Fig. 19.6. The moon at maximum northern declination setting first into the nearby horizon, then re-gleaming at the base of a cliff (forming one side of a notch), indicated by the flat north-east face of stone 8 at the north end of the avenue, Site I. This is one of the three astronomical events viewed from, and indicated from, the north end of the avenue.

Secondly, it was Somerville (1912) who first noticed the anomalous stones 9 and 34 to the south-west and north-east of the circle which did not fit into the arrangement of circle and alignments, but which acted as markers for the maximum northern declination moonrise. This was the first confirmation anywhere that prehistoric man had made lunar observatories - i.e. accurate observatories, not ritual temples. From stone 9 an observer would see the broad slab of stone 34 end-on, acting as the foresight and set in a viewing window across the circle (Fig. 19.5). These stones were needed to define the alignment unequivocally because of the lack of a foresight on the horizon.

Thirdly, if the observer then moved to the north end of the avenue, he could see the moon, at the same lunation, set about 20 hours later into a nearby ridge, and then a few minutes later, *reappear* briefly at a small cliff at the end of a ridge (Fig. 19.6). The end stone (stone 8) of the east side of the avenue is set askew, its flat north-east face indicating this cliff.

Fourthly, my theodolite work has also disclosed a third astronomical event which could be seen from the north end of the avenue, but *not* from the circle. The midwinter sun set, then briefly reappeared in a notch, during a few days around the solstice (Fig. 19.7). No pair of stones indicates this today, but documentary research (Ponting & Ponting 1979) and the excavation in 1980 by Patrick Ashmore (1983) revealed a socket hole (stone 18A). It is not impossible that the stone which once stood here acted as a nearby foresight for the midwinter sunset and accompanying re-flash as seen from the end stone (stone 8) of the east side of the avenue.

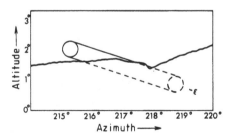

Fig. 19.7. The sun at maximum, southern declination (midwinter sunset) seen from stone 8 at the north end of the avenue (Site I) setting first, then re-flashing in a notch. Stone 18A, for which the socket hole has been found, may have been the indicator for this line. This is one of the three astronomical events viewed from, and indicated from, the north end of the avenue.

Discussion: North end of avenue

A limited area of ground exists by stone 8 at the north end of the avenue where dramatic lunar and solar events were observed, events which could not be seen from the circle. Stone 8 was a backsight for three astronomical sets, each with a reappearance: the annual solar set and re-flash to the south-west; the rarer but predictable lunar set and re-gleam to the north-west; and the lunar southern set followed by a re-gleam within the circle, all capable of being observed within a single period of 24 hours once every 18.6, 37.3 or 56 years. The Callanish main site must therefore be regarded as a prehistoric observatory of special importance, as Professor Thom has shown that single astronomical alignments are common, double alignments occur, but treble alignments are very rare.

Thom (1971: 83-90), Wood (1978: 114-39) and others have proposed that lunar observatories would have required extrapolation devices. It has been shown (Ponting & Ponting 1981a: 78 and Fig 2.2c) how part of the avenue and the east row could have been used as such, and also how the spacing of stones along the east side of the avenue (whose end stones are connected with the maximum north lunar declination) could have been used as a timing device (Ponting & Ponting 1981a: 82 and Fig. 2.2d), apparently, so far, unique to Callanish.

Geometry and the megalithic yard

All the Thomian geometrical shapes of true circles, flattened circles and ellipses have been found in the Callanish area (Curtis 1979; Ponting & Ponting 1981a). Nevertheless, the method of setting out these shapes remains unresolved. Excavations by Patrick Ashmore (pers. comm.) in 1980 and 1981 did not reveal any special features to support Professor Thom's proposed 'geometrical centres' in the way that Jack Scott and J.C. Orkney found three markers at temple Wood (Thom, Thom & Burl 1980: 145).

As far as the main site is concerned, I remain unconvinced of the use of the megalithic yard for fixing the intervals between the stones. Other criteria may have dictated the arrangement of stones in the avenue, criteria such as an extrapolation device and a timing device referred to above. Geometrical solutions based directly on integrals of the megalithic yard have been proposed for all the Callanish circles by Curtis in Chapter 16 of this volume.

Although multiples and subdivisions of the megalithic yard are not immediately obvious at Callanish, it is from the habitation site at Dalmore, 12 km away, that I believe there is tangible evidence for a prehistoric unit of length.

The Dalmore bone

Location and context

Dalmore is a beautiful sandy Hebridean beach 12 km north of Callanish. In the autumn of 1982, part of the sea wall collapsed and access was gained behind it to *in situ* prehistoric stone structures. The Western Isles Islands Council allowed investigation and co-operated by altering its work schedule and by providing a mechanical excavator. Initially, there was no possibility of an official team carrying out the excavation, so the various stone structures were recorded and vast quantities of artefacts were retrieved by the author (Ponting and Ponting 1984b).

During the previous decade, Bronze Age material had been obtained from the area on the shore where spoil had been redeposited as a result of construction of the sea wall and stream realignment. Some of these shore finds, however, were in humus-rich chocolate coloured sand which was almost certainly an *in situ* deposit, and in which the small bone artefact discussed here was found on 2 September 1982.

The artefacts recovered included thousands of pieces of pottery, about half of them Beaker decorated sherds; at least 50 bone tools; and dozens of stone tools, including barbed and tangled arrowheads, made from quartz, flint and baked shale. Incidentally, the baked shale appears to come only from the Isle of Skye. There were also numerous bones of sheep, cattle, deer, pig and fish, as well as sea shells. Three quernstones and some carbonised grain indicated a settled agricultural community.

Later, the Scottish Development Department sent a team to record the site in more detail than was possible for one person working alone, literally against time and tide. The site has since been backfilled and will remain buried unless another freak storm breaches the local authority's sea defence work.

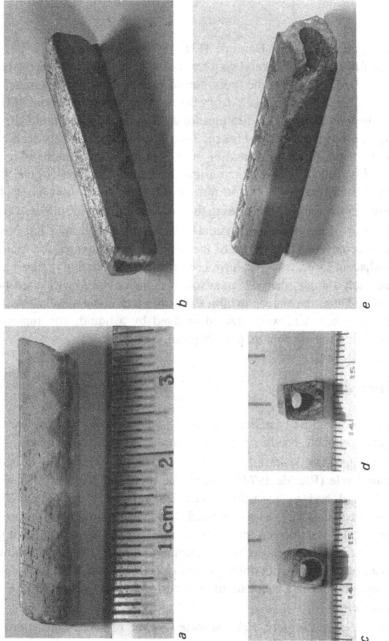

Fig. 19.8. The Dalmore Bone. Length 34 mm (Colin Ramsay)

Analysis

The bone artefact found at Dalmore (Fig. 19.8) is 34 mm long. It has been made carefully from a sheep, goat or possibly even deer, metatarsal or metacarpal. Dr Barbara Noddle (pers. comm.) states that the maximum length of artefact which could be made from these bones would be about 100 mm, based on the overall length of 125 mm for a Soay sheep. One end of the bone is cut square, but the other is broken. As a brittle stick is likely to break at some point between about $\frac{1}{3}$ and $\frac{2}{3}$ of its length, the Dalmore bone may represent either about $\frac{1}{3}$ or $\frac{2}{3}$ of the original artefact. Thus the original length might have been between about 50 mm or 100 mm, the shorter length being less likely and the longer length being the same as the above limit of 100 mm.

The interior of the bone is its natural hollow, almost circular but possibly with ridges removed. The exterior of the bone has been squared and polished. On two adjacent faces, triangles have been made by scratching or cutting lines to outline each triangle, then the areas roughed and coloured with a red-brown stain or dye. These dark areas exhibit extra iron and manganese as shown by X-ray fluorescence, and were stained or dyed by a liquid, not impregnated with a powder (Andrew Foxon, pers. comm.).

Close parallels

There are some close parallels to the Dalmore bone. Single square or rectangular sectioned 'beads' have been recorded on three occasions:

1. *Balbirnie, Fife, Scotland.* In a cist dug into the natural gravel inside a stone circle (Ritchie 1974: 6) were the cremated remains of a woman and child and a 'square-sectioned bone bead' (Fig. 19.9). This is described (*ibid.*: 21) as a 'calcined bone bead or toggle, 28 mm long, rectangular section up to 7 by 6 mm., with an oval longitudinal perforation'. The bead is white, calcined and very fragile. It shows no traces of colouring, but does appear to have been artificially squared off, though the edges are not sharp. The perforation is oval. (Trevor Cowie, pers. comm.)

2. *Barns Farm, Dalgety, Fife, Scotland.* One of three cremations in a single grave in an Early Bronze Age cemetery appeared to have been placed in a bag. It was removed *en bloc* and analyis suggested that the remains may have been of a woman at least in her late teens. 'In

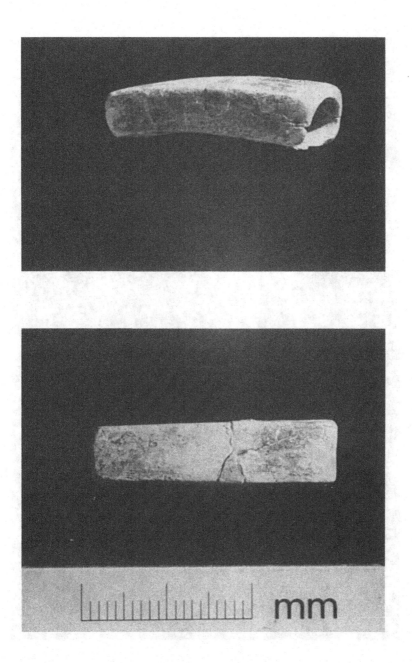

Fig. 19.9. The Balbirnie Bone 'Bead'. Length 29 mm (National Museums of Scotland).

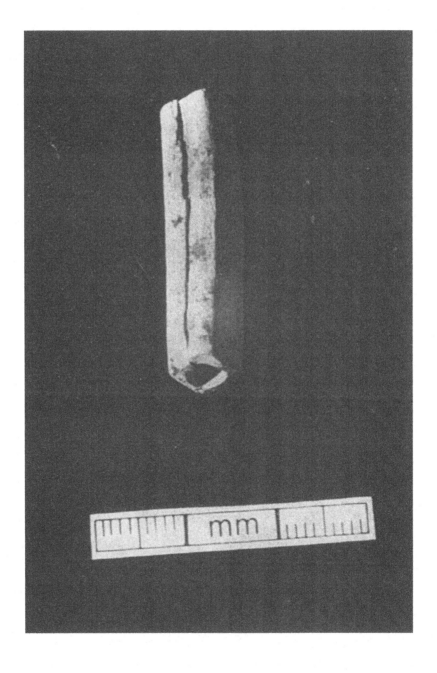

Fig. 19.10. The Dalgety Bone 'Bead'. Length 32 mm (after Trevor Watkins).

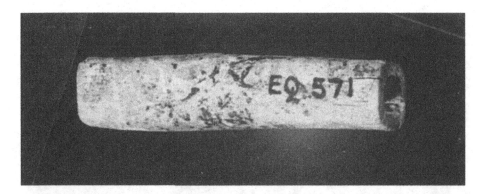

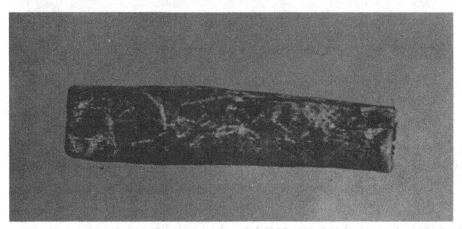

Fig. 19.11. The Patrickholm Bone 'Bead'. Length 33 mm (National Museums of Scotland).

amongst the solid mass of cremated bone debris was a bone bead which had also been burnt' (Fig. 19.10). It is described as '32 mm long, 7 mm wide and the perforation is 4 mm in diameter' (Watkins 1982: 72, 93-4, 108).

3. *Patrickholm, Lanarkshire, Scotland.* In one of the cists in the Bronze Age cemetery excavated by Maxwell, were the cremated remains of at least four individuals. Among the 'mass of bones' were found some stone artefacts and 'three and a half bone beads and about five bone fragments that may have been parts of beads' (Fig. 19.11) which were 'possibly to a certain extent calcined'. The whole 'beads' measure 1^3_8 in, 1 in and 1 in long, with a perforation of about 2-3 mm (Maxwell 1949: 210, 220). Only one of these friable bones has a squarish cross section and is possibly shaped artificially. Length: 33.1 mm; cross sections: 6.6×6.6 mm and 7.2×7.2 mm (Trevor Cowie, pers. comm.).

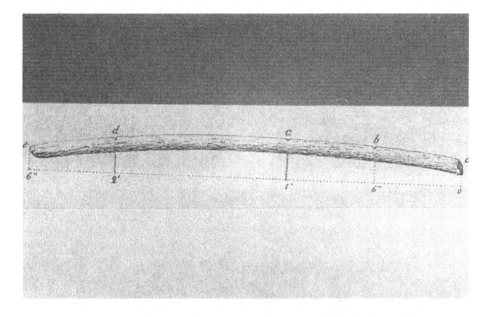

Fig. 19.12. The Borum Eshøj hazel rod. Overall length approximately 82 cm (1 MY) divided into ^{1}s, ^{2}s, ^{1}s, ^{1}s MY. From original drawing by I.M. Petersen (Danish National Museum).

If indeed, any of these are parallels, they are limited to one each from Balbirnie and Dalgety, and a less certain one at Patrickholm. In each case, there is good reason to believe that these artefacts existed singly rather than as the sole surviving piece from a group of similar bones. In each case, as the 'beads' have been calcined, we cannot know the exact original length of each bead because the degree of oxidisation is not known.

If these items were necklace beads, it seems strange that in these finite cremation contexts, only single ones were found. The Dalmore bone has no sign of preferential wear from a cord through it, or from an adjacent bead. It does not have holes through it as in spacers in jet bead necklaces. Its arrangement of coloured triangles forming a zigzag decoration on only two faces seems strange for a decorative bead, or for a die for playing dice, where a greater variety of symbols on the four long faces would be expected. End blown flutes are usually larger than the Dalmore specimen and have holes along one side. If it were part of a set through which a fastening pin could pass, decoration along only two faces would be understandable, but not the lack of internal wear or associated fasteners. If it were a tally stick, the marks

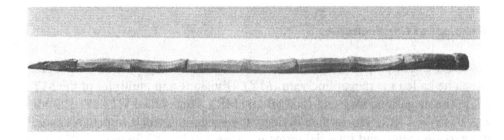

Fig. 19.13. The Borre Fen oak rod. Overall length 135 cm (1^3s MY plus end knob), divided into eight units of ^{1}s MY each (Danish National Museum).

would be less regular. If it were a personal ornament worn through an ear lobe or the nose, wear would be unlikely and the decoration might be understandable.

Zigzag decorations are common on Early Bronze Age items such as gold lunulae, pots or stone carvings, but there is no parallel to the Dalmore artefact.

Longer rods; general parallels

A hazel rod from an Early Bronze Age burial mound at Borum Eshøj in East Jutland, Denmark (Boye 1896) had notches dividing its overall length of 78.55 cm into ^{1}s, ^{1}s, ^{2}s and ^{1}s (Fig. 19.12). Euan Mackie (1977a) gives reasons for preferring an overall length of 81.3 cm. Thus the divisions were 15.71 cm or 16.26 cm long. Whether or not these measurements allow for the curvature is unknown. There is also the possibility of the rod having altered length during 3000 years in a waterlogged grave. Glob (1974: 38) states, 'There can be little doubt that it was a Bronze Age measuring rod marked out in feet ...'.

An oak rod found at an Iron Age stronghold in Borre Fen, Denmark (Glob 1967: 261) was 135 cm long (Fig. 19.13). It 'has a button at one end and is pointed at the other. It is divided into eight equal measuring lengths of 16.5

centimetres, cut along one edge in the form of alternately convex and concave curves'. Thus five of these parts total 82.5 cm, which is close to the megalithic yard of 82.96 cm (see below).

It is interesting to note that both of these measuring rods are divided into units equal to one fifth of a megalithic yard, or eight megalithic inches, measuring 16.60 cm.

Metrological Considerations

Professor Thom proposed a unit of measurement used by prehistoric people, which he called the megalithic yard (MY). Divisions and multiples of the MY led to the megalithic inch and rod (MI and MR). Thus 40 MI = 1 MY; 100 MI = $2^1\!/_2$ MY = 1 MR. He further divided the megalithic inch into halves and quarters. In metric and imperial terms we have 1 MY = 82.96 cm = 2.722 ft = 32.64 in; also 1 MI = 2.075 cm = 0.82 in.

The markings on the Dalmore bone have been measured photogrammetrically by Dr P.J. Scott (pers. comm.) and the distances between all the points of the triangles determined as in Table 19.1.

If the bone is a rule, it is to be expected that the edge figures would be more accurate than the face figures. Taking the average of all edge figures, we get 5.104 mm. Now Professor Thom (Thom & Thom 1978a: 49) quotes 1 MI

Table 19.1. *Measurements of the markings of the Dalmore Bone*

Values in mm	Upper face	Upper edge	Lower edge	Lower face
	4.59	5.01	5.99	5.93
	5.85	5.57	4.87	5.12
	4.88	4.67	4.66	4.85
	4.68	4.78	4.76	4.73
	4.62	5.50	5.23	5.28
Total	24.62	25.53	25.51	25.91
Average	4.924	5.106	5.102	5.182

equal to 20.725 mm, and $^1\!4$ MI equal to 5.18 mm. Therefore four spacings on the Dalmore bone, which average 20.416 mm, closely represent 1 MI. In fact, the error of 0.309 mm is within 2%, which, for the size of ruler being considered, is negligible.

Discussion

There are several ways in which megalithic man could have used the Dalmore bone rule with its $^1\!4$ MI marks. With a fine twig bone splinter or flaked stone point held vertically on a rock surface and a thong looped around the twig and through the bone rule, the rule could be rotated around it and marks made on the rock using one mark to draw a circle or successive marks to draw spirals. The existence of such a scale has been presupposed by both Professor Thom (Thom & Thom 1978a: 45) in the construction of rock carvings involving $^1\!4$, $^1\!2$ and $^3\!4$ MI units, and by Ronald W.B. Morris (1979: 21, 22, 24, 28) in the setting out of megalithic designs.

If, as suggested by Dr J.C. Orkney (pers. comm.), this method were used on leather, geometrical patterns could have been worked out to a conveniently small scale prior to setting them out full size on the ground. For example, if each $^1\!4$ MI on the drawing represented 1 MY on the ground, the scale would be 1:160. Thus the sophisticated geometrical constructions proposed by Professor Thom for flattened circles etc. might have originated on a leather 'drawing board'.

Three rules of different lengths on a fine thong could have been used to study Pythagorean triangles, or a number of shortish scales on a thong could have been used to study perimeters. The famous Gourock golf course cupmarks in the form of two Pythagorean triangles (Thom & Thom 1978a: 49) could have been set up with three rules 5 MI long, two rules 3 MI long, two rules 4 MI long, and grass or leather cords. Alternatively, these cup-marks could have been laid out with the use of seven rules each of them 5 MI long.

Furthermore, the bone item could have been used in a liquid as a dip-stick, or as a pipette for administering medicine (Derrick Lees, pers. comm.)

If indeed the bone artefact from Dalmore is a megalithic rule divided into quarters of a megalithic inch, it is perhaps not suprising that it is, so far, unique.

I believe that I have put forward a case for re-examining any other square-sectioned 'beads' for traces of 'decoration'; for vigilance with respect to traces of colouration, natural, accidental or deliberate; and also for comparing any such traces with the Dalmore 'bead' and the megalithic inch.

20

The Thom paradigm in the Americas: The case of the cross-circle designs

ANTHONY AVENI

Introduction

Alexander Thom's work has influenced scholars far beyond the confines of his own homeland. As Baity's (1973) survey of the literature has demonstrated, once the results of his investigations began to appear at regular intervals in the *Journal for the History of Astronomy* and in his textbooks, investigators all over the world took seriously the possibility that astronomy might have been a motivating factor in the placement and orientation of ancient ceremonial architecture. Nowhere has the impact of the Thom methodology been more deeply felt than in the Americas, where alignment studies have been conducted with varying degrees of success in cultures ranging from the prehistoric period to the Spanish Conquest and from New England and south-west Canada to Peru. Indeed, archaeoastronomy, as astro-archaeology came to be known in the Americas, became something of a fad indulged in by many members of the scholarly set as a leisure weekend pastime in some cases.

In this essay I shall not attempt to survey the literature on the subject, but I do intend to tell the story of how archaeoastronomical studies took hold on our side of the Atlantic and especially how they have developed over the past two decades during which they have been practised. I believe I can best achieve this goal by referring to specific case studies; therefore I shall employ as my subject matter a set of artefacts of assumed astronomical importance, the cross-circle designs (formerly called pecked circles) of Mesoamerica. I choose these curious petroglyphic carvings for at least three reasons: they have been under investigation at least since the time of Thom's first major work; I have had enough familiarity with them to be able to perceive an evolution of their interpretation that offers us the opportunity to assess both the short- and long-term effects of the Thom legacy upon the histories of science and culture in the Americas; and finally, the present study offers me the opportunity to present a status report on our continuing study of these fascinating symbols - a study which grows ever more complex with each discovery of a new 'specimen' and with each new idea proposed to account for it.

Just as Thom had his archaic predecessors in Lockyer, Somerville, and Trotter, so too had Mesoamericanists proposed prior to the 1960s that ancient Mexican buildings were astronomically oriented. Zelia Nuttall (1906) was among the first to suggest that astronomical orientations existed in Mexico. She used several pictures from the Mixtec codices to suggest that people were aligning their structures by looking over notched sticks. One of the earliest archaeoastronomical arguments employing the passage of the sun in the zenith was offered by Marquina & Ruíz (1932), who claimed that the 17° north of west orientation of the Pyramid of Tenayuca could be attributed to sunset at zenith passage - an idea that seems to have been propagated through the literature to its ultimate source at Teotihuacan. As it turned out, both the measurement and its interpretation are false. At about the same time Blom (1924) first proposed the idea that the Group *E* structure at Uaxactún, an Early Classic Maya site in the Petén rain forest, incorporated solstitial and equinoctial alignments; and shortly thereafter Ruppert (1935) hypothesized that the Caracol of Chichén Itzá was an astronomical observatory. Students of Thom will be interested to learn that Ruppert had suggested that two of the windows incorporated the lunar extrema. And in South America, Paul Kosok (1965) proposed that the Nazca lines of Peru constituted the 'largest astronomy book in the world'. All of these opinions were espoused well before the rejuvenation of alignment studies, which occurred with the publication, wide dissemination, and popularization of Hawkins' study of Stonehenge in the early 1960s. To many outsiders, Hawkins' publications opened the floodgates to the receipt of Thom's numerous papers on the other megalithic sites that had already appeared in some of the more remote scholarly publications.

Kuhn (1962) has argued that revolutions in the conduct of science are usually characterized by a pattern of behaviour that takes on essentially the same qualities across the disciplines. The first stage of these revolutionary developments is characterized by the presentation of a new way of doing things, a new set of questions being asked, usually by someone outside or on the fringe of the discipline. Though at first perceived as radical, the new paradigm, once offered up in its nascent and pristine form as an original idea detached from the usual set of disciplinary suppositions, becomes articulated through the practice of a new method or process by converts to the new approach. The paradigm articulation process results in the gathering of large quantities of new information and it gradually becomes transformed into the business of ordinary science which is installed as normal practice by an ever-increasing number of followers of the new leadership.

There are some striking parallels between Kuhn's model and the history of archaeoastronomy. Utilizing the British megalithic corpus of material remains, Thom demonstrated the power of quantifiable knowledge residing in alignments gleaned from highly accurate measurements made in the field. The very same methodology was then tried out on Native American archaeological remains. Thus Hawkins (1975) himself subjected the lines of Nazca to the same analysis he had executed at Stonehenge. Eddy (1974) studied the alignments of certain Medicine Wheels which, with their roundness, radial quality, and lithic composition, bore more than a casual structural relationship to the megalithic stone circles. Indeed, one pair of authors (Ovenden & Rodger 1981) went so far as to postulate that the ovoid forms of Medicine Wheels were derived from megalithic circles and that the megalithic yard was represented in their plan. In the American south-west, the search for solar alignments was conducted with a fervent passion not exceeded anywhere in the world. Many a weekend was spent by individuals in hot pursuit of daggers of light entering cracks and openings in caves and buildings - especially light patterns that crept across petroglyphs. In New England, a group of investigators undertook to determine the orientation of certain stone chambers (which more likely were colonial root cellars) and pairs of standing stones supposed by some to have been descended from the megalithic builders of Great Britain.

Thom's idea that ancient man could record 'unwritten' astronomical knowledge caught hold of the imagination of the astronomer, engineer, and lay scientist, but (perhaps fortunately) not the somewhat more sceptical archaeologist. With all of the objectivity advertised to be a virtue of their new-found profession, budding archaeoastronomers followed the order of enquiry as if it were a recipe in a cookbook: find first the solstices, then, if successful, look for the lunar extremes. Having recently surveyed the Native American archaeoastronomical literature,[*] I was surprised to find the degree of parallelism between the US south-west and megalithic case studies. I attribute the purported discovery of common astronomical concepts in the

---

[*]   A.F. Aveni, Archaeoastronomy in the U.S. south-west: A neighbor's eye view. *In* Carlson, J.B. & Judge, J. (eds.), *Astronomy and ceremony in the American southwest*, Albuquerque. In press.

alignments to the absence of a written record, a body of evidence which I shall suggest below provides a source for archaeoastronomical hypotheses in studies of other cultures.

I do not mean to negate the valuable insight and useful knowledge about ancient astronomical practice that has emanated from the study of alignments alone. However, I must hazard a few critical comments in order to stress two central points that cannot be overlooked: that often these studies have been conducted without reference to the pre-existing data base relating to the cultural remains; and that too often the central goal of a field investigator seems to consist of finding an alignment for the sake of alignment without proceeding to the far more important questions of what was the purpose of the alignment, how and why were the astronomy and calendar developed, and how do such matters fit into what we know about these societies in general? I believe we are now just beginning to correct this monolineal methodology in American archaeoastronomy because we realise that we have, on this continent, one commodity that does not exist (or exists only minimally) in the area that served as parent to the paradigm we have defined: a rich cultural record. The success of our studies, therefore, will rest upon our ability to modify and adapt the Thom Paradigm to a more complex set of data.

We shall judge the first chapter in our success in handling this transformation by discussing the study of the astronomical meaning of the cross-circles as it has evolved over the past two decades.

Cross-circles: the data base

The earliest case studies for astronomical orientation in ancient Mesoamerica that can be attributed directly to the Thom paradigm deal with Teotihuacan, the great ceremonial centre located immediately north-east of Mexico City. Unlike the studies in the British Isles, those undertaken at Teotihuacan involved the archaeologists - in fact, they were conducted initially by members of the investigative team that had been actively excavating the site during the late 1950s and early 1960s. Several Teotihuacan orientation hypotheses are discussed in a short article by Dow (1967) and some of the data upon which they are based are presented in greater detail in the site excavation report (Millon 1973a).

Two vital points are stressed in the argument:
(a) that Teotihuacan is peculiarly laid out with respect to the local landscape according to a pre-ordained plan; and

(b)   that a pair of permanent markers, one etched into the floor of a building and the other carved on a rock outcrop, may have figured in that plan.

Since that time, similar markers have been discovered at Teotihuacan and elsewhere. These discoveries, as well as the recognition that the marker had been mentioned in earlier published sources, made it necessary to reassess its meaning at several different levels. Now we realise that the use of this symbol passes far beyond the realm of archaeoastronomy, but nevertheless, most explanations of it still embrace the orientation question.

Designs consisting of holes, pecked with some sort of percussive device, appear in ceremonial buildings and on outcrops of rock all over Mesoamerica. Enough such symbols have been located and photographed that one can recognize them as consituting a class of symbols by virtue of their similarity. The design usually consists of single, double, or triple concentric circles centred on a cross. Occasionally, a square or a set of concentric squares replaces the circle. Rarely, circle and square appear in the same design and in at least one instance the concentric design resembles a 'Maltese Cross'. A sample of representative cases is given in Fig. 20.1.

For want of space, we are unable to assemble all the data here. Instead, we must refer the reader to publications in which the relevant information is presented. The data pertaining to the 29 examples of cross-circles known up to 1977 are laid out in tabular form in Aveni, Hartung & Buckingham (1978)*. In a later paper (Aveni & Hartung 1982a) we reported the existence of three designs (TEO 13, 14 & 15) that accompany TEO 7 on rock outcrops at Cerro Chiconautla, a mountain located 14 km WSW of the pyramids. These, like some of the other symbols cited in our 1977 paper, had already been pointed out by Gaitan *et al.*† Of them, only TEO 13 and TEO 14 bear a close resemblance to the cross-circle form, the other two consisting only of curved axes that intersect in a non-right angle.

---

*   All notations for the naming of the symbols follow the system developed in this reference. Since the publication of this information, the authors have determined by close examination that the two designs, TEO 6 and TEO 11, which were thought to have been carved (i.e. formed by a continuous line), were indeed pecked and later embellished, probably in modern times. Of the other designs mentioned in that paper, we should add that we have never been able to locate TEO 9 and TEO 10. M. Wallrath has offered the more descriptive term 'pecked cross-circle' to replace 'pecked circle.' As this paper was going to press yet another pecked circle discovery was reported, the first in the region of Oaxaca (near the Pacific coast) (Zarate 1986). The carving is especially interesting because the interior portion of it bears a distinct resemblance to the 'Maltese Cross' pattern evident in TEO 2 (Fig. 1b).

†   Gaitan, M.A. Morales, Harleston, H. Jr. & Baker, G., La Triple Cruz Astronómica de Teotihuacan. Paper presented at the International Congress of Americanists, Mexico City 1974. Manuscript unpublished.

Fig. 20.1. Representative photographs and tracings of various types of pecked cross design. All notations are after Aveni *et al.* (1978).

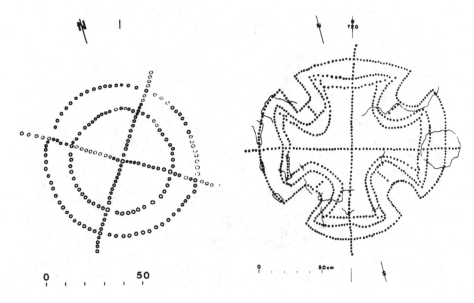

(a)   TEO 1; carved in stucco floor (Horst Hartung).   (b)   TEO 2; as TEO 1.

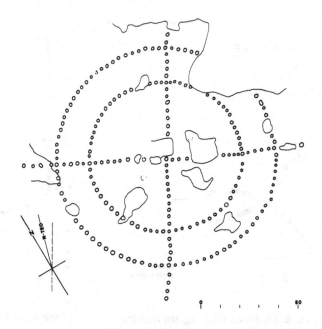

(c)   TEO 17; as TEO 1.

Fig. 20.1 (continued).

(d)   TEO 5; carved in rock (Horst Hartung).

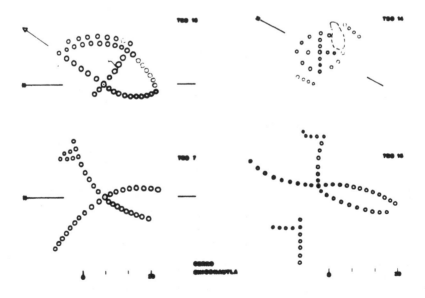

(e)   TEO 7, 13, 14, & 15; forms retaining some characteristics of the cross-circle design, all carved in rock; vicinity of Teotihuacan.

Fig. 20.1 (continued).

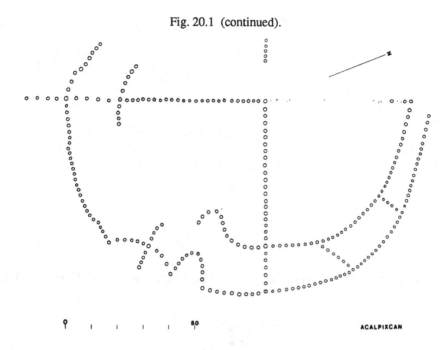

ACALPIXCAN

(f) ACA, from Acalpixcán, near Teotihuacan; carved in rock, only a part of the design remains (Horst Hartung).

(g) CEC1, from Cerro de la Campaña, State of México; carved in rock (Horst Hartung).

Fig. 20.1 (continued).

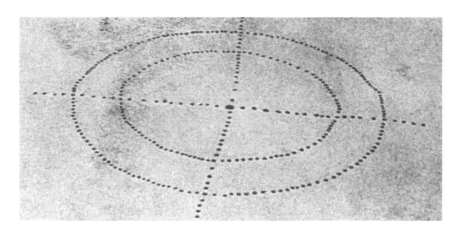

(h)   UAX 1 from the Maya city of Uaxactún; carved in stucco floor (Smith 1950: Fig. 15a).

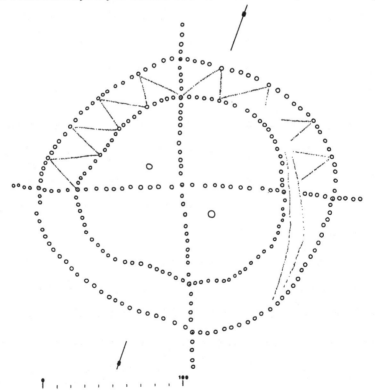

(i)   CHA 1, near the Tropic of Cancer, north-west Mexico; carved in stone (Horst Hartung).

Fig. 20.1 (continued).

(j) BOC 1, State of México; pecked square with flying corners, carved in stone.

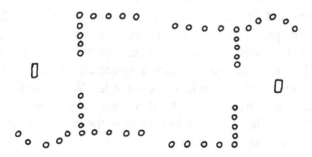

(k) Quinze board after Bennett & Zingg (1935: 343) (redrawn and photographed by Horst Hartung). See also Fig. 20.6.

In 1982, we published data (Aveni & Hartung 1982b) on two new cross-circles that had been discovered by R. Cabrera C., one in the Ciudadela of Teotihuacan (TEO 17) and one adjacent to the ruins of Purepero, in the Mexican state of Michoacán. In the same publication we reported the existence of two additional petroglyphs at Tepeapulco (more appropriately, Xihuingo), another carved on a loose stone now residing in the Museo Coatelelco, in the state of Morelos, and a third (actually a square composed of seven large pecked holes) at Chalcatzingo (CHL). Iwaniszewski (1982) summarized data relating to four more designs discovered by archaeologists from the Instituto Nacional de Antropología y Historia (INAH) in the Ciudadela and vicinity (TEO 18, 19, 20, 21). Of them, only TEO 19 can be classified as a true cross-circle design. TEO 18 and TEO 21, like TEO 7 on Chiconautla, are made up only of rows of peck marks or intersecting axes. TEO 20, the most curious of all the newly discovered designs, consists of a pattern reminiscent of a pair of concentric 'Gaussian curves' that seem to comprise half of a quadripartite design. It lies almost flush against the outer wall of the Ciudadela on the side fronting the Street of the Dead. Another design, TEO 22 (unpublished), was recently discovered in the Ciudadela.

Folan (1978) and Folan & Ruíz (1980) have reported several pecked designs at Cerrito de la Campaña (CEC) and Boctó (BOC) in the north of the state of México. The symbol CEC1 is a double pecked circle carved in the vertical position on a large, loose boulder that now forms part of a wall, while at least eight other designs appear as pecked squares or portions thereof.

Recently, Grijalba (1984) has listed additional petroglyphs at Xihuingo, which he has located on a map. He also referred to the continuing work there of Wallrath, which was summarized in a paper read at a meeting on Archaeoastronomy and Ethnoastronomy held at Mexico City in September 1984. Following his presentation, Wallrath was kind enough to lead me on a detailed guided tour of Xihuingo, where he disclosed the existence of at least 40 designs, confined to an area of less than a square kilometre, all of which fit the standard definition of a pecked cross-circle. We eagerly await the publication of these new data, which, by their sheer abundance and exacting detail, are certain to offer us fresh insight into the nature of the symbol, especially in the context of these ruins.

To the north, Zeilik (1985) has reported the existence of cross-circle patterns in pictographic form in the vicinity of Chaco Canyon, New Mexico, USA (NMX). It remains to be seen whether these examples fit the true category of definition of pecked circle.

Finally, to complete a listing of the basic literature on the data base relating to this widespread symbol, we should report that in other publications we

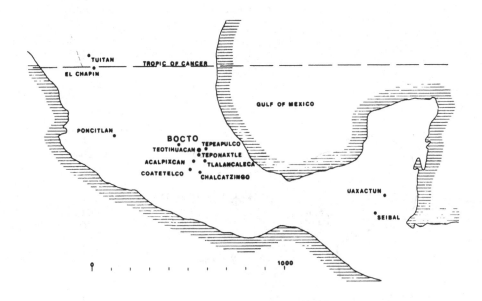

Fig. 20.2. Map showing the widespread locations of pecked cross circles in Mesoamerica (Horst Hartung).

have interpreted data on many of the crosses already known to exist (cf. also Aveni & Gibbs 1976; Aveni & Hartung 1979; Aveni, Hartung & Kelley 1982; and Aveni 1980 for further discussion and interpretation). Also, Coggins (1980) has discussed and interpreted the cross-circles in the broader context of quadripartite symbolism in Mesoamerica.

In Fig. 20.2 we present an updated map showing the location of the designs we now believe to comprise the category of pecked cross-circle. In Fig. 20.3 we show the locations of clusters of these symbols at Teotihuacan and in its vicinity. Unfortunately, the state of preservation of many of these petroglyphs is extremely poor. TEO 1, for example, is now practically destroyed.

As one can see, the literature on this subject has increased enormously. Now that there is an abundance of them, we find it useful to classify these designs on the basis of the general configuration of their component parts and the medium in which the symbol is etched or carved. Thus, we employ the notation given in Table 20.1.

Fig. 20.3. Pecked crosses at Teotihuacan. (All parts by Horst Hartung.)

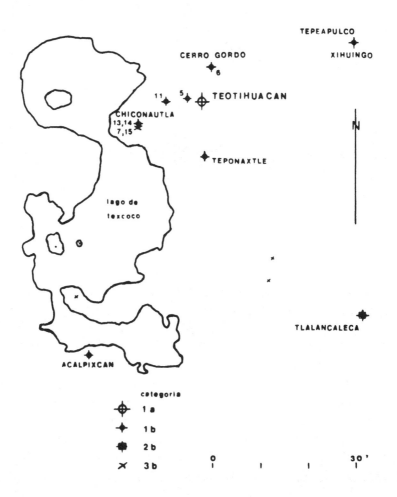

(a) Map showing pecked crosses in the vicinity of the ceremonial centre of Teotihuacan.

Fig. 20.3 (continued).

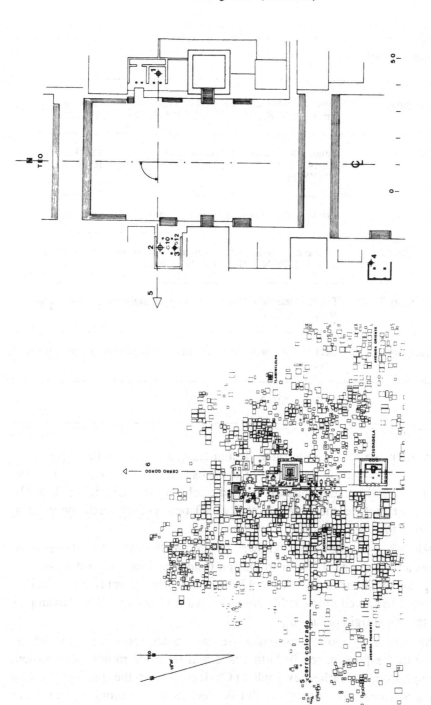

(b) Map showing pecked crosses in the ceremonial centre proper.

(c) An enlargement of the central region shown in (b).

Table 20.1. *Classification of cross-circle designs by form and medium in which they are presented*

Design Category 1	Single, double or triple pecked cross-circle in floor of building: TEO 1, 3, 4, 8?, 9?, 10, 11; UAX 1-3; TEO 17, 19, 22.
Design Category 2	Single, double or triple pecked cross-circle on rock: TEO 5, 6, 13, 14; ACA; TEP 1-5; TEX; CHA 1, 2; TUI; RIV?; SEI; COA; NMX?; PUR.
Exception	Triple pecked Maltese Cross in floor of building: TEO 2.
Design Category 2	Single or double pecked squares with or without diagonal: CC; TLA 1-3; PON; CHL; BOC 1-7 (always carved in rock).
Design Category 3	Pecked line or intersecting lines, straight or curved, carved in rock: TEO 7, 15.
Design Category 3	Pecked line or intersecting lines, straight or curved, carved in floor: - - - -
Inadmissable	TEO 18, 20?, 21.

Hereinafter, we refer to the three design categories as labelled above, the first being the most rigid definition of the cross-circle. The map includes the newly-discovered but unpublished pecked circle in the Ciudadela (TEO 22), but does not include the (approximately) 40 newly-discovered symbols at Xihuingo or the one in Oaxaca.

Of the 93 examples now known (counting the 40 recently-reported examples at Xihuingo), 14 refer to Design Category 1 carved in the floors of buildings at Teotihuacan (eleven) and Uaxactún (three). Twenty more designs of this type are carved on rock. Nearly all of Wallrath's unpublished examples also fit this most rigid definition.

Of the twelve Category 2 examples we have noted, both the forms of the designs and the provenance in both space and time are rather more varied; these range from the earliest example at Chalcatzingo, to the (probably) rather late ones at Boctó. The other handful of designs of Category 3, which are composed of simply straight or curved intersecting lines, perhaps ought not even be considered as members of the class. Even with these restrictive definitions, the pecked cross-circles of Category 1 emerge as a common widespread symbol, of which we can cite at least 60 examples (if we include the Wallrath data), that likely persisted throughout a long period of Mesoamerican history.

Does the development and diffusion of this perplexing design indicate a continuity of purpose in Mesoamerican culture? And does astronomy fit into the more expanded and complicated body of data on cross-circles? To answer these questions we must review the various explanations that have been offered for the meaning and function of this symbol, and in particular the evidence supporting each explanation.

Pecked cross-circles: a discussion of the hypotheses

The calendar hypothesis

The earliest explanation for the use of the pecked cross-circle was that suggested by A. Chavero (1886) who, to my knowledge, was actually the first to report the existence of the symbol in the literature. He suggested that the RIV petroglyph, which has never been located since the original citation, but which I suspect is likely identified with CHA 1, was used as a calendric counter. Chavero's hypothesis seems to have been based on the fact that one could count 20 points along each axis of the cross as well as in each quadrant of the inner circle. Twenty is a very common Mesoamerican calendrical number.

More recently Worthy & Dickens (1983) developed a somewhat complex method for marking the 365-day year through the use of the 260-point form of the pecked circle. With some evidence in our 1978 paper (Fig. 6) that numbers of calendric significance were present on many pecked crosses, Aveni, Hartung & Kelley (1982) explored this idea in further detail for the two petroglyphs located on the summit of Cerro El Chapín overlooking the ruins of Alta Vista (Chalchihuites), very close to the Tropic of Cancer. Both these petroglyphs contain nearly 260 holes, as best we can count them given their state of preservation. Not counting the centre hole, the pattern consists of 20 holes on the axes, 20 on each quadrant of the inner circles and 25 on each quadrant of the outer circles.* The pattern of 20 holes on the axes, arranged in

---

* This breakdown into units of 20 works only if we include the point of intersection of circle and axes on the *axial* count, an observation which might suggest that in the

a 10-5-5 sequence as counted from the centre outward, is practically universal. This fact alone, given the importance of the number 20 in Mesoamerican mathematics and the calendar, is very strong evidence in support of the idea that pecked circles were used as calendrical counting devices.

The counts on the circles also appear to be grouped around multiples of 20, namely 60 (six examples), 80 (twelve examples), and 100 (twelve examples).†

TEO 2 is a petroglyph that bears a number of calendrical properties. The count on the outer circle of the design is exactly 260, the same as that on the periphery of the 'World diagram' shown on page 1 of the Codex Fejéváry-Mayer (1971), which it also resembles in its general 'Maltese cross' form. Moreover, the diagram in the codex contains specific reference to calendrical day names. On the contrary, there are some counts of elements in the pecked circles that appear to have nothing to do with calendars, at least as we know them; for example the outer circle of TEP 5 consists of 130 peck marks while the same portion of UAX 1 contains 156 distributed very unevenly in each quadrant (43-45-32-36). Often the distance between the pecked holes varies, the points sometimes appearing too crowded (cf. the north-eastern outer quadrant of UAX 1, the north-western outer quadrant of CHA 1, and the south-eastern outer quadrant of TEO 1), as if the maker had not paid much attention either to executing a precise count or to making a precise design. We must conclude that while it is very likely that some of the designs were intended to count the days in groups of 20 and its multiples, perhaps even in totals of 260, still not all of them could have functioned that way. Though it is possible to hypothesize that variation in the count could be attributed to the need to use different agricultural calendars in different altitudes and climate zones (a suggesion actually made by Chavero), we can offer no concrete evidence that such was the case. Still, we must admit that over the long period of Teotihuacan influence in Mesoamerica, one certainly would have had need of controlling and universalizing the calendar over a wide domain.

---

pecking order the axes were made first. We have found a number of designs that consist only of the axes or crossing lines. These may constitute examples of designs that had never been completed.

† These estimates are based upon a probable error in the total hole count of 7 which results from the fact that many of the samples are substantially eroded.

The architectural marker and astronomical orientation hypotheses

Though these two hypotheses for the function of the pecked cross symbol are separable, we consider them together for historical reasons. Stated simply, the first hypothesis proposes that the markers are related to the layout of ceremonial centres and the second that lines between pairs of markers and/or axes of markers point to the rising/setting positions of astronomical bodies used to mark events in the calendar. Conjoined, the two hypotheses imply that astronomical considerations were a part of the planning and orientation of ceremonial centres.

The arguments interrelating the astronomical function of the Category 1 markers TEO 1, 5, 6, 11 in the plan of Teotihuacan have been laid out fully in the literature (Dow 1967; Millon 1973b; Aveni, Hartung & Buckingham 1978; and Aveni & Hartung 1982a). As hinted earlier, the motivating factor for proposing these relations had been the recognition of the precise, likely pre-planned, $15^{1}2°$ east of north skew of the axial grid of the city, a misalignment from the cardinal directions that runs against the general trend of the landscape. This odd skew appears to have been imitated all over the Valley of Mexico in later times (Aveni & Gibbs 1976). A number of astronomical bodies have been suggested to account for this skewed orientation. In our judgment, the one that makes most sense is the setting point of the Pleiades, which we know played an important role in Mesoamerican cosmology and astronomy. Also, this prominent star group crossed the overhead point of Teotihuacan at about the time the city was built, and furthermore, the first annual appearance of the Pleiades in the pre-dawn sky of Teotihuacan occurred on the same day as the passage of the sun overhead. This solar-stellar event pairing suggests to us that the Teotihuacanos may have employed some sort of functional calendrical relationship as a year marker, with for example the first appearance of the Pleiades providing a warning of the most important solar event in the annual calendar. That the Teotihuacanos would incorporate a calendrical function into their ceremonial architecture does not seem unreasonable. Many of the developers of early cities may have shared what Loew (1967: 5, 13) calls a 'cosmological conviction', i.e. that there is a cosmic order that binds man, society, and the heavens. As part of this conviction, man adopts the attitude that the structure and dynamics of society are discernible in the patterns of movements of heavenly bodies. Often, there exists the belief that human society should be a microcosm of the divine society reflected in the heavens. It is the chief responsibility of the priests to adjust the human order on earth to fit the divine order manifest in the heavens. Therefore, it seems reasonable to believe that among societies of the past, the structure of the urban centre was deeply influenced by religious systems of belief which, in turn, were tied intimately to the stars.

Given the Teotihuacan astronomical orientation hypothesis, we must be careful not to anticipate, as some investigators have (e.g. Chiu & Morrison 1980), that every new petroglyph that is discovered must fit into a system. Some designs do fall into an astronomical orientation pattern: thus TEO 13-15 point to the pyramids from a distance of 14 km and indicate the general direction of sunrise at summer solstice as well. However, some of them do not: the axes of ACA, TEX, and TEP 1, for example, point in the general direction of Teotihuacan (as seen from three different directions), but have no apparent astronomical function that we can discern.

Ruggles & Saunders (1984) have attempted to apply statistical significance tests to determine whether the correspondence of alignments between the positions of pecked cross-circles in the vicinity of Teotihuacan and any deliberate orientational scheme might be purely accidental. Their method of testing can be criticised on the grounds that their choice of which symbols to include is, itself, arbitrary and subjective. They also fail to sort out pecked circles into different design categories, instead treating all of them the same way. One cannot simply assume that a marker is a marker and that all markers within an arbitrarily chosen radial distance from the pyramid will be included and those outside that radius excluded from a test. All we can conclude from these tests is what we had already suspected: all the petroglyphs taken together do not constitute an astronomical system. Hawkins (1975) drew the same conclusion about the Nazca lines but, as we have argued, astronomy still could have been a part of the picture (Aveni 1986).

No discussion of astronomical orientation hypotheses involving pecked circles is complete without a discussion of the Alta Vista markers, two remarkable cross-circles located near the Tropic of Cancer which we believe provide the strongest argument for relating pecked circles, architecture, and horizon astronomy (Aveni, Hartung & Kelly 1982). Viewed from a pair of petroglyphs (CHA 1, 2) carved on the eastern edge of an isolated plateau (Cerro El Chapín), the summer solstice sunrise occurs precisely over the top of Picacho El Pelón, a prominent peak to the north-east, at the base of which certain blue stones (Chalchihuites) were mined. The ruins of Alta Vista, a site which the archaeological record indicates was very strongly influenced during the Tlamimilolpa-Xolalpan phase of Teotihuacan, lie a few kilometres to the north. As seen from these ruins, sunrise at the equinoxes occurs precisely over the same peak. These facts led us to posit that the architecture of Alta Vista and the pecked circles were associated with the mountain because, taken together, they produced a double astronomical alignment that marks important points in the year count. J.C. Kelley has excavated a possible ceremonial walkway at the ruins that leads eastwards in the direction of the prominent peak. It is likely that the petroglyphs and the site were deliberately placed at the Tropic in order to mark the place in the Teotihuacan sphere of influence where the sun 'turned around'. Here, and only here, it stands at the overhead point on the longest day of the year - the June solstice - when it also stands still on the horizon in its annual course (Fig. 20.4 illustrates the details).

The scheme at Alta Vista reinforces the connection between the alignment and calendar hypotheses for the use of pecked circles. As indicated in the previous section, both Cerro El Chapín petroglyphs are composed of nearly 260 peck marks; therefore, it is possible that at the Tropic, in the northern extremes of their vast empire, the Teotihuacanos had attempted to co-ordinate the spatial and temporal aspects of their calendar by both counting the days and fixing the important directions in space to mark the movement of the sun.

At Uaxactún in Guatemala, an Early Classic Maya site with established Teotihuacan influence (Smith 1950), we find another example of a design (Fig. 20.1(h)) that falls in precisely the same group (Category 1) as the pair at the Tropic of Cancer and the two petroglyphs involved in the Teotihuacan east-west orientation scheme. Recently, Aveni & Hartung have shown* that the positioning of this marker may have been related through physical alignments to the Group *E* structures at Uaxactún, for which an astronomical function has long been argued.

All of the aforementioned examples appear to exhibit unusually strong evidence of association with alignment schemes but, as in the case of the calendrical hypothesis, we must be careful about anticipating unwarranted precision, for the evidence argues that if the alignments were astronomical, they were not executed with the same order of precision that seems to have captivated Thom and his followers, but which may be unrealistic (cf. Ruggles 1984b). The temptation to over-interpret these data may be a part of our western mental make-up. The specification of alignments to fractions of a minute of arc is, in my opinion, no more justified in the American case studies than it is in the British. Notice, for example, in Fig. 20.1 that the axes of many of the pecked circles are often not perfectly straight nor do they usually cross at right angles. Even the examples that seem at first glance to be more

---

*   Aveni, A.F. & Hartung, H., Uaxactún, Group *E*: An archaeoastronomical reconsideration. *In* A.F. Aveni (ed.), *World Archaeoastronomy*, Cambridge. In press.

Fig. 20.4. Alta Vista. (All parts by Horst Hartung.)

(a)   Labyrinth with axis of walkway to Picacho (arrowed).

(b)   The same view from a lower level.

Fig. 20.4 (continued).

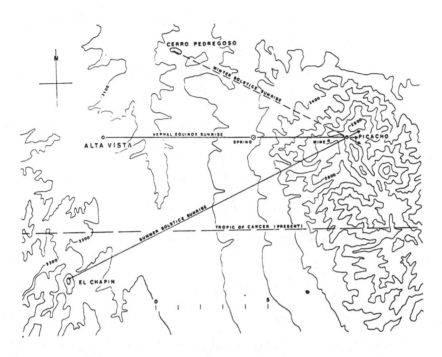

(c)   Map showing the double alignment (solid lines) and a tentative third alignment (dotted line).

carefully constructed are not actually very precise. For example, in TEO 17, the axes deviate from a right angle by ±3° and the ends of the north-east/south-west axis are considerably bent out of line. These linear deviations are reminiscent of the variation in the count of marks in different quadrants of the circles that comprise a given design. To imagine that all of these variations are subtle and deliberate and that they indicate an order of precision beyond our expectation is, in this writer's view, untenable.

To summarize the distribution of alignments of axes of the pecked cross-circle design, we refer to Fig. 20.5, in which we display the averaged direction of alignment, folded about true north, of the axes of each petroglyph. In Fig. 20.5(a) we plot only the petroglyphs within the ceremonial centre (Category 1, Teotihuacan only); in Fig. 20.5(b) we plot those in the Teotihuacan environment, defined as those places from which the pyramids are directly visible (Category 1, carved on rock). Finally, in Fig. 20.5(c), we plot all the other axial directions indicated (all categories, other locations).

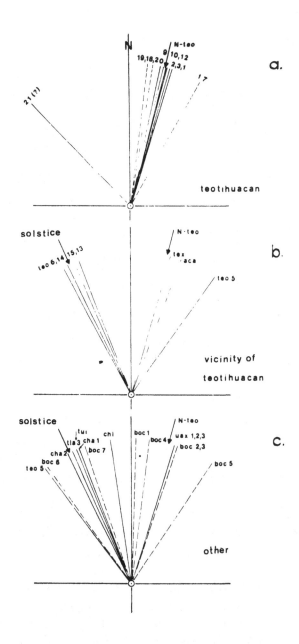

Fig. 20.5. Graphs showing the alignment of the axes of the pecked crosses.

These three plots, formulated on the basis of a consideration of the distribution of the petroglyph in the landscape, differ remarkably from one another. Clearly, nearly all the petroglyphs carved in the floors of Teotihuacan buildings adjacent to the Street of the Dead (TEO 1, 2, 3, 9, 10, 12, 20) are oriented almost precisely in line with the grid of the city, the direction of which is indicated by an arrow. The most notable exceptions are TEO 17 and 21, but recall that we had already questioned admitting the latter to our first design category before we ever looked at how it was oriented. TEO 17 (like TEO 19, which also deviates somewhat, but by a lesser amount, from the TEO grid) is among those designs located farthest from the Street of the Dead. These results offer us a motive for associating the Category 1 symbols along the Street of the Dead with the orientation plan of the city. Indeed, TEO 1 had long ago been implicated in such a scheme (Dow 1967). But what of the large number of designs in the building on the western side of the street opposite TEO 1? We would hypothesize an eastward alignment or orientation by prolonging the $15\frac{1}{2}°$ (south of east) cardinally deviated axis to the horizon and then searching that vicinity for a matching petroglyph, similar to TEO 5 on the western horizon. But this is the work of the future.

In the second plot virtually none of the axial directions fits the Teotihuacan grid, but five of the seven petroglyphs point toward the pyramids and four of them align (approximately) with an astronomical event - the June solstice sunrise (indicated by an arrow). Thus the Design Category 1 petroglyphs found on the periphery of the TEO valley point either to the ceremonial centre or to an astronomically significant event, or to *both*. Perhaps this reveals the intentions of the builders of the great city: to associate cosmic principles directly with the design and layout of the city's ceremonial frame of reference.

On the contrary, the directions of the axes of all the designs that are territorially remote from TEO seem to follow a random pattern, as Fig. 20.5(c) indicates. Perhaps a few of these could have been intended to indicate the sanctity of the TEO grid alignment; e.g. UAX 1, 2, & 3 line up in the direction of the Teotihuacan grid, though they are located hundreds of kilometers from Teotihuacan. On the other hand, these petroglyphs may simply incorporate the common, generally east of north alignment that can be ascribed to nearly all Mesoamerican sites (Aveni 1975b; see also Aveni & Hartung, 'The Maya City and the Sky', *Transactions of the American Philosophical Society*, in press). Four designs may be solstice indicators: the pair of petroglyphs near the Tropic at Alta Vista (CHA 1 & 2) and the TUI petroglyph located 132 km north of the Tropic near Alta Vista. Might this have represented a failed attempt on the part of the Teotihuacan polity to determine the place where the sun turned around in their regime? The fourth solstice example is the pecked cross-square (TLA 3) at Tlalancaleca in the state of Puebla, which is located on a very prominent rock shelf that offers an expansive overlook to the east - a setting rather like that of CHA 1 & 2. In both cases, the place could have served as a convenient observation point from which to witness an astronomical event. Site occupation has been dated to the pre-Teotihuacan period.

The axes of the Boctó petroglyphs appear to spread out in all directions, but as we shall argue in the next section, these probably were game boards and, therefore, their layout may well have necessitated little or no directional significance. Of the TEP (Xihuingo) petroglyphs for which we have collected (but not yet plotted) data, the axes spread out in various directions and appear to follow no particular pattern. TEP 1 points in the direction of Cerro Gordo, the prominent mountain overlooking Teotihuacan from the north, which appears from the location of the petroglyph to be neatly framed in a saddle between two hills that dominate the landscape. Thus, the general setting of this petroglyph accords quite well with other suspected observations points, but in this case the object sighted may not be astronomical; more likely it is the great mountain at the northern end of the TEO valley. Also, Xihuingo lay on the obsidian trade route out of Teotihuacan, a fact we cannot rule out when we refer to the positioning of the markers and their capacity to point out places in the terrestrial rather than the celestial landscape.

The 'game' hypothesis

Aveni, Hartung & Buckingham (1978: 278) summarized the evidence relating cross-circle petroglyphs to *patolli*, a Native American board game. Among the common qualities are the general cruciform design, the occurrence of such designs in the floors of buildings, and the prominence of the number 5 in the counting scheme (the number 5 also figures prominently in the published descriptions of the rules of the game). Except for the isolated examples of the two Huastec patollis (*ibid.*: Fig. 9), the shape of all patolli boards described in the codices or incised at the archaeological sites is more like our Design Category 2, i.e. rectangular rather than circular. Nearly all the nine examples depicted by Smith* indicate a five count from one intersection to the next.

The Boctó petroglyphs particularly resemble native board games (see also Folan & Ruíz 1980). J.C. Kelley (pers. comm.) has brought to our attention

---

*   Smith, J. *The forms of Patollis: Their interrelationships and some symbolic associations.* Unpublished manuscript.

several descriptions of a Tarahumaran game called 'quince' or 'quinze', which was played on a board that is practically identical with the Boctó petroglyphs. Lumholtz (1902) (cf. also diagram in Bennett & Zingg 1935: Fig. 5), includes the 'flying corners' that extend outside the figure, which one also finds on the Boctó petroglyphs (see Fig. 20.1(j)). Players sat opposite one another and were said to begin the count on the extension of the diagonal axis which always extends four or five points beyond the corner. The count continues into the square. According to Bennett and Zingg (*ibid.*: 343) 'Nine holes are made at each corner, namely one at the corner and four running along each side' (see also our Fig. 20.6). Now, if we compare the count on the board for the game of quince with the form of the Boctó petroglyphs (cf. Folan & Ruíz: Figs. 4 & 5), we find an interesting similarity. In each case, the breakdown of the count into units of five is reminiscent of what we find on the pecked circles, both in the divisibility of counts on the circles into families

Fig. 20.6. The board game of quinze as it was observed by Lumholtz (1902: Vol. 1, 279) to have been played. Compare Fig. 20.1(k).

of five and particularly in the count on the axis which almost always proceeds (intersection in parentheses):

4-(1)-4-(1)-4-(1)-10-1-10-(1)-4-(1)-4-(1)-4.

The foregoing comparison makes it clear that a number of the Design Category 2 petroglyphs were very likely employed as game boards. Furthermore, the similarity in the number structure offers the distinct possiblity that some of the Category 1 designs may have had a similar function. However, we should point out that many Category 1 petroglyphs are situated in rather difficult places to have served as game boards (i.e. on slopes or places inaccessible for the purpose of seating even two players). In a personal communication J.C. Kelley notes with interest that in many cases, quinze boards were simply dug in the dirt rather than in stone or plaster, as Lumholtz witnessed (Fig. 20.6). If cross-circles also were dug in the dirt, archaeologists would, of course, have little chance of rediscovering them, a fact which may explain in part the peculiarity of distribution of the design.

Conclusions

I began this essay by suggesting that, for the Americanist, the legacy of Professor Thom's work is two-fold. It has consisted of the offering of an enquiry into the archaeological remains that hitherto had been neglected. For us, Thom raised the engaging question: do the remains of ancient civilizations reflect a knowledge of astronomy by virtue of the way they are laid out in the landscape? Secondly, Thom offered us a methodology for seeking the answer to this question. His method has consisted of making precise measurements at the site and expressing the resulting alignments in the framework of an astronomical reference system. As I have shown (Aveni 1975b), one cannot determine even approximate alignments by using extant Mesoamerican site maps or relying too heavily upon magnetic compass measurements. Thanks to Thom, we have now begun to reassess old arguments that were based on imprecise data.

By reviewing the data on the Mesoamerican pecked circles, I have attempted to show that the enquiry and methodology that make up the Thom paradigm can be misleading. As it stands, it is essentially incomplete as a valid mode of enquiry unless one also employs relevant data about the culture concerning the nature of astronomical and calendrical practice. The concept of quadripartition and quadripartite designs was a part of Mesoamerican cultural *insistence*, the sum total of artistic, religious, and intellectual endeavour that make one culture recognizable and distinct from others. We know that the four-fold division of things was both widespread and common in the Mesoamerican world view and that it is present in the city, cosmological diagrams, codices, games, and even the forms of hieroglyphs. As one expression of this four-fold way of thinking, we find the pecked cross-circle symbol in a variety of contexts all across Mesoamerica.

I have tried to demonstrate that no *single* explanation can account for the occurrence of the cross-circle symbol in its many different forms. I have suggested that for certain examples, e.g. CHA 1 & 2 at the Tropic, a multivalent function was intended; specifically, one could have used these markers as counting devices and astronomical directional pointers at the same time. I have presented evidence that supports a variety of hypotheses relating to use of these symbols in general.

Precision and rigour are two characteristics of good science, for ultimately they relate to the issue of provability. But the use of astronomy among ancient cultures is a socially-based enquiry and when we deal with questions about human behaviour, we must be careful about how we interpret and apply our scientifically-based values.

Given what we have observed of cross-circle construction properties, I think it unwise to assume that the symbol was intended to be precise either in the count of the component holes, in the placement of the symbol in the landscape, or in the orientation of the axes. Perhaps local variations and experimentation with the orientation and/or counting scheme led to an evolution in the structure of the symbol. This may be why the hole counts and orientation directions differ among the samples investigators have recovered to date. We must realise that any suppositions about the relation between precision and function could be a by-product of western European thinking and might not necessarily be applicable in the context of another culture. Thus, it might be incorrect to assume that because Teotihuacan is precisely laid out astronomically, its builders were scientifically minded, like us. It is not fair to assume they sought knowledge for the sake of knowledge, an assumption which one can actually trace in the foundation of Thom's writings.

The role of astrology, religion, and the rituals that attend astronomical pursuits cannot be underplayed in Mesoamerican studies. If one wishes to understand the real meaning of this exotic astronomy, one cannot do it with theodolite alone. One must also engage the formidable, unfamiliar corpus of native writing, iconography, ethnohistory, and even current ethnology, for the survivors of many of these ancient cultures still exist and their ideologies, in many cases, can be shown to have remained to a degree intact. Archaeo-astronomical studies ought not to consist simply of consulting the 'experts'.

One must become familiar enough with relevant materials to employ them as a means of generating, rather than only responding to, astronomical hypotheses.

From the study of the Maya written calendar, we learn that precision - that admirable intellectual quality present in all of our scientific endeavours - can be commensurate with purely religious motives and that one does not need to view nature detached from society in order to pursue its workings. The written record also teaches us that there is no single evolutionary course to all astronomies. All cultures do not simply proceed up an observational ladder consisting of the serial recognition of astronomical phenomena of increasing levels of difficulty beginning with the solstices, proceeding to the lunar standstills, and culminating perhaps with the precession of the equinoxes. Furthermore, the tropical sky of Mesoamerica was substantially different with respect to its phenomenal content, arrangement, and motion from the high latitude celestial environment under which the Thom paradigm originated. We must be prepared to believe that certain sky phenomena recognized by tropical astronomers were of no concern to our European ancestors and vice-versa.

And, finally, as far as alignments are concerned, in our attempts to judge which astronomical targets ought to be included in a test, we would do well to question the standard assumption that all phenomena be accorded equal weight. Are the lunar limits as basic or important as the solar? In Mesoamerica the zenith sunrise-sunset positions ought perhaps to receive added weight because other evidence suggests they were important.

The variety of evidence bearing upon ancient astronomical practice and the additional complicating factors I have tried to point out must not be construed as an invitation to abandon all attempts at making one's studies rigorous. Rigorous intentions, while helpful in setting limits on conclusions we draw, can be a bit misleading if applied too unilaterally.

Like the statistical tests applied to the movement of stars or atomic particles, statistical enquiries in archaeoastronomy serve to provide blanket results for a set of objects taken as a class. They say nothing about the subtle, but nevertheless discernible, nuances that characterise the unpredictable behaviour of people as opposed to things. The use, for example, of the centre of a pyramid as a marker might not be considered valid in such statistical tests unless it possesses a visible marker, yet there is evidence that the location of such a structure has a place in the astronomical framework. Indeed, the astronomical orientation of the Aztec Templo Mayor, which is referred to specifically in the ethnohistoric literature, could never be revealed by techniques such as Patrick's (see, e.g., Patrick & Freeman's paper (Chapter

10) in this volume). The temple possesses an equinoctial orientation (normally azimuth 90°) to azimuth 97°, which was transferred from the open environment into the closed (three-dimensional) space of the ceremonial centre.

Techniques that emphasize extreme rigour can indeed be helpful by revealing whether earlier data have been assessed with a sufficient degree of objectivity. Their proponents have already begun to alter our perspectives about the degree of precision that might have concerned our ancestors. But these techniques are meant to be applied in pristine form only in those cases in which one has nothing more to work with than identical standing stones, all of the same chronological period, accompanied by a horizon along which one can objectively choose natural peaks and notches of predetermined angular dimensions. Is there such a utopian archaeoastronomical world?

Finally, there has developed recently in the archaeoastronomical literature a strategy of requiring that the only valid hypothesis is that which is announced prior to the examination of the data. Perhaps originating out of twentieth century positivism coupled with a reaction to the bizarre conclusions of the first generation of twentieth century archaeoastronomers, this strategy suggests that the only valid way to conduct research is to proceed from hypothesis to test and then back to hypothesis again, according to an orderly scientific method. To me, such a process seems desirable when it can be achieved but in practice it is rather unrealistic. At least our process of discovery of the pair of cross-circle petroglyphs at the Tropic of Cancer followed no such course.

Of the three hypotheses we have entertained to explain the meaning and function of pecked cross-circles, we have tried to suggest that some petroglyphs fit one or another the best, while others agree with more than one hypothesis equally well. The idea that different levels of explanation and multiple functions can co-exist seems an anathema to the nature of scientific enquiry wherein one seeks to isolate paired causes and effects. But in archaeoastronomy we must force ourselves to think somewhat differently from the hard scientist, for we are most likely to be dealing with people who did not think and act as we do. Perhaps once we release from our thoughts the notion that a grand design or single purpose must be accorded all the pecked cross-circle symbols, we shall begin to arrive at a set of explanations, couched at different levels, that makes sense in the context of what we know about the Mesoamerican mentality. Only then will we have accorded the Thom legacy its appropriate place in our studies.

Acknowledgments

Portions of this paper were drawn, modified, and updated from an earlier draft of a paper by Aveni & Hartung (1985). I am grateful to Horst Hartung for his comments and illustrations. I also acknowledge the National Science Foundation (Grant No. BNS-8319920) for supporting this research.

21

Light in the temples

ED KRUPP

Unlike Sir J. Norman Lockyer, who surveyed megalithic monuments in Britain and ancient temples in Egypt with an eye out for astronomical alignment, Alexander Thom restricted his inquiries to standing stones and their kin and left Egypt alone. He sailed the Hebrides, but not the Nile, and the closest he got to the Great Temple of Amun-Re at Karnak in Egypt was the stone alignments of Carnac in Brittany. Despite his scrupulous avoidance of astronomically-oriented antiquities beyond the borders of north-western Europe, Alexander Thom was, indirectly at least, responsible for the field research on Egyptian temples reported here.

Although I, as many others, was entertained and stimulated by the astronomical interpretations of Stonehenge and other monuments detailed by Gerald Hawkins in the 1960s, it was Alexander Thom's work that prompted me to invest time and effort into the idea of ancient astronomy. When, by chance, I discovered the title *Megalithic Lunar Observatories* (Thom 1971) in a Blackwell's catalogue in 1972, I decided I should like to know what besides Stonehenge, and perhaps Callanish, could have some connection with the sky. Shortly after the book arrived in the mail, the comprehensive, systematic approach taken by Alexander Thom convinced me that I should have a first-hand look at these faraway places with strange sounding names - Long Meg and Her Daughters, Castle Rigg, Ballochroy, and Kintraw. The importance of making reliable measurements in the field was underlined by the care and discipline that marked Thom's data. However his hypotheses about prehistoric astronomy, geometry, and measure may be judged, his effort was serious and substantive. The decades he invested and the field data he obtained persuaded others and me that ancient and prehistoric astronomy might yet comprise a respectable area of study.

In *Megalithic Sites in Britain*, Alexander Thom summarised the practical motivations behind an astronomical sightline to the horizon: 'Apart from ritualistic purposes there are three - time indication, calendar purposes, and studying the moon's movements' (Thom 1967: 93). Some of the stone rings in Britain didn't seem to meet these goals and yet still were oriented celestially. Contrasting those he identified as 'functional, scientific' (Thom 1971) with the ones that are less precise, he tagged the latter 'symbolic, mystical.' The 'symbolic' and the 'mystical' were not really the targets of Thom's research. He was seeking in prehistory the surviving artifacts of a developed practical astronomy. That, however, is not what lights up in the Egyptian temples. The astronomy that beamed into an Egyptian sanctuary probably was 'symbolic, mystical'. But it was Alexander Thom's awesome body of work that sent me down the alignments to find whatever might be lodged within them.

The dawn of archaeoastronomy

Claims for astronomical orientation of Egyptian temples were announced by Lockyer, in November, 1890, in a series of lectures, later published in *Nature* (Lockyer 1891). Lockyer's survey and analysis of sites in Egypt were detailed in *The Dawn of Astronomy* (Lockyer 1894), and in the preface to his book Lockyer mentioned that in Germany Professor H. Nissen had proposed the idea of astronomical temple orientation as early as 1885. Nissen's papers did not reach Lockyer until after he had presented and submitted his own work. Lockyer generously acknowledged Nissen's precedence despite the much greater investment his own field study represented.

After Lockyer made his ideas public, Egyptologists ignored or disputed most of what he had said, and *The Dawn of Astronomy*, after a hesitant appearance upon the horizon of Egyptology, slipped back down into that last perilous hour the sun god Re fords before daybreak. And the idea of astronomical temple alignment remained stranded and out-of-sight for 80 years.

Lockyer believed that many Egyptian temples had been oriented on the rising and setting points of certain stars or on the sun. Much of his effort centered on the Great Temple of Amun-Re at Karnak, near Luxor (Fig. 21.1). This extensive zone of religious architecture is really part of a larger, complicated assemblage of ceremonial lakes, processional avenues, courtyards, obelisks, and more than a dozen temples, all within an area roughly one mile by half a mile. Most of Karnak was a product of the New Kingdom period - 1567-1085 BC, the time of the eighteenth, nineteenth, and

Fig. 21.1. According to Lockyer, the Great Temple of Amun-Re at Karnak was originally oriented to the north-west and the summer solstice sunset. This permitted the sun's last light to stream down the temple axis and into the sanctuary, in the south-east. This access is framed here by the temple's first pylon. The view is to the south-east. (E.C. Krupp)

twentieth dynasties. Amun-Re at this time was the pre-eminent god, and he embodied the power of the sun.

The main axis of Amun-Re's temple at Karnak - all 1996 ft of it - was thought by Lockyer to have aligned with the summer solstice sunset to the north-west and the hills of Thebes, when the temple was first laid out (Fig. 21.2). Having convinced himself that most of the gods of the Egyptian pantheon were personifications of celestial phenomena, Lockyer was not particularly surprised about the approximate orientation of the Great Temple of Amun-Re. Dedicated to the solarised and elevated patron god of Thebes,

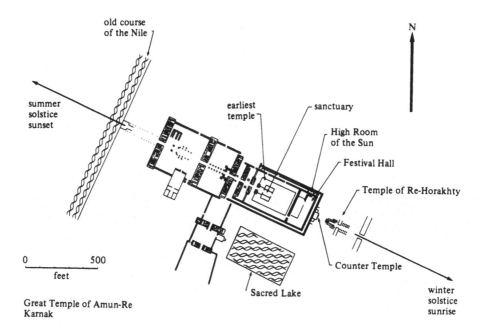

Fig. 21.2. Although Lockyer thought that the main axis of the Great Temple of Amun-Re at Karnak was oriented to the north-west and summer solstice sunset, he recognized that the Temple of Re-Horakhty, built by Ramesses II, opened to the south-east and the winter solstice sunrise. Tuthmosis III interrupted the main axis of the temple when he built his Festival Hall, but a concern for maintaining the link to the winter solstice sunrise may be indicated by the small Counter Temple, built on the south-east side of the temple's main enclosure wall. The Sanctuary of Philip Arrhidaeus was built late in the history of Karnak but is inside the zone of the earliest temple. It is open on both the north-west and south-east sides. (Griffith Observatory: Joseph Bieniasz)

the Great Temple of Karnak opened, more or less, towards the year's northernmost sunset. Lockyer realised, however, that the exact orientation of the axis and the height of the horizon across the river probably would prohibit the sun from entering the conduit of temple pylons and from lighting the darkened sanctuary deep with the temple. Slow change in the obliquity of the ecliptic would, he argued, inevitably reorient the summer solstice sunset with respect to the temple, and accordingly Lockyer arranged for an on-site observation of the 1891 summer solstice sunset by P.J.G. Wakefield, an officer of the Public Works Department in Egypt. Wakefield confirmed the sun's inability to set along the temple axis (Lockyer 1894: 117-118). From the contemporary position of summer solstice sunset, Lockyer calculated how much change in the obliquity of the ecliptic would drop the sun back into place, and with an estimate of the rate of change of the obliquity, he judged the first sun temple at Karnak to have been founded in 3700 BC. There is, however, no archaeological evidence or documentary evidence to carry the temple and the cult of Amun-Re so far back beyond the first dynasties of Egypt. The earliest evidence for activity at Karnak is architectural and textual and dates to about 1900 BC. The ruined area between the sanctuary and the Festival Hall of Tuthmosis III belongs to this period and seems to share the same alignment as the later buildings.

Lockyer mentioned several other possibilities for solar alignment at Thebes. A temple south-east of the main temple shares the same general axis but opens to the south-east, and Lockyer concluded it was oriented to the rising winter solstice sun. This temple was built by Ramesses II (1279-1212 BC) and was named 'Ramesses beloved of Amun is he who listens to the prayers at the main door of the kingdom of Amun'. It is generally known as the 'Enclosure of Amun-Re Horakhty Who Hears Petitions' (Fig. 21.3). An obelisk once stood at the temple's heart and, as a solar symbol, indicated the temple's solar theme. It was, in fact, dedicated to Re-Horakhty, the rising sun. Its obelisk is now in the Piazza di San Giovanni in Laterno, Rome. Just behind this temple, against the main temple's enclosure wall, are the remains of another sanctuary that opened to the south-east. Built by Tuthmosis III (1504-1450 BC), it is sometimes called the Counter Temple, and its alabaster statues, probably representations of Tuthmosis III and Amun, were once flanked by two obelisks. Now only their bases remain.

Two statues on the west bank of the Nile, across the river from Karnak, also face south-east. The Colossi of Memnon are each almost 65 ft tall. Sculpted from sandstone conglomerate, they weigh about a thousand tons apiece, and they once fronted the mortuary temple of Amenophis III (1386-1349 BC). The Colossi are both seated figures of this pharaoh who, according to Lockyer, faced the winter solstice sunrise while its light streamed into the sanctuary of his long-vanished temple.

After Lockyer proposed these and and other Egyptian temple alignments, his work went dormant. Scholars of Egyptian tradition were repelled by the intrusion of an outsider into their realm, and Lockyer's extravagant dates and fast-and-loose mythology didn't draw many of the sceptics back to his bosom. No one paid much attention to temple alignments again until Gerald S.

Fig. 21.3. The main axis of the Temple of Re-Horakhty at Karnak passes south-east through the gate of Nectanebo I. Today this portal frames winter solstice sunrise. (E.C. Krupp)

Hawkins, already having trailed Lockyer to Stonehenge, followed Lockyer's cues in Egypt, too, and described the results of his own analysis of several Egyptian New Kingdom temples (Hawkins 1973; 1974; 1975).

Stalking Lockyer

Relying on the 1932 survey of Karnak by F.S. Richards, Hawkins re-examined the astronomical potential of Karnak's main axis and concluded that Lockyer was wrong.

Sir Norman Lockyer had worked at Stonehenge before me, and also at Karnak. His measurements of the Heel Stone were exact, verifiable, constructive. His survey of the Amon-Re axis was not.

G.S. Hawkins (1973)

It was not Lockyer's survey that was at fault. His azimuth for the main axis agreed with the other determinations. But his interpretation of the axis was flawed. Although Hawkins implied that Wakefield's 1891 attempt to observe summer solstice sunset contradicted Lockyer, Lockyer didn't look at it that way at the time. The sun's failure to align with the axis of the temple in his era convinced him that the first structure of Karnak was aligned with the summer solstice sunset about 56 centuries earlier.

Solar connotations at Karnak

Paul Barguet, a French researcher, studied the Great Temple of Amun-Re with great care, and his report (Barguet 1962) spotlighted the numerous solar and celestial connotations of the temple's architecture and iconography. Barguet was convinced that the axis was really oriented to the south-east, away from the Nile. Hawkins' calculations provided further support: the main axis of the great temple of Amun-Re coincided with the direction of winter solstice sunrise during the eighteenth and nineteenth dynasties.

Hawkins agreed with Lockyer when it came to Ramesses II's Temple of Re-Horakhty: it points to winter solstice sunrise. And Hawkins saw winter solstice significance in an open sanctuary Barguet had called the High Room of the Rising Sun (Fig 21.4). Just north of the Festival Hall of Tuthmosis III, at the south-eastern end of the Great Temple, the High Room of the Sun, as Hawkins called it, appears to have the same orientation as the main temple, and a window on the south-eastern wall permits the winter solstice sunrise to be seen in the chapel. A piece of ritual furniture still sits in the room: a massive alabaster altar in what Barguet called the 'Heliopolitan' style. Its shape looks at first like an eight-pointed star, but actually four *hetep* ('offering') glyphs combine to form the square altar. The four *hetep* points and the four corners create the eight-pointed star (Fig 21.5).

The cult of Re, the primordial solar god, was centered at Heliopolis, in what is today el-Matariya, a suburb on the north side of Cairo. Open temples built for solar service included *hetep* altars and obelisks, solar symbols related to the pyramid and derived from the idea of the primordial mound of creation.

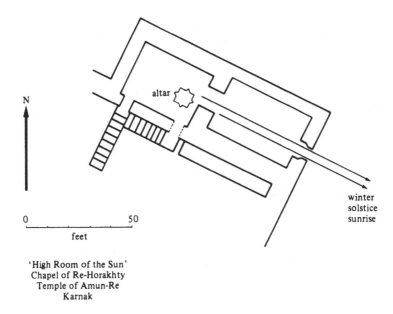

Fig. 21.4. Unconvinced that the Great Temple of Amun-Re was originally targeted on summer solsice sunset, Gerald S. Hawkins suggested that this open-air solar chapel near the Festival Hall of Tuthmosis III was the site for ritual observation of *winter* solstice sunrise. He called this the High Room of the Sun. (Griffith Observatory: Joseph Bieniasz)

Barguet noticed that Karnak's High Room of the Sun is elevated upon a massive pyramid-like construction that was added against the north enclosure wall near the Festival Hall. Texts and reliefs in the High Room allude specifically to the rising sun, and another inscription, in the Festival Hall, also suggested that the High Room might have something to do with the sun:

> One makes for the Festival Hall, horizon of the sky, and climbs there to the 'Place of Combat', secluded place of the majestic spirit, high room of the Ram that crosses the sky; there for him are opened the doors of the horizon [the sanctuary] of the primordial Lord of the Two Lands to see the mystery of Horus shining.

Fig. 21.5. Evidence for a window in the High Room of the Sun, open to the south-east, remains in the ruined frame in the middle of this picture. The *hetep*-glyphs altar is in the foreground. The view is to the south-east, and the gate of Nectanebo I protrudes above the horizon on the right. (Robin Rector Krupp)

Barguet became convinced that the 'secluded place of the majestic spirit, high room of the Ram that crosses the sky' was the High Room of the Sun and thought that the references to combat had something to do with a conflict between the sun and the darkness. In this connection, he considered the presence of a suite dedicated to Sokar, the funerary god originally linked with Memphis, the capital of Egypt during the Old Kingdom (2818-2360 BC). Associated with death, decay, and the darkness of the tomb, Sokar could, in a sense, be considered the night-time, underworld opposite of the reborn rising sun. Barguet saw these two aspects of the cycle of cosmic renewal in the sanctuaries of Karnak's Festival Hall and quoted a text from the Theban tomb of Kheru-ef, steward to Queen Tiy in the eighteenth dynasty. The tomb's wall murals include scenes of the Sed festival. This celebration of the rejuvenation of the pharaoh in his jubilee year took place on the first day of the first month of the first season. In this New Year ritual, the pharaoh demonstrated his physical vigour. By passing the Sed tests, he was re-energised for another bout at kingship. Through his renewal, the world order was re-established and the land itself was reinvigorated. The inscription in the tomb of Kheru-ef suggests that Sokar - the personification of death - was himself revived,

Fig. 21.6. Tuthmosis III's Festival Hall, south-east of the main sanctuary of Amun at Karnak, was probably a mortuary temple. It also commemorates his Jubilee Sed festival and expresses his gratitude to the gods of Egypt for the power and success they had conferred upon him. Its long axis cuts across the main axis of Karnak. The Sed references and the High Room of the Sun highlight the theme of renewal. In this sense, it is similar to Hatshepsut's temple at Deir el-Bahri and to the vanished temple of Amenophis III that once stood behind the Colossi of Memnon. It had a series of Sed festival reliefs, and opened to the winter solstice sunrise. (E.C. Krupp)

resurrected as a reborn rising sun, renewed like the pharaoh on his Jubilee. Tuthmosis III built his Festival Hall (Fig 21.6), with its High Room of the Sun and Sokar chapel, as a mortuary temple that celebrates his Sed Jubilee. The theme of solar renewal is at home there. Hawkins embraced Barguet's interpretation of the High Room of the Sun and saw in its orientation an allusion to the annual renewal of the sun at the winter solstice - another kind of defeat for darkness and chaos. This idea of cosmic restoration is a central theme in Egyptian ritual and belief.

Solar sanctuaries at Abu Simbel

Although Hawkins dismissed nearly all of Lockyer's efforts in Egypt, the idea of solar-aligned temples survived, at least in modified form. Hawkins confirmed the winter solstice orientation of the Colossi of Memnon and verified the solar significance of the rock-hewn temple of Ramesses II at Abu Simbel. There the main axis is not targeted on a solstice sunrise - or on an equinox sunrise either, though this is often claimed. In fact, the temple faces the rising sun when its declination is -9°.6. In our era, this would occur on 18 October and 22 February. At the time of Ramesses II, the October date could have coincided with New Year's Day in the Jubilee year (c. 1250 BC) of Ramesses II (Fig 21.7).

Even though the temples of Abu Simbel were moved to higher ground, above the water level of Lake Nasser, newly formed behind the Aswan High Dam, the original alignments were carefully preserved. Sunlight, then, still beams straight into the temple's inner sanctuary on nearly the same days as it did in the New Kingdom and warms the faces of three of the four statues seated there: Amun-Re, Ramesses II, and Re-Horakhty. On the far left, Ptah, the god who dwells in darkness, is not touched by the sun.

Hawkins noticed another modest chapel on the north end of the temple's main facade. With an altar-platform inside, reached by a short stairway, the chapel is skewed from the main axis of Abu Simbel and faces the winter solstice sunrise. It appears, then, to be a small, open-air solar sanctuary like the High Room of the Sun at Karnak.

Another look at the sun

With these results in mind, I thought it might be fruitful to identify similar architectural components in other Egyptian monuments and to measure their

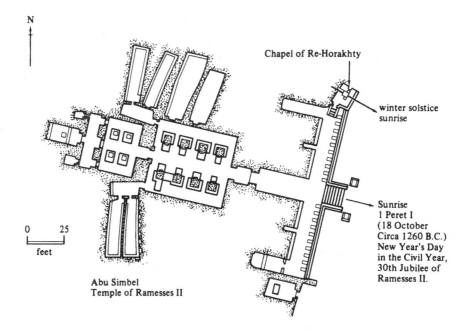

Fig. 21.7. Although the main axis of Ramesses II's rock-cut temple at Abu Simbel is not oriented to the winter solstice, the little Chapel of Re-Horakhty on the north side is. The main axis seems to be linked to the Jubilee celebration of the temple's builder, for light enters its sanctuary on what was probably New Year's Day in that key year. (Griffith Observatory: Joseph Bieniasz)

orientations in the field. Through such a program, other winter solstice chapels might turn up and shed, perhaps, more light on the nature and function of the sanctuaries Hawkins had examined. Because Hawkins relied on Francis Richards' determination of the main axis at Karnak and assumed that the other elements there were set out on the same bearing, I decided to check those orientations in the field as well. Contradictory and incomplete information on the view along the main axis to the north-west prompted me to consider redetermining the azimuth of the main axis and mapping, in detail, the horizon profile visible from the north-western entrance of the inner sanctuary.

On my third visit to Egypt, in August-September 1982, I obtained field data on six structures with potentially astronomical alignments. Axis orientations were measured with a Zeiss theodolite readable to 10 arc seconds. Each alignment was calibrated in the field by making direct observations of the relative azimuth and zenith distance of the sun. Site latitudes and longitudes were obtained from Baines & Málek (1980). A battery-powered digital watch, referenced to WWV time signals in the United States, provided the correct time and permitted reduction of the relative position of the sun to an absolute azimuth through Aveni's (1980) 'azimuth-finder' program modified for the Griffith Observatory Apple II personal computer by Mr Peter Scott. True azimuths, measured from monument foundations or walls and corrected for refraction and for assumed horizon elevation, were given by the same program. In all these cases the true horizon was obstructed by nearby objects (trees or walls). Hawkins estimated Karnak's south-eastern horizon to be $0°.58$ high. On-site inspection showed that the horizons in question are within half a degree of $0°.5$, and a half-degree error in horizon elevation does not affect the broader conclusions of the analysis.

The azimuths were converted to astronomical declinations with a similarly modified version of Aveni's (1980) 'azimuth-declination predictor' program. Finally, the declinations implied by the temple axes were compared to the declination of the sun for the era of each temple's construction. These solar declinations were given by the de Sitter formula for the change in the obliquity of the ecliptic over time. Alignment results are summarised in Table 21.1.

An analysis of the level of error associated with the tabulated results involves consideration of:

(1) original error in the building alignment itself due to errors in sighting the intended celestial target and builders' errors in layout and construction;

(2) error in measuring the sun's position to calibrate measured azimuths - a combination of clock error, sighting error, timing error, scale reading error, and latitude and longitude error;

(3) error in establishing the building's intended axis;

(4) error in measuring the established axis;

(5) error due to uncertainty about horizon elevation; and

(6) error due to uncertainty in the date of the building's construction.

Repeated sun sightings permitted an estimate of the measurement errors involved in calibration of the azimuth. In practice, it was possible to duplicate the calibration of an observed azimuth to ± 16 arc seconds $(= \pm 0°.004)$ at

Table 21.1.

Site and alignment	Assumed date of construction	Azimuth	Declination (assuming 0°.5 horizon elevation, full orb)	Astron-omical event	Error
Karnak Great Temple of Amun-Re main axis	1450 BC	117°.0	-23°.7	winter solstice sunrise (-23° 52′)	+0°.2
Karnak Great Temple of Amun-Re High Room of the sun	1450 BC Tuthmosis III	116°.3	-23°.0	winter solstice sunrise (-23° 52′)	+0°.9
Karnak Great Temple of Amun-Re Counter Temple	1450 BC Tuthmosis III	117°.3	-24°.0	winter solstice sunrise (-23° 52′)	-0°.1
Karnak Great Temple of Amun-Re Temple of Re-Horakhty	1300 BC Ramesses III	117°.5	-24°.0	winter solstice sunrise (-23° 52′)	-0°.1
Deir el-Bahri Mortuary Temple of Hatshepsut Solar Sanctuary	1450 BC Hatshepsut	115°.8	-22°.5	winter solstice sunrise (-23° 52′)	+1°.4
Abu Ghurab Sun Temple of Neuserre' main axis	2400 BC Neuserre'	91°.8	-1°.5	'equinox' sunrise (0°)	-1°.5

worst. Error in the estimate of horizon elevation was no greater than 0°.25. At the general latitude in question for most alignments considered here, such an error is equivalent to ± 0°.1 error in the declination determined for the alignment. Although uncertainty and controversy characterize the dating of

Egyptian kings and dynasties, an error of two centuries generates an error of only 0°.02 in declination. All of these sources of error then are relatively small, and the chief reservoir of error includes establishing and measuring the original axis intended by the builders. The magnitude of these errors varies with each site. For example, the main axis of the Great Temple of Amun-Re at Karnak was fixed with eight points obtained by bisecting the openings of pylons III, IV, V, and VI, and the width of the sanctuary. Pins from the earlier survey remain in place at Karnak, and their positions, too, were incorporated into the analysis. The standard deviation of the azimuths established by all of these points is 0°.1. But at the Sun Temple of Neuserre' at Abu Ghurab, the azimuth of two separate components - the south edge of altar's foundation and the south foundation wall of the obelisk platform - differed by a degree. Greater dispersion than at Karnak, by a factor of 2 to 3, also implies that the true line there is less easy to obtain.

The Great Temple of Amun-Re at Karnak

Within the errors of measurement, my determination of the orientation of Karnak's main axis, 117°.0 ± 0°.1, is the same as Richards' (116°.9). Wakefield's earlier survey (116°.4) came close to the same value. Toward the south-east, then, the main axis coincides reasonably well with winter solstice sunrise in the eighteenth dynasty. In the opposite direction, to the north-west, the azimuth is 297°, and the measured horizon elevation is 2°.6, a value also reported by Lockyer (c. 2°.5). The declination given by the line to the north-west misses the summer solstice sun now. It missed it in 1891, too, as Lockyer reported, and it missed it in 1450 BC, a time of extensive construction at Karnak. The summer solstice sun could then have been found at declination ± 23°.86. The line matches declination +25°.8, about 2 degrees too far north for full orb and nearly 1.5 degrees too far north for last gleam. I measured the width of the aperture, defined by the most distant pylon, to be 50 arc minutes (0°.85) at the inner sanctuary. This is equivalent to 45 arc minutes (0°.75) in declination. Accordingly, the left side (south) of the opening is still more than a degree too far north to accommodate the last gleam of the summer solstice (Fig. 21.8).

This outermost pylon, the Great Pylon, was built, however, in the Late Period, perhaps in the thirtieth dynasty, and according to Wakefield it does not line up well with the main axis. Dropping back to the second pylon, which was built during the New Kingdom, by Haremhab and Ramesses I (1321-1291 BC), puts the south edge of the 'window' only 0°.07 (4 arc minutes)

Fig. 21.8. Pylons and columns of the Great Temple of Amun-Re at Karnak direct the view from the Sanctuary of Philip Arrhidaeus to the hills of Thebes in the north-west. At no time since the construction of buildings here during the Middle Kingdom would the setting sun manage to slip into the gap. The left side of the frame is too far north. (E.C. Krupp)

closer to the summer solstice - not enough to hit it. Summer solstice sunset never touched the Great Temple's sanctuary. Lockyer, of course, knew this, but his certainty about the solar meaning of the main axis led him to conclude that the temple was first laid out when summer solstice sunset did match Karnak's main avenue, in 3700 BC. The date makes no sense at Karnak, and that is one reason why Egyptologists had no time for him.

Hawkins measured the height of the north-western horizon from Karnak and estimated the height of the south-eastern horizon from contour maps. He adopted Richards' azimuth, and his examination of a photogrammetric survey conducted by the Franco-Egyptian Research Centre led him to conclude that the High Room of the Sun and the Temple of Re-Horakhty share the same orientation as the main axis of the temple. My own measurements of these

Fig. 21.9. Hatshepsut's mortuary temple is built against the vertical cliffs of Deir el-Bahri. Its solar sanctuary is on the upper level, on the right, behind the scaffolding. (E.C. Krupp)

components of Karnak and of the Counter Temple indicate agreement with the main axis to within about ± 0°.6. Variation from the main axis like this is not surprising. None of the alignments needs greater accuracy than it has to funnel symbolic light into the sanctuaries.

Hawkins, however, found an astronomical reason to support Barguet's contention that the temple's orientation was really towards the south-east, away from the Nile, and on the rising sun. Ritual observation of winter

Fig. 21.10. Inside the solar chapel at Deir el-Bahri, a large altar occupies most of the available space. This altar differs from the heliopolitan alabaster altars in the High Room of the Sun at Karnak and in the Sun Temple of Neuserre' at Abu Ghurab, but it is similar to the altar in the solar sanctuary in the mortuary temple of Sethos I at Qurna. The two styles may be related to the primary function of the structures in which they are located. Solar chapels in the mortuary temples perhaps served a different purpose from the shrines dedicated to the service of Re. The niche on the back (north-west) wall is on the same axis as the altar. (Robin Rector Krupp)

solstice sunrise is consistent with the measured orientations of the High Room of the Sun, the Counter Temple, and the Temple of Re-Horakhty. Inscriptions and architectural iconography also support that interpretation.

The solar sanctuary of Hatshepsut's mortuary temple at Deir El-Bahri

Across the river from Karnak, about a kilometre south - and one ridge over - from the Valley of the Kings, Egypt's first woman pharaoh, Hatshepsut, built her mortuary temple (Fig. 21.9). This monument, in the bay of Deir el-Bahri,

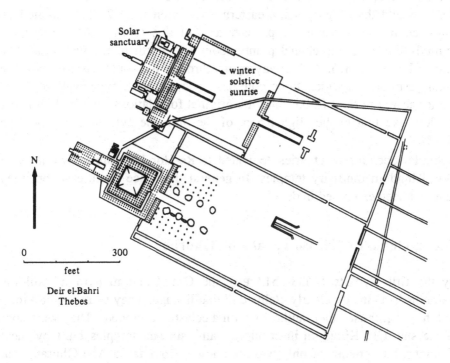

Fig. 21.11. Oriented toward the south-east, Hatshepsut's mortuary temple seems to have been aimed toward the winter solstice sunrise. The Middle Kingdom temple of Mentuhotpe II flanks Hatshepsut's temple on the south and may have been oriented toward winter solstice sunrise in that era. (Griffith Observatory: Joseph Bieniasz)

is regarded as a jewel of New Kingdom religious architecture, and its design appears to have been inspired, at least in part, by the mortuary temple of eleventh dynasty, Middle Kingdom pharaoh Mentuhotpe II (Nebhepetre'). Once crowned with a shrine-shaped building, these ruins lie next to Hatshepsut's temple.

Identification of an open chapel on the north side of the upper terrace as a solar sanctuary caught my eye (Fig. 21.10). An inscription on the altar dedicates the room to Re-Horakhty of Heliopolis, and a text above its doorway declares 'Amun is holy in the horizon'. Ten steps on the north-western side of the altar pointed the priest who climbed them toward the realm of the rising sun. On the north-western wall, behind the altar and on its axis, a small niche may still be seen, and it is likely it once held a statue of Hatshepsut. With the altar in the centre of the room, the chapel resembles, on a larger scale, the winter solstice shrine at Abu Simbel. The various solar elements and the roughly south-eastern orientation (Fig. 21.11) revealed by maps led me to measure the primary axis of this room. With a measured azimuth of 115°.8, the chapel points about 1°.5 north of the winter solstice sunrise line (full orb). A doorway opens through the south-eastern wall to a small vestibule, in which three columns once stood. A final wall seals off further access to the SE, but only the original foundations remain. If it had a window like that in the High Room of the Sun, the evidence is gone (Fig 21.12).

Similar open-air sanctuaries dedicated to Re-Horakhty are known at other New Kingdom mortuary temples. In general they face south-east, but they have not been examined in detail.

The sun temple of Neuserre' at Abu Ghurab

By the fifth dynasty (2638-2513 BC), the Old Kingdom pyramid-building pharaohs had already clearly claimed divine lineage. They were sacred kings, and they legitimised their authority with a celestial argument. They were sons of the sun. Old Kingdom inscriptions name six sun temples built by these pharaohs, but remains of only two are known. Both are at Abu Ghurab, near the four pyramids of Abusir and about 18 miles south of Cairo.

The one that is best preserved was constructed in stone by Neuserre' (2591-2580 BC), and the base of the thick, massive solar obelisk that dominated the temple still stands in the main court, as does an altar just east of the base (Fig. 21.13). This altar, like the obelisk, derives from the solar cult of Heliopolis, and the four-sided altar is really a set of four *hetep* glyphs with a stone disc at

Fig. 21.12. Although the solar sanctuary at Deir el-Bahri is solstitially oriented, the view to the south-east is blocked by a modern wall. The base of the reconstructed original wall still remains and shows no sign of a door. Whether it held a window is conjectural. The altar is in the foreground. (Robin Rector Krupp)

Fig. 21.13. Only the base of the obelisk of the Sun Temple of Neuserre' at Abu Ghurab remains today. The view is west, from the temple's main entrance. Visitors are standing by the altar in the main courtyard. (Robin Rector Krupp)

the centre (Fig. 21.14). The altar in Karnak's High Room of the Sun extends this same tradition a thousand years in time to the New Kingdom. A solar boat built of brick and chapel reliefs depicting Nile Valley seasons further emphasizes the solar character of the temple. Its connections with the sun inspired on-site measurements of its orientation. Two components of the temple were surveyed - the south edge of the *hetep* altar and a foundation wall on the south side of the obelisk's base. For a full orb sunrise, the altar's azimuth (92°.5) indicated a declination of -2°.5. The obelisk platform, oriented toward azimuth 91°.2, gives a declination of -1°.7. Both results indicate nearly cardinal alignment, and this is consistent with the Old Kingdom emphasis on cardinal orientation (Fig. 21.15). East and the sunrise are, by symbol and direction, the targets of the main temple at Abu Ghurab. A

Fig. 21.14. The Fifth Dynasty pharaoh Neuserre' sponsored this sanctuary dedicated to the sun or Re. Such temples are said to follow the pattern established at Heliopolis, the original centre of the cult of Re. For all practical purposes, the main component of the temple is cardinally aligned, like most Old Kingdom religious and funerary architecture. A square altar in the main courtyard and east of the huge, stocky obelisk aligns with the main entrance and suggests that the entire temple was oriented to the east. (Griffith Observatory: Joseph Bieniasz)

causeway led north-east from the temple entrance, on the east, down to the valley-portal, nearer to and oriented to the river, in contrast with the court of the sun on the desert plateau above.

Shedding light on the temples

Orthodox interpretation of Egyptian temple architecture sees the Nile River as the motivating factor for most temple orientation. Some temples are also

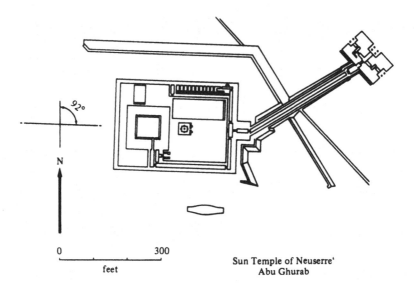

Fig. 21.15. The monumental, cardinally oriented alabaster altar in the main courtyard of the Sun Temple of Neuserre' at Abu Ghurab includes a central disc which is surrounded by four *hetep*, or 'offering', glyphs. The view is east, from the remains of the obelisk and through the temple's main entrance. (Robin Rector Krupp)

oriented to other temples. The solar connotations of certain temples are, however, accepted, and the idea of a general alignment towards the rising sun for some of these temples is not particularly controversial. Textual references to a particular sunrise, a calendrically and seasonally significant sunrise, remain clouded, however.

Karnak, on the other hand, does appear to confirm in stone what is absent among the inscriptions. Certainly it opens to the north-west and to the Nile. Subsequent building extended the temple farther towards the north-west, and the Great Temple of Amun-Re at Karnak is, in that sense, oriented towards the Nile. But its general alignment seems to have been inspired by the winter solstice sunrise, in the opposite direction, south-east. The sanctuary has doorways on both ends and could accommodate both orientations. Winter solstice sunrise is consistent with the iconography and texts that are present despite the absence of explicit references.

In temperate latitudes, the winter solstice is frequently the time of chaos and death that marks the end of one cycle and the start of another. The Egyptians, however, do not seem to have thought of the winter solstice as the time of the year's rebirth. Egypt's seasons are the seasons of the Nile, and the winter solstice plays no obvious role in them. In fact, Parker (1950) has argued that the last month in the Egyptian lunar calendar was related by a variant name to 'the birth of Re'. This times the sun's rebirth with the heliacal rising of Sirius and, perhaps, the *summer* solstice.

Ptolemaic *Building Texts* at Edfu refer to foundation and construction of the mythical first solar temple and call it the 'Place-for-Piercing'. The name comes from the story of the defeat of the serpent enemy of the sun by the primordial Falcon Sun God. The temple, then, commemorates the world's creation and ordering through the snake's defeat in combat with the sun. The Karnak text Barguet associated with the High Room of the Sun refers to the 'Place of Combat' and looks like a reference to the sun's battle.

The piercing or crushing of the serpent has an analogue in the New Kingdom pharaonic tomb texts sometimes called *The Book of What Is in the Netherworld*. In the seventh of the night's twelve hours, the serpent Apep is sliced and pierced and defeated in the Netherworld by Re and his allies. Schematically, the seventh hour is the first realm past the turning point, the halfway mark between the sixth and the seventh. The sun enjoyed a victory against chaos and death in a battle at the very deepest and most remote darkness. In that sense, the seventh hour is a cyclic return like the winter solstice. It is a combat zone in which the sun emerges victorious, a major battle on the road back to the eastern horizon and sunrise.

Alexandre Piankoff (1954a; 1954b) recognized a cycle of birth, death, and rebirth in the symbolism of mythological papyri. He equated sunrise with the east and sunset with the west, of course, but north and south played a part in this scheme as well. North is identified with ascent and the sky. South means descent and the Netherworld and in particular may be bonded to the midpoint of the journey through that realm. South is equivalent to the transition from the sixth to the seventh hour, which was linked to the idea of the transformation from death to life. Sokar was revivified at this junction.

Solar chapels are consistently found on the north sides of the temples. Mortuary cult or Sokar chapels in general occur on the south. The directional placement of these features may reflect a broader use of distinctions between north and south. The two turning points of the sunrise in the north-east and in the south-east would fit comfortably into this picture. If the north-east be a place of renewal, the south-east may represent a place of light's triumph over darkness. This victory, commemorated in what appears to be the symbolic solstitial alignment of New Kingdom solar chapels, reaffirms the pharaoh's close bond with Re. In the mortuary temples, these shrines seem to have something to do with the transformation of the deceased pharaoh's soul. The Netherworld journey of the sun represents the same risky passage undertaken by the soul as it joins the sun and in the sun's company helps defeat the forces of darkness. Royal mortuary temples and the rituals they housed helped sustain order in a world threatened by chaos. If the winter solstice symbolized the crucial battle in that eternal conflict, its incorporation into temple architecture is less mysterious.

The Great Temple of Amun-Re harboured daily and seasonal rituals performed by the Theban priesthood for a similar reason. Their magical renewal of the god's vitality preserved the world's order. The pharaoh, too, provided stability and life for the land. Winter solstice sunrise at Karnak put chaos on notice that Re had defeated Apep.

If the solstitial component of New Kingdom mortuary and cult temples is real, it sets them apart from the Old Kingdom sun temples, which were probably intended to be cardinal. Each morning Re emerged from a nightly ordeal in the Netherworld on the eastern horizon, but the winter solstice played no part in the myth, or at least in the architecture. Something important changed between the Old Kingdom and the New.

Solar alignments in Egyptian temples still give us some problems. But the connection between sunlight and sanctuary is clear enough. Even the pylon - the massive wall that fronts the temple - reiterates the sun's role. In Egyptian temples, the pylon was called the 'Luminous Mountain Horizon of Heaven,' and its two towers made a notched skyline out of the pylon profile. The winged sun emblem occupied the spot above the main door, on the main axis, between the two 'mountain peaks' created by the pylon towers. The sun, then, sat in a notch upon the 'luminous mountain horizon in heaven' and solidified in a monumental symbol what was probably an original source of Egyptian concepts of celestial order: practical observation of the sun on the horizon. Professor Thom argued that the horizon was the fundamental astronomical tool in prehistoric Britain. The horizon-lodged sun, or *akhet*, that shows up in Egyptian hieroglyphics and on temple façades is a hint that someone did watch that horizon - at least in Egypt - for signs of cosmic order. And they put those signs into 'symbolic, mystical' temples whose purpose - the visible display of social and political order - is as practical as the calendar.

Acknowledgements

I am grateful to the Egyptian Department of Antiquities for its co-operation and permission to measure sites in upper Egypt. The American Research Center in Egypt was also extremely helpful. A.R.C.E. co-ordinated necessary connections and permitted me to use its theodolite. Dr James Allen, at that time Executive Director of the A.R.C.E., took a personal interest in the work and made my task easier. Field measurements were undertaken in conjunction with a University of California, Los Angeles, University Extension field study tour I led to Egypt in 1982, and tour members Mr Robert A. Iehl and Dr David S. Gold assisted with some of the surveys. Ethan Hembree Krupp and Robin Rector Krupp also participated in the field surveys. Finally, Mr Peter Scott played the major role in the reduction of the data.

Abbreviations

The following abbreviations are used in the bibliography that follows, in the list of publications by Alexander Thom which appears in Chapter 3, and elsewhere in the volume.

AA	*Archaeoastronomy* (supplement to *JHA*)
AAB	*Archaeoastronomy* (Bulletin, later Journal, of the Center for Archaeoastronomy, University of Maryland)
Am. Ant.	*American Antiquity*
Ant.	*Antiquity*
Ant. J.	*Antiquaries Journal*
Arch. Exc.	*Archaeological Excavations*
Arch. J.	*Archaeological Journal*
ARCR	*Aeronautical Research Council Report*
BAR	British Archaeological Report
CA	*Current Anthropology*
CBA	Council for British Archaeology
CFCB	*Caithness Field Club Bulletin*
CIWP	Carnegie Institute of Washington Publication
Comp. J.	*Computer Journal*
Curr. Arch.	*Current Archaeology*
DASP	*Devon Archaeological Society Proceedings*
D & E	*Discovery and Excavation in Scotland*
GAJ	*Glasgow Archaeological Journal*
Heb. Nat.	*Hebridean Naturalist*
HMSO	Her Majesty's Stationery Office
Ind.	*Indiana*
JBAA	*Journal of the British Astronomical Association*

JHA	*Journal for the History of Astronomy*
JKHAS	*Journal of the Kerry Historical and Archaeological Society*
JRAI	*Journal of the Royal Anthropological Institute*
JRSS	*Journal of the Royal Statistical Society*
Lat. Stud.	*Lateinamerika Studien*
MG	*Mathematical Gazette*
NGR	National Grid Reference
NMRS	National Monuments Record of Scotland
NRRS	*Notes and Records of the Royal Society of London*
Phil. J.	*Philosophical Journal*
PPS	*Proceedings of the Prehistoric Society*
PRS	*Proceedings of the Royal Society*
PSA	*Proceedings of the Society of Antiquaries of London*
PSAS	*Proceedings of the Society of Antiquaries of Scotland*
PTRS	*Philosophical Transactions of the Royal Society of London*
R.A.E.	Royal Aircraft Establishment
RCAHMS	Royal Commission on the Ancient and Historical Monuments of Scotland
RILKO	Research into Lost Knowledge Organisation
RMARC	*Reports and Memoranda of the Aeronautical Research Council*
SAF	*Scottish Archaeological Forum*
SAR	*Scottish Archaeological Review*
Sci. Arch.	*Science and Archaeology*
TAMS	*Transactions of the Ancient Monuments Society*
TCWAAS	*Transactions of the Cumberland and Westmorland Antiquarian and Archaeological Society*
TDGNHAS	*Transactions of the Dumfries and Galloway Natural History and Antiquarian Society*
TGAS	*Transactions of the Glasgow Archaeological Society*
TISS	*Transactions of the Inverness Scientific Society*
UJ Arch.	*Ulster Journal of Archaeology*
VA	*Vistas in Astronomy*

Bibliography

Anderson, J.

(1886). *Scotland in pagan times*. Edinburgh.

Anderson, W.D.

(1915). Some recent observations at the Keswick stone circle. *TCWAAS* **15**, 99-112.

Angell, I.O.

(1976). Stone circles: Megalithic mathematics or neolithic nonsense? *MG* **60**, 189-93.

(1977). Are stone circles circles? *Sci. Arch.* **19**, 16-19.

Ashmore, P.J.

(1983). *Excavations at Callanish Stone Setting, May 1980*. Scottish Development Department (Historic Buildings and Monuments Commisssion, Scotland). Edinburgh.

Atkinson, R.J.C.

(1966). Moonshine on Stonehenge. *Ant.* **40**, 212-16.

(1968). *Review of* Thom (1967). *Ant.* **42**, 77-8.

(1972). Burial and population in the British Bronze Age. *In* Lynch & Burgess (1972), 107-16.

(1981). Comments on the archaeological status of some of the sites. *Appendix 4.2 to* Ruggles (1981a).

Aubrey, J.

(1665-93). *Monumenta Britannica* (2 vols.). Repr. 1980, 1982, annotated by R. Legg, ed. J. Fowles, Sherborne.

Aveni, A.F.

(1975a) (ed.). *Archaeoastronomy in pre-Columbian America*. Austin.

(1975b). Possible astronomical orientations in ancient Mesoamerica. *In* Aveni (1975a), 163-90.

(1980). *Skywatchers of ancient Mexico.* Austin & London.

(1982) (ed.). *Archaeoastronomy in the New World.* Cambridge.

(1986). The Nazca lines: Patterns in the desert. *Archaeology* **39**, no. 4, 32-9.

Aveni, A.F. & Gibb, S.

(1976). On the orientation of pre-Columbian buildings in central Mexico. *Am. Ant.* **41**, 510-17.

Aveni, A.F. & Hartung, H.

(1979). The cross petroglyph: An ancient Mesoamerican astronomical-calendrical symbol. *Ind.* **6**, 37-54.

(1982a). New observations of the pecked cross petroglyph. *Lat. Stud.* **10**, 25-41.

(1982b). Note on the discovery of two new pecked cross petroglyphs. *AAB* **5**, 21-3.

(1985). Las Cruces Punteadas en Mesoamérica. *Cuadernos de Arquitectura Mesoamericana* **4**, 3-12.

Aveni, A.F., Hartung, H. & Buckingham, B.

(1978). The pecked cross symbol in ancient Mesoamerica. *Science* **202**, 267-79.

Aveni, A., Hartung, H. & Kelley, J.C.

(1982). Alta Vista (Chalchihuites), Astronomical implications of a Mesoamerican ceremonial outpost at the Tropic of Cancer. *Am. Ant.* **47**, 316-35.

Bailey, M.E., Cooke, J.A., Few, R.W., Morgan, J.G. & Ruggles, C.L.N.

(1975). Survey of three megalithic sites in Argyllshire. *Nature* **253**, 431-33.

Baines, J. & Málek, J.

(1980). *Atlas of ancient Egypt.* New York.

Baity, E.C.

(1973). Archaeoastronomy and ethnoastronomy so far. *CA* **14**, 389-449.

Barber, J.W.

(1973). The orientation of the recumbent-stone circles of the south-west of Ireland. *JKHAS* **6**, 26-39.

(1978). The excavation of the holed-stone at Ballymeanoch, Kilmartin, Argyll. *PSAS* **109**, 104-11.

Barguet, P.

(1962). *Le Temple d'Amon-Rê à Karnak.* Cairo.

Barnatt, J. & Moir, G.

(1984). Stone circles and megalithic mathematics. *PPS* **50**, 197-216.

Barnatt, J. & Pierpoint, S.

(1983). Stone circles: observatories or ceremonial centres? *SAR* **2**, 101-15.

Barnett, V.

(1982). *Comparative statistical inference*. Chichester & New York.

Bartholomew

(1972). *Gazetteer of the British Isles*. Edinburgh.

Beckensall, S.

(1983). *Northumberland's prehistoric rock carvings*. Rothbury.

Bennett, W. & Zingg, R.

(1935). *The Tarahumara, an Indian tribe of northern Mexico*. Chicago.

Bersu, G.

(1940). King Arthur's Round Table. Final report. *TCWAAS* **40**, 169-206.

Blom, F.

(1924). Archaeology: Report of Mr. Frans Blom on the preliminary work at Uaxactún. *Carnegie Institute of Washington Yearbook* **23**, 217-19.

Bowen, E.G.

(1972). *Britain and the western seaways*. London.

Boye, V.

(1896). *Fund af egekister fra bronzealderen i Danmark*. Copenhagen.

Branigan, K.

(1976). *Prehistoric Britain: An illustrated survey*. Bourne End.

Brennan, M.

(1983). *The stars and the stones: Ancient art and astronomy in Ireland*. London.

Broadbent, S.R.

(1955). Quantum hypotheses. *Biometrika* **42**, 45-57.

Bunch, B. & Fell, C.I.

(1949). A stone axe factory at Pike of Stickle, Great Langdale, Westmorland. *PPS* **15**, 1-20.

Burl, H.A.W.

(1976a). *The stone circles of the British Isles*. New Haven & London.

(1976b). Intimations of numeracy in the Neolithic and Bronze Age societies of the British Isles (c. 3200-1200 BC). *Arch. J.* **133**, 9-32.

(1979).   *Prehistoric Avebury*. New Haven.

(1981b).   *Rites of the Gods*. London.

(1983).   *Prehistoric astronomy and ritual*. Aylesbury.

(1985a).   Stone circles: The Welsh problem. *CBA Report* **35**, 72-82.

(1985b).   *Megalithic Brittany: A guide*. London.

Burl, H.A.W., MacKie, E.W., & Selkirk, A.

(1970).   Stone circles again. *Curr. Arch*. **2** (no. 12), 27-8.

Camden, W.

(1695).   *Britannia*. London.

Campbell, M. & Sandeman, M.

(1961).   Mid-Argyll: A field survey of the historic and prehistoric monuments. *PSAS* **95**, 1-125.

Chart, D.A.

(1940). (ed.). *A preliminary survey of the ancient monuments of Northern Ireland*. Belfast.

Chavero, A.

(1886).   La Piedra del Sol. *Annales Museo Nacional de México* **3**, 1-26.

Childe, V.G.

(1940).   *Prehistoric communities of the British Isles*. 1st edn. London.

Chiu, B.C. & Morrison, P.

(1980).   Astronomical origin of the offset street grid at Teotihuacan. *AA* **2**, S55-64.

Clare, T.

(1975).   Some Cumbrian stone circles in perspective. *TCWAAS* **75**, 1-16.

Clark, J.G.D.

(1948).   *Prehistoric England*. 4th edn. London.

Clarke, D.L.

(1970).   *Beaker pottery of Great Britain and Ireland*. 2 vols. Cambridge.

Clough, T.H. McK.

(1968).   The Beaker period in Cumbria. *TCWAAS* **68**, 1-21.

Codex Fejérváry-Mayer. Free Public Museum, Liverpool. 12014 (HMAI Census No. 118). Codices Select: XXVI, Graz (1971).

Coggins, C.

(1980).   Some political implications of a four-part figure. *Am. Ant*. **45**, 727-39.

Collingwood, R.G.

(1933).   An introduction to the prehistory of Cumberland, Westmorland and Lancashire north of the Sands. *TCWAAS* **33**, 163-200.

(1938).   King Arthur's Round Table. Interim report on the excavations of 1937. *TCWAAS* **38**, 1-31.

Cooke, J.A., Few, R.W., Morgan, J.G. & Ruggles, C.L.N.

(1977).   Indicated declinations at the Callanish megalithic sites. *JHA* **8**, 113-33.

Cowan, T.M.

(1970).   Megalithic rings: Their design construction. *Science* **168**, 321-5.

Cowling, E.T.

(1940).   A classification of west Yorkshire 'cup and ring' stones. *Arch. J.* **97**, 115-24.

Cowper, H.S.

(1934).   Unrecorded and unusual types of stone implements. *TCWAAS* **34**, 91-100.

Critchlow, K.

(1979).   *Time stands still.* London.

Crone, A.

(1983).   The Clochmabanestane, Gretna. *TDGNHAS* **58**, 16-20.

Cummins, W.A.

(1974).   The Neolithic stone axe trade in Britain. *Ant.* **68**, 201-5.

(1980).   Stone axes as a guide to neolithic communications and boundaries in England and Wales. *PPS* **46**, 45-60.

Curtis, G.R.

(1979).   Some geometry associated with the standing stones of Callanish. *Heb. Nat.* **3**, 29-40.

Davidson, D.A.

(1979).   The Orcadian environment and cairn location. *In* Renfrew (1979), 7-20.

Davidson, D.A. & Jones, R.L.

(1985).   The environment of Orkney. *In* Renfrew (1985), 10-35.

Davies, M.

(1945).   Types of megalithic monuments of the Irish Sea and north Channel coastlands: A study in distributions. *Ant. J.* **25**, 125-44.

(1946). The diffusion and distribution pattern of the megalithic monuments of the Irish Sea and north Channel coastlands. *Ant. J.* **26**, 38-60.

Davis, A.

(1983). The metrology of cup and ring carvings near Ilkley in Yorkshire. *Sci. Arch.* **25**, 13-30.

Defoe, D.

(c. 1700). *A tour through the whole island of Britain.* London.

Dover, W.K.

(1882). Excursions and proceedings, 5 October, 1882. *TCWAAS* **6** (old series), 505.

Dow, J.

(1967). Astronomical orientations at Teotihuacan: A case study in Astroarchaeology. *Am. Ant.* **32**, 326-34.

Dymond, C.W.

(1880). Gunnerkeld stone circle. *TCWAAS* **4** (old series), 537-40.

(1881). A group of Cumberland megaliths. *TCWAAS* **5** (old series), 39-57.

(1891). Mayburgh and King Arthur's Round Table. *TCWAAS* **11** (old series), 187-219.

(1902). An exploration of *Sunken Kirk*, Swinside, Cumberland, with incidental researches in its neighborhood. *TCWAAS* **2**, 53-76.

Eddy, J.

(1974). Astronomical alignment of Big-Horn medicine wheel. *Science* **188**, 1035-43.

Evans, I.H.

(1977). *Brewer's dictionary of phrase and fable.* London.

Fell, C.I.

(1972). *Early settlement in the Lake Counties.* Clapham, Yorks.

Fenton, A. & Pálsson, H.

(1984) (eds.). *The northern and western Isles in the Viking world.* Edinburgh.

Fletcher, W.

(1958). Grey Croft stone circle, Seascale, Cumberland. *TCWAAS* **57**, 1-8.

Folan, W.

(1978). San Miguel de Huamango: Un Centro Tolteca-Otomi. *Boletín de la Escuela de Ciencia y Antropología, Univ. Yucatán (Méx.)* **32**, 326-34.

Folan, W. & Ruíz, A.

(1980).   The diffusion of astronomical knowledge in greater Mesoamerica: The Teotihuacan-Cerrito de las Campaña-Chalchihuites-southwest connection, *AAB* **3**, 20-5.

Fraser, D.

(1983).   *Land and society in neolithic Orkney.* Oxford (BAR 117) (2 vols.).

Freeman, P.R.

(1976).   A Bayesian analysis of the megalithic yard. *JRSS* **A139**, 20-55.

Freeman, P.R. & Elmore, W.

(1979).   A test for the significance of astronomical alignments. *AA* **1**, S86-96.

Freer, R. & Myatt, L.J.

(1982).   The multiple stone rows of Caithness and Sutherland, I. *CFCB* **3**, 58-67.

(1983).   The multiple stone rows of Caithness and Sutherland, II. *CFCB* **3**, 120-34.

Freer, R. & Quinio, J.L.

(1977).   The Kerlescan Alignments. *JHA* **8**, 52-4.

Gelling, M.

(1978).   *Signposts to the past. Place-names and the history of England.* London.

Giot, P.R.

(1976).   Dolmens et menhirs, le phénomène mégalithique en France. *In* Guilaine (1976), 202-10.

(1979).   La vie spirituelle au Néolithique. *In* Giot, L'Helgouach, & Monier (1979), 375-440.

(1983a).   *Les alignements de Carnac.* Rennes.

(1983b).   The megaliths of France. *In* Renfrew (1983), 18-29.

Giot, P.R., L'Helgouach, J. & Monier, J.L.

(1979) (eds.).   *Préhistoire de la Bretagne.* Rennes.

Gladwin, P.F.

(1978).   Discoveries at Brainport Bay, Minard, Argyll; an interim report. *The Kist* **16**, 1-5.

(1985).   *The solar alignment at Brainport Bay, mid Argyll.* Ardrishaig.

Glob, P.V.

(1967).   *Danish prehistoric monuments.* London.

(1974). *The mound people*. London.

Gomme, G.L.

(1886) (ed.). *The Gentleman's Magazine library. Archaeology* (2 parts). London.

Gourlay, R.

(1975). *A field survey in the Loch Rimsdale area*. Glasgow.

Grijalba, V.

(1984). Tepepulco. *Cuadernos de Arquitectura Mesoamericana* **2**, 41-6.

Grossman, N.

(1970). Megalithic rings. *Science* **169**, 1228-9.

Guilaine, J.

(1976) (ed.). *La préhistoire française. Tome II. Les civilisations Néolithiques et protohistoriques de la France*. Paris.

Gunn, G.

(1915). The standing stones of Caithness. *TISS* **7**, 337-60.

Hadingham, E.

(1975). *Circles and standing stones*. London.

Hawkes, J.

(1967). God in the machine. *Ant*. **41**, 174-80.

Hawkins, G.S.

(1966). *Astro-Archaeology*. Cambridge, Massachusetts.

(1973). *Beyond Stonehenge*. New York.

(1974). Astronomical alignments in Britain, Egypt, and Peru. *PTRS* **A276**, 157-67.

(1975). Astroarchaeology: The unwritten evidence. *In* Aveni (1975a), 131-62.

Hawkins, G.S. & White, J.B.

(1965). *Stonehenge decoded*. London.

Heggie, D.C.

(1972). Megalithic lunar observatories: An astronomer's view. *Ant*. **46**, 43-8.

(1981). *Megalithic science*. London.

(1982) (ed.). *Archaeoastronomy in the Old World*. Cambridge.

Henshall, A.S.

(1963). *The chambered tombs of Scotland, 1*. Edinburgh.

(1972). *The chambered tombs of Scotland, 2*. Edinburgh.

HMSO

(1983). *The astronomical almanac for the year 1983*. London.

Hogg, R.

(1972). Factors which have affected the spread of early settlement in the Lake counties. *TCWAAS* **72**, 1-35.

Hole, C.

(1978). *A dictionary of British folk customs*. London.

Hoyle, F.

(1966). Speculations on Stonehenge. *Ant.* **40**, 262-76.

Iwaniszewski, S.

(1982). New pecked cross designs discovered at Teotihuacan. *AAB* **5**, 22-3.

Jope, E.M.

(1966) (ed.). *An archaeological survey of Co. Down*. Belfast.

Jope, E.M. & Preston, J.

(1953). An axe of stone from Great Langdale, Lake District, found in County Antrim. *UJ Arch.* **16**, 31-6.

Kendall, D.G.

(1974). Hunting quanta. *PTRS* **A276**, 231-66.

Kermode, P.M.C. & Herdman, W.A.

(1914). *Manks antiquities*. Liverpool.

Kosok, P.

(1965). *Life, land & water in ancient Peru*. New York.

Krupp, E.C.

(1983). *Echoes of the ancient skies: The astronomy of lost civilizations*. New York.

Kuhn, T.

(1962). *The structure of scientific revolutions*. Chicago.

Lambrick, G.

(1983). *The Rollright Stones*. Oxford.

Laufer, B.

(1906) (ed.). *Anthropological papers written in honor of Franz Boas*. New York.

Lees, D.

(1984). The Sanctuary: A Neolithic calendar? *Bulletin of the Institute of Mathematics and its Applications* **20**, 109-14.

Le Roux, C.T.

(1979). Informations. *Gallia Préhistoire* **22**, 526-29.

(1981). Informations. *Gallia Préhistoire* **24**, 395-9.

(1985). New excavations at Gavrinis. *Ant.* **59**, 183-7.

Lewis, A.L.

(1886). On three stone circles in Cumberland. *JRAI* **15**, 471-81.

L'Helgouach, J.

(1983). Les idoles qu'on abat. *Archéologie Armoricaine* **110**, 57-68.

Liestøl, A.

(1984). Runes. *In* Fenton & Pálsson (1984), 224-38.

Lockyer, J.N.

(1891). On some points in the early history of astronomy. *Nature* **43**, 559-63; **44**, 8-11; **44**, 57-60; **44**, 107-10; **44**, 199-202.

(1894). *The dawn of astronomy.* London.

(1906). *Stonehenge and other British stone monuments astronomically considered.* 1st ed. London.

Loew, C.

(1967). *Myth, sacred history & philosophy.* New York.

Lumholtz, C.

(1902). *Unknown Mexico.* New York.

Lynch, F. & Burgess, C.

(1972) (eds.). *Prehistoric man in Wales and the west.* Bath.

Lysaght, A.M.

(1974). Joseph Banks at Skara Brae and Stennis, Orkney, 1772. *NRRS* **28**, 221-34.

MacCana, P.

(1970). *Celtic mythology.* London.

McCreery, T., Hastie, A.J. & Moulds, T.

(1982). Observations at Kintraw. *In* Heggie (1982), 183-90.

MacKie, E.W.

(1969). Stone circles: For savages or savants? *Curr. Arch.* **1** (no. 11), 279-83.

(1974). Archaeological tests on supposed astronomical sites in Scotland. *PTRS* **A276**, 169-94.

(1976). The Glasgow conference on ceremonial and science in prehistoric Britain. *Ant.* **50**, 136-38.

(1977a). *Science and society in prehistoric Britain.* London.

(1977b). *The megalith builders.* Oxford.

(1981). Wise men in antiquity? *In* Ruggles & Whittle (1981), 111-52.

(1986) *Review of* Ruggles (1984a), *AAB* **7** (dated 1984), 144-50.

MacKie, E.W., Gladwin, P.F. & Roy, A.E.

(1985). A prehistoric calendrical site in Argyll? *Nature* **314**, 158-61.

(1986). Brainport Bay: A prehistoric calendrical site in Argyll. *AAB* **8**, in press.

MacKie, E.W. & Roy, A.E.

(1985). Prehistoric calendar. *Nature* **316**, 671.

Mair, G.R.

(1921). *Translation of The Phaenomena of Aratus.* London.

Manby, T.G.

(1970). Long barrows of northern England; structural and dating evidence. *SAF* **2**, 1-27.

Marquina, I. & Ruíz, L.

(1932). La Orientación de las Pyramides Prehispánicas. *Revista de la Universidad de México* **5**, 25-6.

Marshall, D.N.

(1977). Carved stone balls. *PSAS* **108**, 40-72.

Martlew, R.D.

(1982). The typological study of the structures of the Scottish Brochs. *PSAS* **112**, 254-76.

Mason, J.R. & Valentine, H.

(1925). Studfold Gate circle and the parallel trenches at Dean. *TCWAAS* **25**, 268-71.

Maxwell, J.H.

(1949). Bronze Age graves at Patrickholm Sand Quarry, Larkhall, Lanarkshire. Excavated during the autumn of 1947. *PSAS* **83**, 207-20.

Megaw, J.V.S. & Simpson, D.D.S.

(1979). *Introduction to British prehistory.* Leicester.

Millon, R.

(1973a) (ed.). *Urbanization at Teotihuacan, Mexico.* Austin & London (2 vols.).

(1973b). The Teotihuacan map, Part 1: Text. *In* Millon (1973a), Vol. 1.

Mohen, J.P.

(1984). Les architectures mégalithiques. *La Récherche* **161**, 1528-38.

Morris, R.W.B.

(1977).   *The prehistoric rock art of Argyll*. Poole.

(1979).   *The prehistoric rock art of Galloway and the Isle of Man*. Poole.

Morrison, L.V.

(1980).   On the analysis of megalithic lunar sightlines in Scotland. *AA* **2**, S65-77.

Morrow, J.

(1909)   Sun and star observations at the stone circles of Keswick and Long Meg. *Proceedings of the University of Durham Philosophical Society* **3**, 71-83.

Musée de Brest

(1982).   *Cartes de côtes de Bretagne du XVIe siècle a nos jours*. Catalogue d'exposition. Brest.

Myatt, L.J.

(1975).   Stone rows. *D & E 1975*, 54-55.

(1977).   Two possible prehistoric quarries in Caithness. *CFCB* **2**, 46-7.

(1985).   A setting of stone rows, Tormsdale, Caithness. *CFCB* **4**, 4-9.

Norris, R.P.

(1983).   A solar calendrical indicator on Dartmoor? *DASP* **41**, 123-5.

Norris, R.P., Appleton, P.N. & Few, R.W.

(1982).   A survey of the Barbrook stone circles and their claimed astronomical alignments. *In* Heggie (1982), 171-81.

Nuttall, Z.

(1906).   The astronomical methods of the ancient Mexicans. *In* Laufer (1906), 290-8.

Ovenden, M.

(1966).   The origin of the constellations. *Phil. J.* **3**, 1-18.

Ovenden, M. & Rodger, D.

(1981).   Megaliths and Medicine Wheels. *In: Megaliths to Medicine Wheels: Boulder Structures in Archaeology* (Proceedings of the Eleventh Chacmool Conference, Calgary University), 371-86.

Parker, R.

(1950).   *The calendars of ancient Egypt*. Chicago.

Patrick, J.D.

(1974).   Midwinter sunrise at Newgrange. *Nature* **249**, 517-19.

(1978).   An information measure comparative analysis of megalithic geometries. Ph.D. Thesis. Monash University.

(1979). A reassessment of the lunar observatory hypothesis for the Kilmartin stones. *AA* **1**, S78-85.

Patrick, J.D. & Freeman, P.R.

(1983). Revised surveys of Cork-Kerry stone circles. *AA* **5**, S50-6.

Pennant, T.

(1774). *A Tour in Scotland, I.* (3rd edn.). Chester.

Piankoff, A.

(1954a). *The tomb of Ramesses VI.* Princeton, New Jersey.

(1954b). *Mythological papyri.* Princeton, New Jersey.

Piggott, S.

(1968). *The Druids.* London.

Ponting, M.R., MacRae, M. & Ponting, R.

(1983). *A mini-guide to Shawbost stone circle, Isle of Lewis.* Callanish.

Ponting, M.R. & Ponting, G.H.

(1979). *Callanish - The documentary record.* Callanish.

(1981a). *Achmore stone circle.* Callanish.

(1981b). Decoding the Callanish complex - some initial results. *In* Ruggles & Whittle (1981), 63-110.

(1982). Decoding the Callanish complex - a progress report. *In* Heggie (1982), 191-203.

(1984a). *The stones around Callanish.* Callanish.

(1984b). Dalmore. *Curr. Arch.* **91**, 230-5.

Powell, T.G.E.

(1972). The tumulus at Skelmore Heads near Ulverston. *TCWAAS* **72**, 53-6.

Rademacher, H. & Töplitz, O.

(1957). *The enjoyment of mathematics.* Princeton.

Ralph, E.K., Michael, H.N. & Han, M.C.

(1973). Radiocarbon dates and reality. *MASCA Newsletter* **9**, 1-20.

RCAHMS

(1911a). *Third report and inventory of monuments and constructions in the county of Caithness.* Edinburgh.

(1911b). *Second report and inventory of monuments and constructions in the county of Sutherland.* Edinburgh.

(1928). *Ninth report with inventory of monuments and constructions in the outer Hebrides, Skye, and the Small Isles.* Edinburgh.

(1946). *Twelfth report with an inventory of the ancient monuments of Orkney & Shetland.* Edinburgh.

(1980). *Argyll: an Inventory of the ancient monuments, 3: Mull, Tiree, Coll and northern Argyll.* Edinburgh.

Renfrew, A.C.

(1979). *Investigations in Orkney* (Society of Antiquaries of London Research Report 38). London.

(1983) (ed.). *The megalithic monuments of western Europe.* London.

(1985) (ed.). *The prehistory of Orkney.* Edinburgh.

Ritchie, J.N.G.

(1974). Excavation of the stone circle and cairn at Balbirnie, Fife. *Arch. J.* **131**, 1-32.

(1976). The Stones of Stenness, Orkney. *PSAS* **107**, 1-60.

(1985). Ritual monuments. *In* Renfrew (1985), 118-30.

Rolleston, T.W.

(1911). *Myths and legends of the Celtic race.* London.

Ross, A.

(1970). *Everyday life of the pagan Celts.* London.

Roy, A.E., McGrail, N. & Carmichael, R.

(1963). A new survey of the Tormore circles. *TGAS n.s.* **15**, 59-67.

Ruggles, C.L.N.

(1981). A critical examination of the megalithic lunar observatories. *In* Ruggles & Whittle (1981), 153-209.

(1982). A reassessment of the high precision megalithic lunar sightlines, 1: Backsights, indicators and the archaeological status of the sightlines. *AA* **4**, S21-40.

(1983). A reassessment of the high precision megalithic lunar sightlines, 2: Foresights and the problem of selection. *AA* **5**, S1-36.

(1984a) (principal author: contributions by P.N. Appleton, S.F. Burch, J.A. Cooke, R.W. Few, J.G. Morgan & R.P. Norris). *Megalithic astronomy: A new archaeological and statistical study of 300 western Scottish sites.* Oxford (BAR 123).

(1984b). Megalithic astronomy: The last five years. *VA* **27**, 231-89.

(1984c). A new study of the Aberdeenshire Recumbent Stone Circles, 1: Site data. *AA* **6**, S55-79.

(1985). The linear settings of Argyll and Mull. *AA* **9**, S105-32.

(1986). 'You can't have one without the other'? I.T. and Bayesian Statistics, and their possible impact within archaeology. *Sci. Arch.* **28**, 8-15.

Ruggles, C.L.N. & Burl, H.A.W.

(1985). A new study of the Aberdeenshire Recumbent Stone Circles, 2: Interpretation. *AA* **8**, S25-60.

Ruggles, C.L.N. & Norris, R.P.

(1980). Megalithic science and some Scottish site plans: Part II. *Ant*. **54**, 40-3.

Ruggles, C.L.N. & Saunders, N.J.

(1984). The interpretation of the pecked cross symbols at Teotihuacan: A methodological note. *AAB* **7**, S101-7.

Ruggles, C.L.N. & Whittle, A.W.R.

(1981) (eds.). *Astronomy and society in Britain during the period 4000- 1500* BC. Oxford (BAR 88).

Ruppert, K.

(1935). *The Caracol at Chichén Itzá, Yucatán, Mexico*. Washington (CIWP 454).

Sage, D.

(1899). *Memorabilia Domestica*. Wick.

Service Historique de l'Armée de Terre, Vincennes.

(1776-1783). Carte topographique et géométrique des côtes de France offrant celles de la Bretagne depuis le Mont-St-Michel jusqu'à l'isle de Noirmoutier, divisée en 61 parties. 1:14,400 ou 6 lignes pour 100 toises. MSS J 10 C. Tracing paper copies in the Bibliothéque Nationale, Département des cartes et plans (dépôt du Service Hydrographique de la Marine), S.H., pf. 43-4.

Siegal, S.

(1956). *Non-parametric statistics for the behavioral sciences*. Tokyo.

Simpson, D.D.A.

(1972). Raigmore, Inverness. *Arch. Exc. 1972*, 14.

(1973). Raigmore, Inverness. *Arch. Exc. 1973*, 109.

Simpson, J.Y.

(1867). *Archaic sculpturings of cups, circles, etc. upon stones and rocks in Scotland, England, and other countries*. Edinburgh.

Smith, A.L.

(1950). *Uaxactún, Guatemala: Excavations of 1931-7*. Washington (CIWP 588).

Smith, G.

(1752, repr. 1886). Long Meg and her daughters. *In* Gomme (1886), part II, 71-72.

Smith, I.F.

(1965). *Windmill Hill and Avebury. Excavations by Alexander Keiller 1925-1939*. Oxford.

Somerville, B.

(1912). Prehistoric monuments in the Outer Hebrides and their astronomical significance. *JRAI* **42**, 23-52.

(1923). Instances of orientation in prehistoric monuments of the British Isles. *Archaeologia* **73**, 193-224.

(1927). Orientation. *Ant*. **1**, 31-41.

Stukeley, W.

(1776). *Itinerarium Curiosum, Vol. II*. London.

Thom, A.

(1954). The solar observatories of megalithic man. *JBAA* **64**, 396-404.

(1955). A statistical examination of the megalithic sites in Britain. *JRSS* **A118**, 275-95.

(1964). The larger units of length of megalithic man. *JRSS* **A127**, 527-33.

(1966). Megalithic astronomy: Indications in standing stones. *VA* **7**, 1-57.

(1967). *Megalithic sites in Britain*. Oxford.

(1968). The metrology and geometry of cup and ring marks. *Systematics* **6**, 173-89.

(1969). The geometry of cup-and-ring marks. *TAMS* **16** (new series), 77-87.

(1971). *Megalithic lunar observatories*. Oxford.

(1984). Moving and erecting the menhirs. *PPS* **50**, 382-4.

Thom, A. & Thom, A.S.

(1973). A megalithic lunar observatory in Orkney: the Ring of Brogar and its cairns. *JHA* **4**, 111-23.

(1975). Further work on the Brogar lunar observatory. *JHA* **6**, 100-14.

(1978a). *Megalithic remains in Britain and Brittany*. Oxford.

(1978b). A reconsideration of the lunar sites in Britain. *JHA* **9**, 170-9.

(1980). A new study of all megalithic lunar lines. *AA* **2**, S78-89.

Thom, A., Thom, A.S. & Burl, H.A.W.

(1980). *Megalithic rings*. Oxford (BAR 81).

Thom, A., Thom, A.S. & Thom, Alexander Strang

(1974). Stonehenge. *JHA* **5**, 71-90.

Thom, A.S.

(1980). The stone rings of Beaghmore: Geometry and astronomy. *UJ Arch.* **43**, 15-19.

(1981). Megalithic lunar observatories: an assessment of 42 lunar alignments. *In* Ruggles & Whittle (1981), 13-61.

Thomas, F.W.L.

(1852). Account of some of the Celtic antiquities of Orkney, including the Stones of Stenness, tumuli, Picts-houses etc. *Archaeologia* **34**, 88-136.

Thorpe, I.J.

(1981). Ethnoastronomy: its patterns and archaeological implications. *In* Ruggles & Whittle (1981), 275-88.

(1983). Prehistoric British astronomy - towards a social context. *SAR* **2**, 2-10.

University of Glasgow

(1978). *A map of the standing stones and circles at Callanish, Isle of Lewis, with a detailed plan of each site*. Glasgow.

Waateringe, W.G.-van & Butler, J.J.

(1976). The Ballynoe stone circle. Excavations by A.E. Van Giffen 1937-1938. *Palaeohistoria* **18**, 73-110.

Wallace, C.S & Boulton, D.M.

(1968). An information measure for classification. *Comp. J.* **11**, 185-94.

Waterhouse, J.

(1985). *The stone circles of Cumbria*. Chichester.

Watkins, T.

(1982). The excavation of an early Bronze Age cemetary at Barns Farm, Dalgety, Fife. *PSAS* **112**, 48-141.

West, J.F.

(1970) (ed.). *The journals of the Stanley expedition to the Faroe Islands and Iceland in 1789: i, Introduction and diary of James Wright*. Tórshavn.

(1975) (ed.). *The journals of the Stanley expedition to the Faroe Islands and Iceland in 1789: ii, Diary of Isaac S. Benners*. Tórshavn.

(1976) (ed.). *The journals of the Stanley expedition to the Faroe Islands and Iceland in 1789: iii, Diary of John Baine*. Tórshavn.

Williams, B.

(1856).   On some ancient monuments in the county of Cumberland and its borders. *PSA* **3**, 224-7.

Williams, J.

(1970).   Neolithic axes in Dumfriess and Galloway. A preliminary list of axes possibly available for thin section analysis. *TDGNHAS* **47**, 111-22.

Williamson, R.A.

(1982).   Casa Rinconada, twelfth century Anasazi kiva. *In* Aveni (1982), 205-19.

(1984).   *Living the sky: The cosmos of the American Indian*. Boston, Massachusetts.

Wood, J.E.

(1978).   *Sun, moon and standing stones*. Oxford.

Worthy, M. & Dickens, R. Jr.

(1983).   The Mesoamerican pecked cross as a calendrical device. *Am. Ant.* **48**, 573-6.

Yaglom, I.M. & Boltyanskii, V. G.

(1961).   *Convex figures*. New York.

Zarate M., R.

(1986).   Tres piedras grabadas en la región Oaxaqueña. *Cuadernos de Arquitectura Mesoamericana* **7**, 75-6.

Zeilik, M.

(1985).   The ethnoastronomy of the historic pueblos, 1: Calendrical sun watching. *AAB* **8**, S1-24.

Printed in the United States
By Bookmasters